Ecosociocentrism

Gopi Upreti

Ecosociocentrism

The Earth First Paradigm for Sustainable Living

Springer

Gopi Upreti
Fairfax, VA, USA

ISBN 978-3-031-41753-5 ISBN 978-3-031-41754-2 (eBook)
https://doi.org/10.1007/978-3-031-41754-2

This Springer imprint is published by the registered company Springer Nature Switzerland AG
The registered company address is: Gewerbestrasse 11, 6330 Cham, Switzerland

Paper in this product is recyclable.

This groundbreaking book by Professor Gopi Upreti delves deep into the complex issues of the global environmental crisis and climate change, highlighting the threats that jeopardize the very continuity of life in our planet. It vividly depicts the Anthropocene, the most perilous epoch in human history, and the stark realities of climate change, environmental destruction, and degradation. The author presents innovative ideas, strategies, and policy recommendations that encourage responsible practices toward sustainable living. It challenges us to reassess our mode of thinking, reform our dominant development model, and reorient our cultural-behavioral patterns toward the ones guided by ecological wisdom consciousness. It is a valuable addition to every environmental scholar's bookshelf. It is recommended for those engaged in environmental conservation and development, including at the UN and similar international organizations, governmental policymakers, civil societies, academics, media, and students.

—Ambika P. Adhikari, *D. Des, Sr. Global Futures Scientist, Arizona State University, Former Nepal Country Representative, International Union for Conservation of Nature (IUCN)*

Professor Gopi Upreti deserves sincere congratulations for his invaluable contribution to the timely book, written amid a global environmental crisis and the looming threat of climate change. The author's keen observation that we live in the most dangerous era, besieged by ecological and environmental crises, is spot-on. In his treatise, *Ecosociocentrism: The Earth First Paradigm for Sustainable Living*, Professor Upreti postulates that our destiny is inextricably linked to Earth's state of functional health. The author's thesis is that we must first secure a functionally healthy Earth to ensure a future for humankind and other living systems. He advocates for an alternative development vision, where policies are guided by ecological principles and wisdom that prioritize ecosystem health and sustainable use of Earth's systems, promote cooperation and equity, and resist competition, control, and domination driven by the insatiable greed of manmade capital accumulation. This book is a compelling narrative that highlights the importance of Earth as our only home and the need to protect it as a fundamental requirement for our safety and survival. It is a must-read book for anyone committed to social and environmental sustainability, nature conservation, and the pursuit of a sustainable human civilization that coexists with the diverse living system on Earth.

—Megha N. Parajulee, *Regents Fellow Professor and Faculty Fellow, Texas A&M University, Texas*

Ecosociocentrism: The Earth First Paradigm for Sustainable Living, by Prof. Gopi Upreti, presents a comprehensive and thought-provoking analysis of the existential crossroads faced by humanity in our current Anthropocene epoch. It pierces through the fog of growth-driven solutions, often misguided and serving only to perpetuate the crisis. It compellingly argues that changes in human consciousness and behavioral patterns are necessary and imperative for the survival and flourishing of humans and Earth's diverse life system. The book's central thesis resonates profoundly—Earth, our only abode in the vast cosmos, requires our first and foremost protection for our safety and security. Anyone concerned about the future of our life-supporting planet and the continuance of human civilization should consider this insightful book a critical read. It serves as an eye-opening exploration of our relationship with Earth and offers an actionable blueprint for a more harmonious and sustainable living on planet Earth.

—Jagadish Timsina, *Editor of Agricultural Systems, Visiting Professor, University of Peradeniya, Sri Lanka*

We are living in a critical period in Earth's history, in which humanity's impact on the environment has escalated to a scale that is impacting not only the health and vitality of the planet Earth but also the existential threats of diverse living system and the species, including our own. Only the functionally healthy Earth's systems can ensure the security of humanity, the living system, and the sustainable living for all, which is possible only if "The Earth First Paradigm" becomes the conscious working algorithm of humanity in the Anthropocene epoch of the twenty-first century. This is not merely a hopeful vision of the future; it is a logical and necessary path that we must embark upon if we wish to exist on this planet.

To the memory of
Professor Nicholas Polunin (26 Jun 1909–8 Dec 1997)

United Nations Environment Programmer's International Sasakawa Environment Prize (1987) United Nations Global 500 Roll of Honor (1991).

I would like to take a moment to honor the legacy of Professor Nicholas Polunin, a truly remarkable individual whose pioneering efforts in the field of environmental conservation continue to inspire and inform our collective efforts to protect our planet. As the founder and dedicated editor of the journal *Environmental Conservation*, Professor Polunin was a true visionary who recognized the urgent need to preserve our planet's invaluable ecosystems and worked tirelessly to promote greater awareness and action in this critical area. Through his unwavering commitment and indomitable spirit, Professor Polunin embodied a profound dedication to preserving our planet and its natural resources. His teachings and inspirations continue to resonate within the pages of his works and in the lives of those fortunate enough to encounter his wisdom. I count myself among these fortunate individuals, having had the privilege of receiving the Best Paper Award in Switzerland in 1995 and the monetary prize for my article published in the journal *Environmental Conservation*, Volume 21(1) of Spring 1994, bestowed upon me by Professor Polunin himself. It marked a pivotal moment in my life, inspiring me to persist in adding chapters to the manuscript of this book.

This book that I now dedicate to Professor Polunin's memory is a small but heartfelt tribute to his enduring legacy. It is my hope that the pages within will not only capture the spirit and vision of this remarkable individual but will also inspire others to carry on his mission and to work tirelessly in defense of our planet's most precious resources. May we collectively strive for a healthier, greener, and more caring Earth in memory of Professor Nicholas Polunin's vision and work.

Gopi Upreti
Emeritus Professor
(IAAS), Tribhuvan University
Kathmandu, Nepal

Foreword

This is a book written by a longtime professor and practitioner in Nepal's leading academic institution, the Agriculture and Forestry University, formerly known as the Institute of Agriculture and Animal Science of Tribhuvan University. The author, a colleague of mine whom I have known for some years, is passionate about the issues of environmental equity, justice, and social inclusion. The book's 13 chapters follow one another logically and each chapter serves to prepare the reader for the topics in the next chapter. Prof. Upreti performs the difficult task of making the book comprehensible to the layperson while maintaining depth and rigor to satisfy the professional scientist. The global topics of climate change and other human-driven deterioration of earth systems are well covered in the book. I find the book to be a useful resource for professional and can be adopted as supplementary resource material for environmental science and studies course at the university level.

Mohan B. Dangi

Jefferson Science Fellow,
Fulbright US Scholar,
Embassy Science Fellow, Professor,
California State University
Fresno, CA, USA

Foreword

In this timely and compelling book, *Ecosociocentrism: The Earth First Paradigm for Sustainable Living*, Professor Gopi Upreti provides a robust examination of the multifaceted issues our world confronts today, spanning ecological, sociological, environmental, and climate change domains. He expounds on the notion that the qualitative enhancement of both human life and the environmental sustainability should take precedence over mere growth-driven development, thereby challenging the current trajectory of global development. Professor Upreti's comprehensive analysis of our complex environmental and social landscapes firmly posits the need for a fundamental reconfiguration of our dominant development paradigm, beliefs, methodologies, consciousness, and moral obligations. He incisively illuminates the principal triggers of our ongoing ecological and climate crisis, ranging from the destruction and degradation of ecosystems and unsustainable exploitation of natural resources to socioeconomic inequities prevalent in developing countries, and most critically, the ecologically detrimental consumption cultural patterns in developed Western societies. To address these pressing environmental and societal challenges, Professor Upreti argues that significant alterations to our current environmental and development policies and their robust implementation are essential in both developed and developing nations. He presents policy instruments and strategies for the protection of functional health and the ecosystem services of the planet, the advancement of social justice and equality, and the cultivation of global environmental and social sustainability.

Professor Upreti critiques the prevailing valuation approach of biodiversity, ecosystem services, and natural capital. He posits that the flaws inherent in the valuation methodologies are primary drivers of global environmental degradation, thereby imperiling the survival of humanity on our home planet. Contrary to the conventional market-based valuation that undermines the value of biodiversity, ecosystem health and services, he advocates for an economic valuation framework that appreciates the life-support services and material contributions of the natural ecosystem. He accentuates the necessity for new development initiatives to transcend the confines of mere economic growth and to incorporate values that promote qualitative enhancements in human life and the environment. Central to his argument is the vital role that *environmental ethics* can play in shaping human behavior and guiding development initiatives fostering an environmentally conscious society capable of mitigating existing environmental challenges, including global environmental and climate crises.

Navigating the complexities of social and environmental sustainability, Professor Upreti introduces the term *Ecosociocentrism*, a neologism derived from the amalgamation of *ecocentrism* and *sociocentrism.* This innovative paradigm emphasizes that our socio-economic system (sociosphere) is subordinate to, and functions within, the planetary ecosystem (ecosphere), and thus, must operate in harmony with Earth's regenerative biocapacity. Professor Upreti poses a critical question: *how can our finite Earth support infinite growth-driven development within its limited biocapacity?* Professor Upreti maintains that achieving sustainable development requires implementing development policies and strategies that enhance ecological resilience while operating within the *biocapacity* of planet Earth. He presents *Ecosociocentrism: The Earth First Paradigm* as a development model that serves as the foundation of ecological and social sustainability safeguarding the continuity of humanity and the wider biotic community on planet Earth.

In this book, Professor Upreti artfully navigates the complex societal, ecological, environmental, and global challenges currently confronting humanity. An invaluable resource for those wishing to understand the theory and practice of ecological and sociological systems, including the current environmental crises and climate change, this book is a seminal reference for students, academics, researchers, ecologists, sociologists, environmentalists, and policymakers alike.

Founder of Asta-Ja Framework, Professor and Coordinator, Environmental Sciences
University of Louisiana at Lafayette
Lafayette, LA, USA

Durga D. Poudel

Preface

This book is the culmination of over three decades of reflection, engagement, and an enduring quest for a meaningful understanding of our complex relationship with our environment and nature. The inception of this journey took place in 1990 during my East West Center (EWC) Doctoral Fellowship at the University of Hawaii. The 1990s was a decade marked by a global awakening toward *environmental conservation* and *sustainable development*. These concepts became central to development discourses among professionals, planners, and institutions involved in shaping policy across the globe. Significantly, this was also the era of *Our Common Future*, the pivotal Brundtland World Commission on Environment and Development (WCED) report published in 1987. The report was the first to offer an official, albeit inadequate, definition of *sustainable development*, which subsequently informed policy directives in nations worldwide. In this era, the environment earned its rightful place within the development policy framework of nations, spawning a movement that continues to this day. The *mantra* of *sustainable development* has since echoed in every corner of national and international forums and discourses, from development experts to politicians. Following my rigorous immersion in the *Our Common Future* report and the previously published *Limits to Growth* report by D.H. Meadow's team, I found myself intrigued by a glaring discrepancy. While *sustainable development* emerged as the touted model for progress, it became apparent to me that the model failed to embrace the foundational ecological principles that underpin genuine environmental sustainability. This revelation made it clear that *sustainable development*, in many ways, was a sophisticated reframing of the existing neoliberal growth model, masked by its appealing novelty.

The current growth economic paradigm has, with striking audacity, succeeded on two fronts: a relentless, unsustainable extraction of Earth's resources to feed the insatiable appetite for growth, and the promotion of ecologically hostile consumerism as a means to perpetuate corporate wealth accumulation. This model's egregious consequences are palpable and far-reaching, ravaging the health and integrity of Earth's systems with an alarming ferocity. The devastating fallout from this model is manifested in the catastrophic breakdown of planetary ecosystems, rapidly accelerating climate change, the extinction and annihilation of millions of species, ocean acidification, destruction of coral reef ecosystem, toxic pollution, and the desertification of previously fertile lands. The continuity of *Homo sapiens* and Earth's living system now hangs in the balance, under an increasingly ominous cloud of uncertainty. We can agree that humanity's destiny is inexorably intertwined with the functionally healthy planet. This forms the crux of this book that to secure our collective future, we must first ensure the functionally healthy and flourishing planet. *Ecosociocentrism: The Earth First Paradigm for Sustainable Living* is a call for a radical change in our dominant mode of thinking and development model seeking a viable path for sustainable living on planet Earth.

Changes in human consciousness and cultural behavioral patterns are not only necessary but also moral imperative for the survival and flourishing of both humans and Earth's diverse life systems. Anyone concerned about the future of our life-supporting planet and the continuance of human civilization should seriously ponder into revisiting the fundamental *assumptions* and *modus operandi* of our current growth model and reconceptualize what we really mean by *development*.

We are living in a critical period in Earth's history, in which humanity's impact on the environment has escalated to a scale that is impacting not only the health and vitality of the planet but also the existential threat of entire living system including our own. This crisis can no longer be addressed solely through technological advancements or cosmetic policy changes. A fundamental shift in our *collective consciousness* and *behavioral patterns* is required—a shift that allows us to view the environment not as a separate entity, but as an integral part of ourselves and of all life. It behooves that Earth's systems be maintained in a functionally healthy and resilient states so that they can continuously generate ecological goods and services across multiple generations for sustainable living.

This journey begins with a profound look at *human consciousness*, which was understood not merely as an isolated phenomenon, but as an evolutionary development with substantial implications for human behavioral change. When *human consciousness* is elevated, we are more likely to perceive and value the interconnectedness of life, recognizing the environment as an organic extension of ourselves. Yet, consciousness alone is insufficient. Sustainable living also requires a profound sense of *spirituality*—a sense of connection that transcends the self and binds us to the planet Earth and the cosmos. Such *spirituality* gives rise to a deep sense of awe and reverence for the natural world, promoting a stewardship-based approach to environmental interactions. Spirituality, in this sense, is not confined to religious or mystical experiences. Instead, *spirituality* can emerge from our everyday encounters with Nature, from the simple act of observing a sunrise or sunset on the snowcapped mountain top to the amazing deep and colorful sea world and to the complex *web of life* of tropical and Amazonian rainforests and, in introspection, realizing of our special existential role within this grand cosmos.

I believe *human consciousness* is a powerful and liberating force of unlocking our full potential as individuals and as a species. It can open a new vista for the *quantum leap* to *eco-cultural enlightenment*, a new milestone in the evolutionary history of *Homo sapiens* liberating humanity from its current delusion. Albert Einstein's following words impeccably evoke this vision:

> A human being is part of the whole called by us "the universe," a part limited in time and space. We experience ourselves, our thoughts, and our feelings as something separate from the rest—a kind of optical illusion of our Consciousness. This delusion is a kind of prison for us, restricting us to our personal desires and affection for a few persons nearest to us. Our task must be to free ourselves from this prison by widening our circle of understanding and compassion to embrace all living creatures and the whole of Nature in its beauty.

No doubt, the biggest challenge of humanity today is how to free itself from the prison of this optical illusion to preserve the web of interconnectedness and live in peaceful coexistence with other living entities in Nature. The interweaving of consciousness, spirituality, and moral imperatives in environmental stewardship presents a holistic approach to addressing our environmental challenges and climate crisis. It goes beyond the traditional, segmented methods, embracing a more comprehensive, integrated, and ecologically guided profound engagement with our planet. Through this integration and engagement, we can hope to bring about a societal transformation that redefines our relationship with the environment and our roles as custodians of the Earth. Only through a transformative process empowered by *collective consciousness* within humanity itself can we secure a sustainable existence for all. To facilitate this transformation, a paradigm that fosters ecological harmony between the *ecosphere* and *sociosphere* is indispensable. Planet Earth is the sole abode for all living entities, including *Homo sapiens*. Only a healthy and functionally nourishing Earth can ensure the security of humanity and sustainable living, which is possible if *The Earth First Paradigm* becomes the conscious working algorithm of humanity in the *Anthropocene epoch* of the twenty-first century. This is not merely a hopeful vision of the future; it is a logical and necessary path that we must embark upon if we wish to prolong sustainable living on planet Earth.

Emeritus Professor
IAAS, Tribhuvan University (TU)
Kathmandu, Nepal

Gopi Upreti

Acknowledgment

It is with profound gratitude and respect that I acknowledge the contributions of several individuals without whom this manuscript would not have come to fruition. Foremost, my earnest appreciation extends to Mr. Richard Morse, Senior Research Fellow of the Program on International Economic and Development Policy, and the Coordinator of the Participatory Development Group at East West Center (EWC). His instrumental role, akin to sowing a seed that subsequently germinated and culminated in this book, cannot be overstated. Equally, my deep sense of indebtedness rests with my Professor Richard W. Hartmann, Tropical Agriculture, University of Hawaii, whose meticulous editing and valuable feedbacks on the initial few chapters enhanced my confidence to pursue it further.

Similarly, I would like to acknowledge Catherine Wilson, Professor of Urban and Regional Planning at the University of Hawaii, and Arthur Getz, Visiting Research Fellow at EWC for intellectually stimulating discussions on various development themes, particularly "environment and sustainable development," which have been instrumental in shaping my thoughts in this book. I greatly appreciate Dr. Uttam Gaulee, Professor of Higher Education Administration and Policy at Morgan State University for his inspirational advice and encouragement.

On a personal note, my gratitude goes out to my wife, Samita Upreti, whose patience and enduring support have been a pillar of strength. Her providing me with nourishing meals and maintaining my wellbeing, while gracefully bearing with my idiosyncrasy, deserves special mention. Similarly, my son Asim, my daughter-in-law Eva, and younger son Bibek, warrant my grateful acknowledgment. Without their support and compassionate care during my periods of illness, surgery, and the subsequent recovery, completing this work would have been impossible.

Finally, I would be remiss if I did not acknowledge the inspiration derived from my beloved granddaughter, Arya Upreti, whose cheerful smile energizes me and her mere presence symbolizes the generation that will become the true custodians of our planet. This book, in many ways, is a testament to their future stewardship of the planet Earth.

Introduction

As we can observe, the fundamental ecological variables—matter, energy, space, time, and diversity—govern all ecological phenomena. The interactions and interplay of these variables determine the behaviors of ecological systems. Chapter 1 begins with a brief discussion of these variables and introduces emerging ecological concepts that enhance our understanding of ecological complexity and behavior, which we discuss in later chapters.

Chapter 2 delves into the importance of biodiversity, ecosystems, and ecosystem services for human survival. It highlights the necessity of preserving these natural systems and ensuring their continued healthy functionality. The chapter provides a critical perspective on how the current economic system recognizes only the tangible, market-driven commodity values, thus neglecting the integral life support ecological services and resources provided by diverse biotic communities and ecosystems. It emphasizes that all human needs are met through materials and ecological services derived from these natural ecosystem processes, for which no comprehensive economic valuation system currently exists.

Chapter 3 critically examines the current state of biodiversity and ecosystem destruction and degradation, attributing this primarily to hyper-anthropogenic causes. It identifies key forces of environmental destruction, primarily ecologically detrimental consumption and production patterns in wealthier, industrialized countries, and population pressure, poverty, and inequitable development in developing nations. The chapter underscores the alarming impacts of unsustainable resource extraction and economic activities, particularly in developed and rapidly developing nations. The chapter concludes that environmental conservation strategies will only succeed if adequate cultural, socio-economic, and political measures are implemented to shift current production and consumption patterns and alleviate human poverty.

Chapter 4 explores the fundamental principles underpinning ecosystem evolution, examining the dynamic processes that have shaped biodiversity and ecosystems over time. Acknowledging the importance of comprehending these principles and mechanisms for interpreting ecosystem responses to disturbances and forecasting future changes, the chapter explicates how anthropogenic and biotic interactions, abiotic factors, and random events instigate long-term alterations in ecosystems. This intricate interplay underscores the complexity of ecosystem evolution, which often involves shifts spurred by climate change, plate tectonics, and species interactions. Attempts have been made to integrate perspectives with insights derived from evolutionary biology, underscoring the instrumental role of adaptation and speciation in shaping ecosystems.

Drawing from biological, evolutionary, and ecological sciences, the chapter outlines the foundational principles of ecosystem evolution, including natural selection, ecosystem succession, coevolution, diversity and stability, interconnectedness, interdependence, mutualism, and system complexity. It positions ecosystems as the basic units of biological organization, emphasizing the symbiotic feedback mechanisms between biotic communities and the physical environment that sustain their structure and function. It highlights how biological subsystems' growth and development influence and impact the physical system. The chapter delves into a more profound understanding of these principles and theories to inform future ecosystem protection, preservation, and biodiversity management efforts.

Chapter 5 brings perspectives on how *autopoiesis*, or the self-organizing property of the living system, is the basis for the emergence of a complex form of ecosystem structure. Autopoiesis provides the fundamental basis for the system view of life. It pervades the biophysical realm of Nature as a ubiquitous phenomenon. Ecosystems can be comprehensively conceived as autopoietic systems that engender and sustain themselves via homeostatic responses to shifting environmental conditions. Ecosystem health encapsulates its capacity for *resilience*, *self-organization*, and preserving the ecosystem's functional integrity. Therefore, the health of ecosystems should occupy the focal point of any policy-making and managerial strategy that aims to safeguard Nature, promote conservation, and guide ecologically informed management and societal values such as human health and wellbeing, which are intrinsically interconnected to the health of ecosystems.

Understanding the concept of *autopoiesis*, *ecosystem health*, and their implications for human health and wellbeing is imperative for nature conservation and sustainable living. Furthermore, there is an urgent need for restoration and ecological reengineering efforts to restore and redevelop degraded ecosystems, reinstating their functional integrity, which is critical to the continuation of the living system on planet Earth.

Chapter 6 critically analyzes the complex task of aligning human needs with the imperatives of nature conservation, protection, and sustainable development, acknowledging the profound influence of basic human needs on the behavioral patterns of people. It draws upon Abraham Maslow's hierarchy of needs, noting the constant nature of fundamental human needs irrespective of cultural or historical contexts. These needs largely rely on ecosystem services, including material inputs, life-supporting resources such as land, air, water, biodiversity, and waste management services. The human economy operates as a subsystem within the planetary ecosystem, emphasizing sustainable development as a multidimensional concept with inherent ecological, economic, and social dimensions. In the *Anthropocene era*, humanity grapples with the monumental task of balancing these dimensions of sustainability. The dominant economic model, characterized by infinite growth, excessive consumerism, and utter disregard for ecosystem health and resilience, is inherently unsustainable. Thus, an ecological value-based development model that promotes ecosystem health and maintenance of ecological processes and operates within the biocapacity of our planet Earth forms the basis for sustainable living.

The chapter argues for creating socio-economic governance systems capable of mitigating poverty, enabling Earth-friendly consumption patterns and growth, and preserving ecosystem health and vital environmental life support services. With the looming threat of infinite economic growth surpassing the Earth's *biocapacity*, there is an urgent need for a global consensus on sustainable production, distribution, and consumption of goods and services. A steady-state economy—marked by circular economy, population stability and reduced per capita consumption—to ensure the sustained functionality of the planetary ecosystem fosters optimism for sustainable living on planet Earth.

As a defining crisis of the *Anthropocene*, climate change has far-reaching impacts on human and Earth systems. Chapter 7 offers a comprehensive examination of climate change's current and projected consequences, investigating global warming scenarios ranging from 1.5 to 4 °C. It underscores the escalating threats facing human societies, including increased extreme weather events, rising sea levels, and high risks to food and water security. Each incremental degree of warming amplifies these risks, potentially leading to unimaginable catastrophic outcomes, should global average temperatures rise by 4 °C above pre-industrial levels. This is crucial to avoid a looming existential threat to humanity. The chapter highlights the role of fossil fuels corporate capitalism in accelerating global warming to alarming levels, pushing Earth's tipping points to potentially irreversible extremes. Scientists have spotlighted the rapid pace of contemporary changes compared to those marking the end of the last glacial period, notably the 52% loss of biodiversity between 1970 and 2010.

The chapter highlights the efforts of investigative journalists and scholars in holding fossil fuel industries accountable for their environmentally destructive acts and their role in promoting stricter adherence to environmental laws and regulations. It posits that through the unified efforts of scientists, policymakers, journalists, environmental activists, NGOs, and grassroots movements, there is potential for a *paradigm shift* in the mindsets of mainstream politicians and the world's corporate leaders. Replacing the current ecologically hostile cultural superstructures with new ones grounded in *ecological wisdom* is essential for securing the future of humanity and planetary Earth systems. The chapter concludes by endorsing the integration of ecology and political economy with ecological facts backed value-based development imperative as the sole viable solution to rectify humanity's unfortunate disconnection from Nature, suggesting the dawn of an *ecological civilization*, a transformative phase in human development that harmoniously aligns our *sociosphere* with the *biosphere*.

Chapter 8 critically evaluates the existing valuation approaches of *biodiversity*, *ecosystem services*, and *natural capital*, positing their inadequacy as a principal driver of environmental destruction and threats to human existence on planet Earth.

The current *neoliberal economic model* recognizes only the tangible benefits and market-determined commodity values, thereby neglecting the invaluable ecological services contributions from diverse ecosystems and biotic communities. Such disregard for the natural ecosystem's life-support services and material inputs, fundamental to producing human-made goods, necessitates revising economic valuation systems, with the principle of opportunity cost applied to maintaining a healthy natural ecosystem.

It argues for a comprehensive valuation of entire ecosystems, recognizing their potential for generating goods and services vital for human wellbeing and happiness. The intertwined nature of the economic system and natural ecosystem signifies the urgent need for our policies and valuation strategies to abide by ecological laws and principles, emphasizing protecting and maintaining healthy ecological systems. The discourse around the valuation of natural systems, encompassing biodiversity and ecosystem processes, has become a pivotal issue among conservationists, environmentalists, and economists. It underscores the urgent need for experts from various disciplines, particularly ecology, biology, agriculture, and economics, to collaborate to develop valuation techniques and approaches that integrate both the instrumental and intrinsic values of biodiversity and ecosystems. The chapter concludes with a call to integrate biodiversity and ecosystem services values into economic valuation systems, stating it as a prerequisite for sustainable development and effective environmental protection and conservation.

Chapter 9 critically assesses the *metaphysical* foundation of the *dominant development paradigm* that has marked the *Anthropocene epoch*, underscoring its inherent tendency to treat Nature as an entity to be dominated, subdued, and exploited. The pervasive ecological crises we face today—climate change, global warming, large-scale species extinction, and desertification of terrestrial and aquatic ecosystems—can be attributed to this worldview, rooted in Rene Descartes' philosophy and the paradigm of modern science. Descartes's acclaimed system envisages the natural world as a *mechanical* and inert system subject to human intervention and manipulation, engendering inflated *anthropocentrism* and consequent instrumental exploitation of Nature.

The chapter advocates for a transformation in our relationship with and understanding of Nature, emphasizing the need to regard Nature as possessing *functionally nurturing* and *survival values* crucial for perpetuating living system including human civilization. This shift requires an *ecological wisdom* that underscores the protection and cultivation of Nature as a living system and guides the development of science and technology toward restoring damaged planetary ecosystems. This critique extends to the current neoliberal market-driven model, which renders sustainable development unthinkable unless radically restructured to internalize *environmental externalities* and costs linked to the depletion of natural capital and ecosystem services. Empirical data on global ecological footprint growth suggest humanity is consuming Nature's services 44% faster than its regenerative capacity, leading to critical overshoots of four out of nine planetary boundaries and severely undermining the Earth's biocapacity.

The chapter underscores the crucial role of *values* in the development model and argues that without restructuring the current economic development model's *assumptions* and integrating *ecological facts* and *values*, humanity will inevitably confront crises in both the socio-economic and planetary ecosystem. Hence, this chapter's key focus is identifying and conceptualizing alternative development paradigms and trajectories that redirect our currently unsustainable development path toward a healthier planet.

Chapter 10 critically analyzes the indispensable role of *environmental ethics* in the context of sustainable development and Nature conservation. The chapter critically reviews the necessity of a foundational shift in our development approach advocating for pragmatic development ethics rooted in preserving and conserving Nature and satisfying basic human needs. It explores the metaphysical underpinnings of *environmental ethics* and their implications for Nature protection, conservation, and sustainable development. Sustainable development is discussed through multiple interconnected dimensions, including ecology, social and economic, and cultural and ethical systems. The term *sustainable development* has been reduced to a mere rhetoric due to its excessive use with little substance, often camouflaging the neoliberal growth model with minor adjustments, likened to repackaging old wine in new attractive bottles.

The chapter insists that development should not solely focus on quantitative metrics like GDP, but instead, it must encompass qualitative improvements in people's lives and their social and environmental relations. Since the publication of the influential report, *Our Common Future* (1987), the concept of *sustainable development* has undergone considerable changes, with scholars adding various social and ecological dimensions. It underscores that genuine sustainable development necessitates preserving and managing environmental resources but also requires profound social, cultural, and institutional transformation. The chapter presents five principles of sustainable development and proposes to reconceptualize *sustainable development.* The diligent implementation of those principles, underpinned by political commitment, could guide nation-states toward realizing sustainable development goals and foster an eco-civilization grounded in social justice and environmental sustainability.

Buddhism, Gaia, and System Theory share a common foundation of *interconnectedness* and *interdependence*. The Gaia hypothesis and System Theory emphasize systems analysis and remind scientists, policy analysts, and others concerned with local environmental problems that their local systems are embedded inside larger systems. Buddhism emphasizes the interrelatedness and interconnectedness of living and non-living systems in Nature through *Dharma* concept and dependent co-origination of all worldly pehnomena. Chapter 11 offers the potential to guide human behavior to build a harmonious relationship with the planetary ecosystem.

The scientific perspectives embodied in the Gaia hypothesis and system theory adopts a processual view of life, affirming the indivisibility of humanity from the intricate network of relationships with other entities in Nature. Analogously, Buddhist *Eco-Dharma* is rooted in the principle of human interconnectedness with Nature, attributed to the condition of *dependent co-origination* also known as

Pratītyasamutpāda in Sanskrit. The chapter argues that our current crises stem from an egocentric and pathologically misconstrued perception of the human self and Nature. Buddhism offers a pragmatic and practical framework for developing coherent *environmental ethics*. Gaia and System theory provide a unifying framework focusing on living systems spanning individual organisms, ecosystems, and human social systems. The convergence of Buddhism, Gaia, and system theory on interconnectedness, feedback mechanisms, interdependence, emergent boundaries, and hierarchies emphasize their combined potential to illuminate the path out of our current predicament. Buddhist ethical tenets of *compassion, non-violence*, and *reverence for life* amplify this potential. The chapter highlights Buddhism's unique perspective on environmentalism and man's relationship with Nature, underlining the doctrine of dependent co-origination *(Pratītyasamutpāda)* and Eco-Dharma, which affirm the interconnectedness and interdependence of all things.

The chapter explores how these fundamental Buddhist teachings, combined with Gaia and system theory, enhance our understanding and practice of an emerging *ecological paradigm*. Collectively, Buddhism, Gaia, and system theory imbue us with an ecological worldview, or *Eco-Dharma* guiding humanity toward a harmonious relationship with the planet, thereby paving the way for sustainable living and coexistence.

Chapter 12 explores the enormous role human *Collective Consciousness* can play in bringing the desired changes in people's behavior and attitude and consequently altering the behavior of political institutions and power centers. With a collective *ecological awakening*, it is possible to bring about desirable political outcomes that align with maintaining the functional integrity and resilience of planetary ecosystem based on ecological laws and scientific epistemology. However, achieving this vision depends on the emergence of informed and environment-friendly politicians and managers who possess a vision of creating a sustainable society that prioritizes meeting the essential needs of all over satisfying the *greed* and *self-aggrandizement* of a few. The failure to do so would result in the inability to protect the planetary ecosystem and maintain its *regenerative* and *resilient capacity*.

Philosophers and scientists have long grappled with defining and understanding *Consciousness* from various perspectives, but recent neuroscience research has provided new insights into this complex subject. The biological evolution of *Homo sapiens* took millions of years. However, the relatively recent evolution of human Consciousness and rationality can resolve the planetary environmental crisis that humanity currently faces. Human collective ecological Consciousness has the power to effect social and cultural changes that benefit humanity and the biotic community in Nature, and it is the only means that can reconnect and reestablish humanity's ruptured relations with Nature, the planetary ecosystem. The development of science and technology should be directed toward the sustainable uses of *Earth's systems* resources while maintaining the *biocapacity* of the *Earth's systems*. Ultimately, a transcendental ecological consciousness that integrates the *ecosphere* and *sociosphere* can give rise to an ecological civilization that is the basis for an ecologically sustainable and equitable global society.

Building upon the previous chapters, Chap. 13 proposes a new paradigm called *Ecosociocentrism: The Earth First Paradigm* to reconcile *instrumental, relational,* and *intrinsic* values in Nature and claims to provide the foundation for *sustainable development* and *eco-civilization.* This paradigm stresses the significance of maintaining social and ecological integrity to ensure the continuation of all species, including *Homo sapiens.* Recognizing and embracing the interconnectedness and interdependence of all life forms, the *ecosociocentric paradigm* embodies the ethical perception that human activities that promote social and ecosystemic health, stability, integrity, and diversity should be deemed right and just, while those that undermine these values are morally wrong and unjust.

The proposed paradigm suggests a shift to a new development ethics that adopts a comprehensive approach to justice, equity, and social and ecosystemic wellbeing. The paradigm integrates the *ecosphere* and *sociosphere* with the ethical view that embodies the concept that human response to Nature must be for the collective needs of humans and other beings, maintaining self-organizing creative processes *(autopoiesis)*, which are intrinsically intertwined in Nature. The *sociosphere* is a subsystem of the *ecosphere* and is always intertwined and entangled with the ecosphere in a *dialectical nexus.* A correct understanding of the Nature of this interaction is necessary to realize sustainable development in the sociosphere and to prolong and maintain environmental sustainability in the ecosphere.

The insurmountable challenges sustainable development faces today are integrating social and ecological integrity recognizing instrumental and some intrinsic values in Nature and placing them at the heart of normative discourse on development. Ecosociocentrism: The Earth First Paradigm seeks to provide a new definition of sustainable development that captures the essence of social and ecological sustainability. The paradigm defines *sustainable development* as "*the development that satisfies human needs of present and future generations while maintaining the resilience and the biocapacity of Earth's systems in such a way that human socio-economic throughputs of sociosphere remain within the biocapacity of the ecosphere, the planetary ecosystem.*" Only such development can sustain the social and ecological integrity to fulfill and actualize human potential and protect the living system on planet Earth.

The proposed paradigm seeks to integrate ecosphere and sociosphere with an ethics-based development approach that entails Nature's instrumental and intrinsic values. Included in the prescription of this paradigm are ten *directive principles* and *six policy strategies* to achieve the underlying objectives and goals of *sustainable living* on planet Earth. A functionally healthy and nourishing Earth ensures the security of humanity and sustainable living if *The Earth First Paradigm* becomes our conscious *working algorithm* in the *Anthropocene epoch* of the twenty-first century.

Contents

About the Author

Gopi Upreti taught at Tribhuvan University (TU) in Nepal for three decades. He is an emeritus professor at the Institute of Agriculture and Animal Sciences (TU). Throughout his tenure, he ascended various academic echelons, culminating in his reception of the esteemed Best Teacher and Academic Administrator (Campus Chief) Award, a commendation bestowed by Tribhuvan University. In addition to his contributions to academia, Prof. Upreti dedicated his expertise to pivotal research administrative roles, notably as the Chief Commissioner of the Nepal Agriculture Research Council (NARC) and as a strategic advisor to the Environmental Protection Council (EPC) under the auspices of the government of Nepal. Prof. Upreti has authored books on Agriculture and Water Resource Development and published over four dozen peer-reviewed research and review articles on Environmental Conservation, Agriculture, and Sustainable Development in referred journals. He was an East-West Center (EWC) Doctoral Fellow from 1988 to 1992 at the University of Hawaii (UH). His foundational academic credentials include a BSc. Ag (Hons). from Punjab Agriculture University (PAU) an MS in Horticulture and an MS in Envt. Management from UH.

As a noted environmentalist, Prof. Upreti received the coveted Best Paper Award from the Switzerland-based International Foundation of Environmental Conservation for his seminal paper, "Environmental Conservation and Sustainable Development Require a

New Development Approach," featured in the *Journal of Environmental Conservation*, Vol. 21(18-29) in 1994. A proponent of secular humanism, Prof. Upreti holds membership in the International Council of Secular Humanism based in New York and is a life member of the American Humanist Association (AHA). He served on the editorial advisory board of *The New Encyclopedia of Unbelief*, published by Prometheus Books, New York. He edited and published the English magazine *Humanist Voice* from Kathmandu, Nepal, and is the foundational president of the Humanist Association of Nepal (HUMAN).

Professor Upreti has been actively involved in many social and professional organizations in USA. He is a member of STAR Scholars, Co-Chair of advisory council of Association of Nepalese Agricultural Professionals in America (NAPA), Chief Advisor of Equality Foundation in USA, Advisor and life member of Blood Donors of America (BDA), life member and Advisor of Association of Nepalese in America (ANA), and life member of America Nepal Society (ANS). He is the Trustee of International Nepali Literary Society (INLS) and the past Chair of the Award Committee of INLS.

Chapter 1
Ecological Variables and Emerging Concepts in Ecology

If global civilization runs out of natural resources, we cannot replace them by investing in commodities through financial markets. People cannot eat money. Substitution, in the long run, may be possible at the micro-level of economic activity, but long-term macro-level substitution is downright wishful thinking.

Erald Kolasi (2017)

1.1 Fundamental Ecological Variables

The biosphere is defined as that part of Earth in which life is permanently possible and which contains all living organisms, their environments, and ecosystems. It consists of the terrestrial biota, oceans, and the surface of the continents, together with the adjacent atmosphere. The ecosphere embraces the upper layers of the lithosphere and the whole of the atmosphere above the troposphere (Ramade, 1984). Ecology investigates the relationship of living organisms with each other and with their environments and provides an essential basis for the rational approach to the study of the biosphere. There are five fundamental ecological variables, namely, matter, energy, space, time, and diversity, which are intrinsically interconnected with each other and are the basis for the evolution of the biosphere. All ecological phenomena can be explained by the interplay of these five variables in the biosphere (Ramade, 1984).

1.1.1 Matter

We know that living organisms are made of a certain number of chemical elements essential for the building of their biological molecules. There are about 40 chemical elements found in living organisms, which play a significant role in their structure

G. Upreti, *Ecosociocentrism*, https://doi.org/10.1007/978-3-031-41754-2_1

and functioning. There are three fundamental principles that govern the ways in which matter is used by living organisms and matter circulates in ecological systems:

1.1.1.1 The Law of Tolerance

This law states that there is a range of concentrations, i.e., the interval of tolerance, for each element in which all physiological processes involving that element can take place normally. For any plant or animal life, there is a range of concentrations of these elements within which the life processes become possible. There is an optimum concentration of the elements within which the metabolic processes occur at maximum speed. The concentration of the elements beyond the lower and upper limits of the tolerance of the organisms causes their death due to the deficiency and excesses of the elements. For example, in soil, nitrate is essential for plant growth because it is a source of inorganic nitrogen; however, the optimum growth of plants can occur only at some definite concentration of the salts in the soil, with other factors being constant. Below a certain concentration, nitrogen deficiency will prevent development of the plant, whereas if it is present beyond the upper limit of tolerance, then it can again inhibit growth and even lead to death due to phytotoxicity (Table 1.1).

Table 1.1 Chemical elements present in living matter and in the ecosphere

Elements	Atomic number	Ecosphere (lithosphere, atmosphere, and hydrosphere; %)	Human body (%)
Hydrogen	1	0.95	9.31
Carbon	6	0.18	19.37
Nitrogen	7	0.03	5.14
Oxygen	8	50.02	62.81
Fluorine	9	0.10	0.009
Sodium	11	2.36	0.26
Magnesium	12	2.08	0.04
Aluminum	13	7.30	0.001
Silicon	14	25.80	Negligible
Phosphorous	15	0.11	0.64
Sulphur	16	0.11	0.63
Chlorine	17	0.20	0.18
Potassium	19	2.28	0.22
Calcium	20	3.22	1.38
Manganese	25	0.08	0.0001
Iron	26	4.18	0.005

Adapted from Ramade (1984). *Ecology of natural resources*, Wiley 1984

1.1.1.2 The Law of Minimum

This law states that the growth of an organism is only possible as long as all essential elements are present in sufficient quantities and that the rate of growth is controlled by the essential element that is present in the lowest concentration. Caseation of growth occurs if only one element is present in an insufficient amount. At the other extreme, an element present in excessive amounts will be enough to prevent growth and development because the threshold for toxicity has already been reached. There is an optimum concentration of an essential element at which the growth of the organism will be maximum.

1.1.1.3 The Law of Conservation of Matter

Matter never becomes a "waste product" in an ecological system. There is always a permanent recycling of matter between organic and inorganic forms, which is brought about by the three categories of living organisms, namely, the primary photosynthetic producers (autotrophic), animal consumers, and animal decomposers, both of which are heterotrophic.

1.2 Energy and the Second Law of Thermodynamics

From the perspective of bioenergetics all living systems, right from a simple cell to the most complex ecological community, can be regarded as energy converters. The biological processes within the systems channel energy into various activities in such a way that the flow of energy is adapted and controlled. It is the energy that drives all the vital processes at every level, from the most elementary cellular mechanism, to the entire biosphere. For example, factors such as the number of organisms populating a particular locale as well as their rate of development and reproduction all depend upon the amount of energy available. This flow of energy in ecological systems is subject to the law of thermodynamics as stated by Carnot (1924). A brief discussion of the law of thermodynamics is imperative to understand their implications in the evolution of ecological systems.

1.2.1 The Principle of Conservation of Energy

This principle states that energy can neither be created nor destroyed but can only be transformed from one form to another. The amount of energy in the universe has been fixed since the beginning of time and will remain fixed till the end (Rifkin, 1989). This law is universally true and applies to all biological processes including photosynthesis (transformation of light energy to biochemical energy), muscular work (transformation of biochemical energy to mechanical energy), and neural conduction (transformation of chemical energy to electrical energy). It equally applies to the processes involving energy flow within an ecosystem, which is flow of energy to the communities of organisms, such as primary and secondary producers.

1.2.2 The Principle of Degradation of Energy (the Entropy Law)

The second law of thermodynamics is also called the entropy law. This law states that no process involving the transformation of energy can take place without partial degradation of energy. Every time energy is transformed from one state to another, there is always a loss in the amount of available energy. This is called "entropy." There is always a loss of some useful energy in the form of heat. This loss of useful energy from a system is proportional to a quantity called entropy, S. This law can be expressed by the equation

$$\Delta G = \Delta H - T\Delta S$$

where G is the change in usable energy (free energy)
H is the exchange of heat with the surroundings
ΔS is the change in entropy
T is the absolute temperature at which the process occurs

Entropy is a measure of the amount of energy that is no longer available to do some kind of work. The total energy content of the universe is constant, but total entropy is continually increasing.

The inescapable implication of this principle is that no process whether biological, physical, or mechanical can take place with 100% energy efficiency. In other words, a biological or any other process cannot convert energy with 100% efficiency from one state to another.

1.2.3 Implications of the Entropy Law

Energy is different from matter in that it can only pass once through any given trophic level of the food chain. During such energy flow within the ecosystem, the second law demonstrates that it is degraded as it progresses so that it is gradually dispersed and lost to the surrounding in a non-usable form called "entropy." There is only one external source of energy upon which all ecological systems depend: input from the Sun (light energy). This implies that the structure and functioning of communities of living organisms inhabiting a particular region on Earth's surface will be determined by the absolute values of the flux of solar radiation, at each point, together with the size of their annual fluctuations. Consequently, the distribution of large ecological units (macroecosystems or biomes) over the biosphere depends on the characteristics of solar radiation at different latitudes (Ramade, 1984).

There exists a law of optimization in the use of energy at the level of species and communities. The species that occupy a particular "ecological niche" use the available energy more efficiently than other species having similar requirements but less well-adapted to the environmental conditions appropriate to that niche. The whole ecosystem tends to evolve toward a structure consisting of communities that use the available energy most efficiently. This is expressed in terms of the ratio of the biomass B of the community to the amount of energy E entering the biotope per unit surface area per unit time. Hence, the biomass supported per unit of the energy flow, B/E, will increase as the ecosystem progresses toward maturity. These B/E ratios are low for cultivated crops and natural grasslands but become greater for shrublands and reach maximum for old-growth forests and coral reefs.

One direct consequence of the second law of thermodynamics (the entropy law) concerns the transfer of energy in an ecosystem: only a fraction of the energy reaching a given trophic level in the community is transmitted to a higher trophic level. This can be illustrated by the following example:

$$\text{Plant} \rightarrow \text{herbivore} \rightarrow \text{carnivore 1} \rightarrow \text{carnivore 2}$$

As a general rule, not more than 15% of the energy contained at the herbivore level passes to carnivore 1 and likewise from carnivore 1 to carnivore 2, and so on. This indicates that in each subsequent phase of the energy transformation, progressively more energy becomes unavailable and ultimately ends up as "entropy."

Our worldview has a strong hold over our perception of reality, which is so overwhelming that we cannot possibly imagine any other way of looking at the world. We are still living under the influence of the seventeenth century "Newtonian paradigm." This paradigm is being challenged in the wake of long-term energy crisis, escalating global warming and climate crisis. These new and dangerous realities have prompted us to question many of the basic tenets and the operating assumptions of the Newtonian paradigm. A new worldview called "*entropy law,*" has emerged that will replace the Newtonian paradigm. "*Entropy law*" will be the ruling paradigm over the "*Anthropocene*" era of human history. It is the supreme "*metaphysical law*" of the entire universe. As Rifkin (1989) argues, many people are still under the delusion that technology can free us from our dependence on the environment. Technology does not create energy; it only uses up the available existing energy. The more efficient the technology, the more available energy it uses up. Technology is only the transformer, nothing more or nothing less. The "entropy law" states that matter and energy can only be changed in one direction, that is, from usable to unusable, from available to unavailable, or from ordered to disordered. According to the entropy law, everything in the entire universe began with structure and value and is irrevocably moving toward the direction of random chaos and waste. "Entropy" is the measure of the extent to which the available energy in any subsystem of the universe is transformed into an unavailable form.

First, all matter and energy in the world are constant. They can neither be created nor destroyed but only be transformed from one state to another. Second, the transformation of energy is always from an available to an unavailable, a dissipated form, or from an ordered to a disordered state. Living systems are not closed systems and can only survive by exchanging energy with the environment. Without constant flowthrough of energy from the environment, living systems will perish within weeks or months. Another misconception is the belief that technology creates greater order in the world. The entropy law tells us that every time available energy is used up, it creates greater disorder somewhere in the environment/universe. The massive flowthrough of energy in modern industrial society is creating massive pollution and waste in the world we live today. This is the sad reality of our time and is the biggest problem and existential threat we are facing.

1.3 Space

Space is a fundamental ecological variable because the total mass of living organisms that can populate an ecosystem directly depends on the space available. First, the available space determines the intensity of the competition within and between the species. The relation between the dry weight p of an organism and the density d (space available to each individual S, where $d=1/S$) is

$$\log p = a - b\log(1/S) = a - b\log d$$

where a and b are constants. From field measurements, the values of a and b can be obtained as

$$p = k' \ d - 3/2 = k' \ S3/2$$

where k' is another constant. There is a certain value of density for which the rate and speed of growth, weight, fertility, and other physiological factors become maximum and on both sides of which they consequently show a decrease. The density of population can become a limiting factor by being either too high or too low. The distribution of the individual in each species over the available space also plays a fundamental ecological role. This distribution may be random, clumped, or uniform.

1.4 Time

This is another essential ecological variable. Its role can be understood in the evolution of every ecological system toward a state of maturity characterized by the optimum accumulation of biomass and therefore of energy per unit surface area. A normally functioning ecosystem harnesses the maximum amount of energy per unit time. As Ramade (1984) states "the presence of any particular species or any particular community in a given biotope depends above all on the length of the favorable period during which sufficient biomass (reserves) can be accumulated for living through the unfavorable season."

Time becomes a limiting factor for the accomplishment of some biological processes, such as germination of seeds, capture of prey by predators, or the meeting of mates for reproduction. If the density of prey diminishes, then the increase in time for their capture causes expenditure of extra energy, which can be fatal for predators. In an arid region, soils contain enough water for an unusually short period of time, which prevents the germination of seeds of annual plants or stops the growth of annual plants. Consequently, only those plant species that can capture this shortly available water resource can survive and leave offspring to maintain the continuity of their gene pools. Thus, in arid regions, we get to see an extremely different kind of plant community with specialized adaptive mechanisms. Their morphology, anatomy, and physiological processes are uniquely different from other types of plant communities and are all geared up to a special kind of adaptive mechanism that evolved through the action of "natural selection" over time in that environment.

1.5 Diversity

This is the most fundamentally important of all ecological variables because it characterizes the whole community of living organisms and ecosystems. Diversity entails the collection of myriads of millions of species, organisms and their ecosystems, renewable biological resources, and itself as a resource. Hence, the more diverse an ecosystem is in terms of its inter-and intraspecies composition, the more resourceful is the ecosystem. The concept of diversity is similar to a naturalist's concept of the richness of a population. Species diversity d can be mathematically expressed in terms of two variables: the number N of individuals present in a given biotope and the total number of species s to which they belong (Menhinick, 1964). Thus, it follows:

$$d = \frac{s}{\sqrt{N}}$$

As to the significance of species diversity, Ramade (1985) puts forward a powerful argument in the following words "… the greater the number of species in a given community, the greater is the degree of saturation of potential niches and thus greater is the structural complexity of the food web. Finally, the amount of information contained in an ecological system becomes larger as its species diversity increases."

From an evolutionary perspective, species diversity is the repository of a "gene pool," the biological information that came into existence as a result of million years of stochastic evolutionary process.

The Shannon–Weaver index for species diversity is given by

$$\bar{H} = \_\Sigma\left(\frac{ni}{N}\right)\log_2\left(\frac{ni}{N}\right)$$

where the ratio *ni*/*N*=pi expresses the probability of encountering the *i*th species in the total population *N*. Three types of diversities can be distinguished by the application of the Shannon index:

(a) *Alpha diversity*: This is also called intrabiotopic or microcosmic diversity and is the diversity that exists within one community.
(b) *Lambda diversity*: This is called sectoral or macrocosmic diversity and is calculated by considering the whole collection of mixed "biotopes" contained in a given geographical region. This is the diversity that exists among communities.
(c) *Beta diversity*: This is diversity that exists between the populations of two adjacent biotopes (biotopes 1 and 2) and can be measured as:

$$\bar{H}\beta = \bar{H}_{\alpha 12} - 0.5\left(\bar{H}_{\alpha 1} + \bar{H}_{\alpha 2}\right)$$

Taking logarithms to the base 2, *H* bar alpha 1 is the Shannon index for population 1, *H* bar alpha 2 for population 2, and *H* bar alpha 12 for both together. Hence, *H* bar beta represents an index of the similarity between biotopes. The species diversity of an ecosystem, as Ramade argues (1985), depends not only on their area but also on the stability of the habitats they contain and on their degree of geographical isolation. From the point of view of stability, it becomes clear that diversity decreases as one passes from an environment with stable conditions with respect to climate, soils, etc. to one where conditions fluctuate or are irregular. The geographical areas that encounter instability or fluctuation in climatic and soil variables are usually low in species diversity. Hence, a high species diversity is the consequence of stability in climatic factors over a long period of time. It is assumed that such stability produces a much wider ecological diversity in the spatial distribution of habitats and species.

There seems to be a strong correlation between the successional stage of an ecosystem and the magnitude of species diversity. For example, compared to an early successional ecosystem, a mature or climax community has always shown a greater degree of species diversity. For the same reason, the coral reefs among oceanic biomes and tropical rainforests among terrestrial ones have the largest species diversities of all ecosystems (Ramade, 1984). The fundamental reason for such a great diversity in these environments is their great age – millions of years – which demonstrates the great stability of the environments in which they were developed. Geographical isolation has been regarded as a mechanism of speciation and of lowering species diversity. Other things being equal, an island harbors fewer plants and animal species than a continental region of the same area. It is important to understand the relationship between diversity and other ecological variables (stability, time, productivity, etc.) to get an idea about why the protection and conservation of biodiversity is so critical for the survival and flourishing of the biotic community, including human beings.

1.5.1 Relationship Between Diversity and Stability

The relationship between diversity and stability is such that the greater the stability of ecological factors, the greater the diversity of the ecosystem. For example, biocoenosis (interacting organisms living together in a specific habitat) coral reefs and tropical rainforests that have evolved in a stable environment, or have suffered only regular and foreseeable fluctuation, show maximum species diversity. Ecologists argue that human activity, because of its unforeseeable and acyclic nature, inevitably decreases species diversity. The capacity of a community for homeostasis grows with its diversity because there will be a larger number of links in the "food web" and more organisms with redundant functions. Contrary to the stable and diversified climax ecosystems, young ecosystems with low diversity possess communities with fewer species. The species in such ecosystems are characterized by an "r-type" strategy and are usually subjected to the fluctuation of population when environmental conditions change. The absence or elimination of a species with equivalent functions may make the homeostatic regulation of a community uncertain or impossible.

1.5.2 Relationship Between Diversity and Time

This relationship relates to the effects of time on the species diversity and is manifested in a fundamental ecological phenomenon called "succession." Ecosystem diversity increases as a function of time from an initial state to a more climax state, or the diversity of an ecosystem increases as a function of time from a minimum to a maximum value. An ecosystem achieves maximum diversity when it reaches the "climax." The application of this relationship can be seen when a given ecosystem is allowed to redevelop naturally after human intervention. For example, the rangelands that have been overgrazed by domestic animals and the moorlands produced by excessive clearing of primitive forest cover, when allowed to redevelop naturally without any human intervention, may evolve continuously but not necessarily toward its earlier state characterized by maximum species diversity.

1.5.3 Relationship Between Biomass Productivity and Diversity

The biomass productivity ratio in an ecosystem is proportional to the diversity available in the system. In essence, diversity is a measure of the degree of the organizational complexity of an ecosystem. In other words, diversity is a measure of the degree of "negentropy." Hence, it is linked to the amount of energy (B) stored in the biomass and to the flow of energy per unit time (P) in the ecosystem. The time during which a given quantity of living matter and energy remains in the system is given by the equation:

$$t = K\frac{B}{P}$$

where k is a constant. The ratio B/P consists of time dimension because, for example, B can be expressed in kcl per hectare and P in kcl per hectare per year. The equation implies that the time taken by the amount of energy to traverse an ecosystem will be greater with greater diversity. Diversity can also be viewed as a measure of the complexity of a food web. The energy path from producers through to super carnivores and detritivores will take longer in a food web than a short linear chain of only producers to herbivores. Hence, diversity D can be linked to the ratio B/P by the equation

$$D = K\frac{B}{P}$$

where K is a constant.

1.5.4 Relationship Between B/P Ratio and Diversity

This principle is essentially derived from the above equation and states that the biomass/productivity (B/P) ratio grows as a linear function of diversity and/or time up to an asymptotic limiting value. The rearrangement of the above equation shows that

$$\frac{B}{P} = \frac{D}{K}$$

so that B/P and D are both proportional.

Empirical observations on ecosystem succession seem to indicate that this principle is valid. Ramadae points out that the B/P ratio is low for a field of cereal or grass land and is maximum in tropical forests. This can be explained by the fact that during ecological succession, the increase in diversity from the initial stage to the "climax" stage is accompanied by a growth in the B/P ratio as a function of time.

1.5.5 Diversified and Less Diversified Ecosystems

Mature climax ecosystems have been observed to remove energy and matter from the less diversified systems that surround them. A classic example is that of forest–savanna interface in tropical regions. Forest animals seek food in the neighboring open space ground where grass is the predominant ground cover. By grazing on this open ground, the animals prevent the savanna from evolving into a more diversified stage: because the removal of seeds by birds, trampling of young trees by the animals prevent the advanced stage of succession from evolving.

1.5.6 Ecological Variables and Resources

A resource, in simple terms, can be defined as a form of energy and/or matter, which is essential for the functioning of organisms, populations, and ecosystems. "In the particular case of humans, a resource is any form of energy or matter essential for the fulfillment of physiological, socioeconomic, and cultural needs, both at individual and community levels." The five fundamental ecological variables that can be regarded as the natural resources, namely, "energy," "matter," "space," "time," and "diversity," all fit the definition of a resource. Hence, all the theories about the use of natural resources should depend on the laws that govern the interaction and variations in these five variables.

1.6 Emerging Concepts in Ecology

Eminent ecologists have advanced some emerging concepts of ecology. Not all these concepts are universally acclaimed and accepted, but I strongly believe that these concepts will help us understand the complex nature of ecological systems and their behavior and why their protection, conservation, and sustainable use is imperative given that it is upon this that lies the very existence of humanity and the biotic community of the planet.

Concept 1. *Ecosystem*: An ecosystem is a thermodynamically open, far-from-equilibrium system. This concept states that input and output environments are an essential part of a system. For example, when considering a forest tract, what is coming in is as important as what is inside the tract. The same holds true for a city. It is not a self-contained unit either ecologically or economically; its future depends as much on the external life support environment as on the activities within city limits (Odum, 1983; Patton, 1972; Prigogine et al., 1972).

Concept 2. *The source–sink concept*: One area or population (the source) exports to another area or population (sink). This statement is a corollary to concept 1. It is applicable at both the ecosystem and population levels. At the ecosystem level, an area of high productivity (a salt marsh, for example) may feed an area of low productivity (adjacent coastal waters). At the population level, a species in one area may have a higher reproductive rate than needed to sustain the population, and surplus individuals may provide recruitment for an adjacent area of low reproduction. Food chains may also involve sources and sinks (Lewin, 1989; Pulliam, 1988).

Concept 3. *Hierarchical organization*: In a hierarchical organization of ecosystems, species interactions that tend to be unstable, in nonequilibrium, or even chaotic are constrained by the slower interactions that characterize a large system. Short-term interactions, such as interspecific competition – the evolutionary arm race between a parasite and its hosts, herbivore–plant interactions, and predator–prey activities – tend to be oscillatory or cyclic. Large, complex systems – such as oceans, the atmosphere, soils, and large forests – tend to go from randomness to order and tend to have more steady-state characteristics, for example, the atmosphere's gaseous balance.

Accordingly, large ecosystems tend to be more homeostatic than their components. This principle may be the most important of all because it warns that what is true at one level may or may not be true at another level of organization. Moreover, if we are serious about sustainability, we must raise our focus to management and planning of large landscapes and beyond (Allen & Starr, 1982; Kauffman, 1990; Ulanowicz, 1986).

Concept 4. *Environmental stress*: The first sign of environmental stress usually occurs at the population level, affecting especially sensitive species. If there is sufficient redundancy, then other species may fill the functional niche occupied by the sensitive species. Even so, this early warning should not be ignored because the backup components may not be as efficient. When stress produces detectable ecosystem-level effects, the health and survival of the whole ecosystem is in jeopardy. This idea is a corollary of concept 3: parts are less stable than the whole (Odum, 1985, 1990; Schindler, 1990).

Concept 5. *Feedback*: Feedback in an ecosystem is internal and has no fixed goal. There are no thermostats, chemostats, or other set point controls in the biosphere. Cybernetics at the ecosystem level thus differs from that at the organism level (body temperature control, for example) or that of the human-made mechanical systems (temperature control of a building, for example) where the control is external with a set point. Ecosystem control, where manifested, is the result of the network of internal feedback processes, as yet little understood – another corollary of concept 3 (Patten & Odum, 1981).

Concept 6. *Natural selection*: Natural selection may occur at more than one level. This idea is another corollary to concept 3. Accordingly, coevolution, group selection, and traditional Darwinism are all part of the hierarchical theory of evolution. Not only is the evolution of a species affected by the evolution of an interacting species, but a species that benefits its community has a survival value greater than a species that does not (Axelrod, 1980, 1984; Axelrod & Hamilton, 1981; Gould, 1982; Wilson, 1976, 1980).

Concept 7. *Competition, mutualism, and adaptation*: There are two kinds of natural selection or two aspects of the struggle for existence: organism versus organism, which leads to competition, and organism versus environment, which leads to mutualism. To survive, an organism does not compete with its environment as it might with another organism, but it must adapt to or modify its environment and its community in a cooperative manner. This concept was first suggested by Peter Kropotkin soon after Darwin (Gould, 1988; Kropotkin, 1902).

Concept 8. *Competition may lead to diversity rather than extinction*: Although competition plays a major role in shaping the species composition of biotic communities, competition exclusion (in which one species eliminates another, as in a flour beetle microcosm) is probably the exception rather than the rule in the open system of nature, where species are often able to shift their functional niches to avoid the deleterious effects of competition (den Boer, 1986).

Concept 9. *Resource scarcity, mutualism, and cooperation*: Evolution of mutualism increases when resources become scarce. Cooperation between species for mutual benefit has a special survival value when the resource gets tied up in the biomass, as in mature forests, or when the soil or water is nutrient-poor, as in some coral reefs or rainforests (Boucher et al., 1982; Odum & Biever, 1984). The recent shift from confrontation to cooperation among the world's superpower nations may be a parallel in societal evolution (Kolodziej, 1991).

Concept 10. *Negative and positive interactions*: Indirect effects may be as important as direct interactions in the "food web" and may contribute to network mutualism. When food chains function in food web networks, organisms at each end of the trophic series (for example, plankton and bass in a pond) do not interact directly but indirectly benefit each other. Bass benefits by eating planktivorous fish supported by plankton, whereas planktons benefit when bass reduce the population of its predators. Accordingly, there are both negative (predator–prey) and positive (mutualistic) interactions in a food web network (Patton, 1991; Wilson, 1986).

Concept 11. *Environmental adaptation and modification*: Since the beginning of life on Earth, organisms have not only adapted to physical conditions but have also modified the environment in ways that have proven to be beneficial to life in general (e.g., increase oxygen (O_2) and reduce carbon dioxide (CO_2)). This modified Gaia hypothesis is now accepted by many scientists. Especially important is the theory that microorganisms play major roles in vital nutrient cycles (especially the nitrogen cycle) and in atmospheric and oceanic homeostasis (Cloud, 1988; Lovelock, 1979, 1988; Kerr, 1988; Margulis & Olendzenski, 1991).

Concept 12. *Energy flow*: Heterotrophs may control energy flow in food webs. For example, in warm waters, bacteria may function as a sink, in that they short-circuit energy flows so that less energy reaches the ocean bottom to support demersal fisheries. In cooler waters, bacteria are less active, allowing more of the fruits of primary production to reach the bottom (Pomeroy, 1974; Pomeroy & Diebel, 1986; Pomeory & Wiebe, 1988). Small heterotrophs may play similar controlling roles in terrestrial ecosystems such as grasslands (Dyer et al., 1982, 1986). This concept is a corollary of concept 11.

Concept 13. *An expanded approach to biodiversity*: An expanded approach to biodiversity should include genetic and landscape diversity, not just species diversity. The focus on preserving biodiversity must be at the landscape level because the variety of the species in any region depends on the size, variety, and dynamics of patches (ecosystems) and corridors (Odum, 1982; Turner, 1988; Wilson, 1998).

Concept 14. *Autogenic succession*: Ecosystem development or autogenic ecological succession is a two-phase process. Early or pioneer stages tend to be stochastic as opportunistic species colonize, but later stages tend to be more self-organized (perhaps another corollary of concept 3; Odum, 1989a).

Concept 15. *Carrying capacity*: Carrying capacity is two-factor concept involving the number of users and the intensity of per capita use of the resource. These characteristics track in a reciprocal manner – as the intensity of the per capita impact goes up, the number of individuals that can be supported by a given resource base goes down (Catton, 1987). Recognition of this principle is important for estimating the carrying capacity of humans at different quality-of-life levels and in determining how much buffer natural environment to set aside in land-use planning. This is equally true in drawing resources by anthropogenic activities from the planetary ecosystem for human consumption.

Concept 16. *Nonpoint pollution*: Input management is the only way to deal with nonpoint pollution. Reducing waste in developed countries by source reduction of the pollutants will not only reduce global-scale pollution but will also spare resources needed to improve the quality of life in underdeveloped countries (Odum, 1987, 1989).

Concept 17. *Energy expenditure*: An expenditure of energy is always required to produce or maintain energy flow or a material cycle. According to this net energy concept, communities, and systems, whether natural or human-made, as they become larger and more complex, require more of the available energy for maintenance (the complexity theory). For example, when a city doubles in size, more than double the energy (and taxes) is required to maintain order (Odum & Odum, 1981; Pipenger, 1978).

Concept 18. *Bridging the gap between unsustainable and sustainable resource use*: There is an urgent need to bridge the gap between human-made and natural life support goods and services and between non-sustainable short-term and sustainable long-term management. Agroecosystems, tropical forests, and cities are of special concern. H.T. Odum's "emergy" concept and Daly and Cobb's index of sustainable economic welfare are examples of recent attempts to bridge these gaps (Daly & Cobb, 1989; Folke & Kaberger, 1991; Holden, 1990; Odum, 1988).

Concept 19. *Transition cost*: Transition costs are always associated with major changes in human affairs. The society has to decide who pays, for example, the cost of new equipment, procedures, and education in changing from high- to low-input farming or when converting from air-polluting to clean power plants (Renner, 1991; Spencer et al., 1986).

Concept 20. *Parasite–host model*: A parasite–host model for humans and the biosphere is a basis for turning from exploiting Earth to taking care of it (going from dominionhood to stewardship, to use a biblical metaphor). Despite or perhaps because of technological achievements, humans remain parasitic on the biosphere for life support. Survival of a parasite depends on reducing virulence and establishing reward feedback that benefits the host (Alexander, 1981; Anderson & May, 1981, 1982; Levin & Pimentel, 1981; Washburn et al., 1991). Similar relationships hold for herbivory predation (Dyer et al., 1986; Lewin, 1989; Owen & Wiegert, 1976). In terms of human affairs, this concept involves reducing wastes and destruction of planetary ecosystem to reduce human virulence, promote the sustainability of renewable resources, and invest more in Earth care.

Chapter 2
Importance of Biodiversity, Ecosystems, and Ecosystem Services

.... It is the self-organizing ability of the system or more particularly the resilience of that self-organization, which determines its capacity to respond to the stresses imposed by predation or pollution from external sources. The importance of biodiversity lies in its role in preserving ecosystem resilience.

Charls Perrings (1992)

2.1 Introduction

Biodiversity (species richness) is Earth's most important living system that has resulted as a consequence of billions of years of the evolutionary process ever since the origin of the first life on planet Earth. Plants, Animals, and a myriad of microbial organisms and species are not only integral parts of natural ecosystems but are also essential components of the economic and cultural life of human beings. Biodiversity has provided the basic foundation for the origin and evolution of human civilization. Thousands of species of plants and animals supported the development of early societies, thus providing the basis for evolution from hunting and gathering to subsistence farming to present-day agricultural and industrial levels of organizations. Ever since humans began living in social groups, right after the hunting and foraging stage, they systematically domesticated both plant and animal species on which depended their very survival. In this process, they discovered and developed agriculture upon which the human civilization was founded and flourished.

When we talk about biodiversity, usually the larger and more conspicuous species of animals and plants attract immediate human attention, but there are thousands or even millions of smaller and less conspicuous species of animals and plants, such as fungi, bacteria, and arthropods, which are equally or perhaps more important than the larger ones when it comes to maintaining the structural integrity and health of an ecosystem (Piementel et al., 1992; Miller, 1991; Perring et al., 1992). It is difficult to directly observe their functions and roles because this requires knowledge and understanding of the structural and functional linkages of all the

G. Upreti, *Ecosociocentrism*, https://doi.org/10.1007/978-3-031-41754-2_2

ecosystem components and the trophic level and food web that relate to the biotic community. For example, the energy and material recycling that is essential for ecosystem functioning is mostly carried out by small microorganisms. The larger species are dependent upon the interrelated service networks of thousands of these smaller species. The smaller species are highly specialized and occupy a specific niche in the system, so if there is a sudden change in their niche environment, they cannot perform their function and may perish or, if they survive, must adapt to a different niche environment. Here, the important question is does this change have any bearing on other species or life-forms in the system? Of course, the answer is "yes." First, there may be some organisms that were directly dependent upon the services of those that perished or adapted to a different niche environment. These organisms then have only two choices, either perish or replace the energy and materials obtained from the lost source by adopting a different survival strategy (Odum, 1969, 1977; Ehrlich and Ehrlich 1992). The same choice must be made by all organisms that were dependent upon those that adopted a different survival strategy, and this continues up to the last trophic level in the pyramid of biotic life.

The pyramid of the biotic community and the associated food web in natural ecosystems is so intricate and complex that a quantitative change at one tropic level may cause both quantitative and qualitative changes at another level. The whole system is so interconnected and interactive that synergistic relationships are widespread and feedback mechanisms maintain the system. For this reason, system ecologists do not accept the reductionist's thesis that the total effect is simply the sum of the effects of its components because the components interact synergistically. It is in this context that the importance of biodiversity becomes apparent. Economists consider only tangible benefits, i.e., the commodity values determined by market forces, and, consequently, often overlook the value of the life support services and materials provided by the diverse biotic community in the ecosystem, even though it is the source of all the material inputs and services for the production of man-made goods and utilities (Gee, 1992; Reid & Miller, 1989; Upreti, 1996). In other words, all material inputs and services for the production of man-made goods are derived from diverse life-forms operating in natural ecosystems for which there is no system of economic valuation. Mankind will enter a new dimension of development, where the life support services and material inputs of natural ecosystems are considered part of the economic production and valuation system and the principle of opportunity cost is applied to maintain a healthy natural ecosystem.

2.2 Biodiversity and Its Importance

Although the term "biodiversity" describes genetic diversity that exists both within and between species indicating the relative number and abundance of species in a given unit area of a natural habitat, Gee (1992) argues that a more useful and meaningful definition conceives biodiversity as the sum of interactions between species rather than a list of species themselves. Biodiversity, as a system of interaction, implies that the biophysical environment is a product of the interactions between diverse life-forms over millions of years from the very first life on Earth. The gaseous atmosphere, soil, and freshwater, the physical basis for our survival, have all evolved over millions of years of synergistic interactions within and between the biotic and abiotic communities. As Gee (1992) eloquently points out, pollution of the atmosphere, denudation of soil, and diminution of supplies of freshwater in many parts of the world could be seen, in part, as consequences of the loss of biodiversity. The World Resources Institute (Reid & Miller, 1989) also conceives biodiversity as a system of interrelated parts and components: "Biodiversity is the variety of the world's organisms, including their genetic diversity and the assemblages they form. It is the blanket term for natural biological wealth that under girds human life and wellbeing. The breadth of the concept reflects the inter relatedness of genes, species, and ecosystems. Because genes are components of the species, species are the components of ecosystems, altering the makeup of any level of this hierarchy can change the others."

Irrespective of the level at which we conceive biodiversity, species constitute the central unit in the concept of biodiversity and community structures within an ecosystem. A species may comprise a number of subspecies, races, and populations, each of which constitutes a distinct reservoir of gene pools upon which natural selection acts to determine the evolutionary course of the organism and the biotic community. When these gene pools are reduced by human intervention in a natural ecosystem, a species or population may be subjected to extinction because it lacks the adequate amount of genetic variability to cope with changes in the environment. Thus, anthropogenic activities cause losses of species through both ecosystem destruction (deforestation, logging, and grazing) and detrimental changes in the environment from air, land, and water as a result of chemical and waste products used in agricultural and other related management operations. Biological diversity not only benefits agriculture, forestry, medicines, and industrial human subsystems but also plays an essential role in maintaining evolutionary processes and ecological integrity (Pimentel et al., 1992; Upreti, 1996). Ecosystem evolution, development (a natural biological process), and maintenance may all be severely constrained in the absence of ample biodiversity, which may be quite antagonistic to humankind's long-term survival and benefits. For example, a desert ecosystem is far less productive than temperate and tropical rainforest ecosystems in terms of providing life support services and materials to its inhabitants.

Many of today's desert landscapes were once believed to be a part of productive natural ecosystems. Anthropocentric activities were partly responsible for degrading and finally converting these productive ecosystems into less productive desert-like ecosystems. The demise of the Mayan civilization in Central America (northern Guatemala), the ancient culture of Easter Island, and a bunch of Oak trees that survived from the old climax vegetation in the Middle East (Lebanon, Syrian) bear a powerful testimonial to this process (Ponting, 1990; Upreti, 1994). Destruction of vegetation accelerates the loss of topsoil, which can sustain regrowth or alter the micro environment to permit the establishment of species and harvesting of crops without replacing the nutrients removed from the soil. This ultimately leads to a less productive and degraded state of land. At present, anthropocentric transformation of rich natural ecosystems into less productive and degraded ecosystems has been occurring on a massive scale all over the world, particularly in Africa, Asia, and South America, primarily arising from the need to meet the basic physiological needs of the people living in these areas. Literatures seem to indicate that the gradual desertification of many parts of these continents is already in progress (Upreti, 1987; Erickholm, 1975; Ponting, 1990). It is in the interest of mankind to prevent these rich ecosystems of diverse life-forms from being destroyed and degraded and gradually lead to the emergence of desert ecosystems.

The importance of biodiversity and ecosystem services to human beings is tremendous, and its true potential for the well-being of human civilization cannot be realized unless it is considered in the right perspective. This requires both quantitative compilation of information and knowledge from different disciplines and qualitative assessment and correct interpretation of such knowledge. This section attempts to provide some perspectives on why biodiversity as a living system, ecosystems, and ecosystem services are so important for the immediate and long-term well-being of humankind and for the biotic community as a whole.

2.2.1 Instrumental Values

Instrumental values include all those values of the direct or indirect use of resources. The opportunity cost of a resource is the measure of its value in a particular use. The more useful or more productive the resource, the greater its value (Perring et al., 1992). Usually, the price of a resource is an approximation of its value, but, as Perring et al. (1992) argue, this is not the case with many organisms because markets for the services produced by these organisms, particularly ecological services, do not exist. This nonexistence of market values for the ecological services provided by diverse life-forms seems to be a major obstacle for the conservation of biodiversity and habitats; nevertheless, the instrumental values (direct use value) of biological diversity can be summarized under the following major categories:

2.2.1.1 Agriculture

In 1989, Reid and Miller of the World Resources Institute reported that approximately 40% of the tropical forests of the world have already been destroyed and 20 million hectares of forest land are being converted to agriculture every year. Biodiversity plays an essential role in the maintenance of a productive agroecosystem. There are diverse forms of soil microorganisms, which are constantly involved in recycling material and energy to enrich the soil and maintain its high productivity. The total nitrogen fixed per year in the soil from the atmosphere by diverse groups of microbes in the United States and the world has been

Table 2.1 Biomass of various organisms per hectare in a temperate region pasture

Organisms	Biomass (kilogram fresh weight)
Plants	20000
Fungi	4000
Bacteria	3000
Arthropods	1000
Annelids	1320
Protozoa	380
Algae	200
Nematodes	120
Mammals	1.2
Birds	0.3

Source: Adapted from Piementel et al. (1992)

estimated to be more than $7 and $50 billion, respectively (Piementel et al., 1992). An agroecosystem that promotes and maintains this diversity will always be at a high state of productivity in terms of total biomass and energy production (Table 2.1).

The plant kingdom is the source of the very roots of the life of human existence, yet only about 20 species provide more than 80% of the world's food; 3 of which, namely, corn, wheat, and rice, contribute about 65% of the total food supply (Carson, 1990). The hazards of depending on such a narrow range of biodiversity were further illustrated by the severe corn blight epidemic in the United States in 1970. More than 70% of the corn grown in the United States were hybrids with the same cytoplasm, called "Texas" male-sterile cytoplasm. When a corn blight pathotype that was especially virulent on plants with this type of cytoplasm arose and weather conditions highly favorable for the disease occurred, more than 15% of the entire crop and as much as half of the crop in some southern states was lost to the disease. The total loss was estimated to be more than $2 billion (Carson, 1990). The next year, another severe loss was averted only through the crash development of other hybrids using other types of cytoplasm (and the fortuitous lack of favorable weather conditions for the disease). There are other similar examples of vulnerability occurring as a direct result of a lack of biodiversity in our crops.

Present-day economically important crops and animals have been genetically improved by plant and animal breeders, so they are greatly superior to their predecessors in terms of yield, quality, and resistance to pests, drought, frost, severe cold, and other factors. The genes for this improvement have come from and will continue to come from the diverse gene pools found in wild or unimproved cultivated populations of the economically important crops and animals. Modern technology cannot supplant the need for the continued availability of a wide range of genetic materials. Thus, it is highly imperative to conserve and protect the habitats of wild relatives of domesticated plants and animals to ensure that a large, diversified gene pool remains available for future work to genetically improve domesticated crops and animals.

2.2.1.2 Medicines

Although biodiversity in natural ecosystems is important for agricultural crops, it is even more important and valuable for medical proposes. It has been estimated that more than 40% of the prescription drugs sold in the United States contain pharmaceuticals and chemicals originally found in wild species: about 25% of these drugs come from plants, another 12% from fungi and bacteria, and 6% from animals. The value of medicinal products derived from biological organisms has been estimated to be $40 billion a year (Myers, 1984). There are still a vast number of plant and animal species whose prospects have not yet been evaluated, which may possess tremendous potential medicinal values. There may be numerous terrestrial plant and animal species whose medicinal and agricultural potential still needs to be explored and tapped. If these unstudied species are not protected and conserved, their potential contributions to human health and well-being may remain undiscovered and be lost forever.

2.2.1.3 Industry

The biotic community provides raw materials for many essential industrial products. Timber and other wood products, including lumber, paper, and wood-based chemicals, are the most economically important category of industrial products derived from living resources. Rubber, another major industrial product, is also derived from trees. Natural rubber derived from plants constitutes one-third of the current world use because of its superior quality. Other examples of plants, animals, and other biotic life that contribute raw materials for industrial products include bamboo, teak, musk deer, and many animal species for their leather and hides. As with medicine, there are a vast number of species whose prospects have not yet been evaluated, which may remain undiscovered and will be lost once and for all along with the loss of the species.

2.2.1.4 Scientific Value

Studies on species diversity, ecosystem functioning and development, and evolution have provided models that seek to test, confirm, and verify the principle of coexistence between man and Nature. It has long been established that the immature or early successional stage in the development of an ecosystem has a higher net productivity for human needs than the late successional or mature stage, in which the net productivity tends to fall to near zero (Odum, 1969). Agriculture and forestry are based on this characteristic of ecosystems. On the other hand, mature ecosystems (though their net productivity is low) are known to serve other functions important to man. Odum (1969) describes them as "protective ecosystems" in which, for example, there is a high rate of carbon dioxide/oxygen (CO_2/O_2) exchange; a closed or nearly closed system of nutrient cycles; more qualitative changes than quantitative; high ecological efficiency in terms of biomass supported per unit of energy flow; maximum information content; and a minimum rate of entropy gain. Our studies on ecosystem behavior and evolutionary processes on a larger spatial and temporal scale are virtually nonexistent. These studies cannot be conducted in a laboratory.

The ultimate verification of ecological theory, ecosystem functioning and behavior, and evolutionary processes must come from studies carried out in Nature. Protected diverse natural ecosystems could play a key role in understanding the ecological and evolutionary processes and ecosystem functioning and behavior. Biological sciences, until now, have identified and studied only a small fraction of the biotic community. Millions of species still remain to be identified, and their potential uses and values are yet to be discovered. We should not risk losing species with potential values that are not quantified and monetized. Who knows what the price and monetary value of a plant or animal or any other living organism that may have the potential to cure cancer and AIDS would be? How will they exist for future generations of mankind and future scientific studies, if they cannot exist now for their own sake?

2.2.1.5 Aesthetic and Spiritual Values

National parks, reserves, protected wilderness areas, or other natural ecosystems provide valuable aesthetic and spiritual services to humanity. It is in the nature of mankind to curiously observe, admire, and commune with Nature. Nature has inspired great thinkers, writers, and artists in many eras and across all cultures. The works of these thinkers and artists are full of wisdom and creative ideas that have enormously impacted and enriched human civilization. It is amazing how the ideas expressed and emphasized in the writings and works of these thinkers from all over the world coincide, regardless of their cultural backgrounds. They all seem to convey the same idea and message, which is the portrayal of a gentle and harmonious relationship between humans and Nature. I do not see any fundamental difference in the ways in which Nature was regarded by the American Indians or by Thoreau. Thoreau wanted to be close to and live with Nature in a way that could nurture mutualism and promote harmony between humans and Nature, which is identical to the way of life the American Indians have followed for centuries, though the dominant American culture does not recognize it. Similar ideas and concepts are found in the Eastern cultures also, particularly in Buddhism, Hinduism, and Taoism.

The aesthetic, recreational, and spiritual values of protected wilderness areas, parks, and Nature reserves are becoming progressively more significant and essential as a result of today's rapid urbanization, hectic timing, and mechanical lifestyles. Human beings can hardly find time to engage in speculative and imaginative activities while adapting to life in overcrowded, noisy, and utterly artificial environments. Thus, they have become alienated from Nature, and, on realizing this, they develop a longing for what they have missed. That is why parks, Nature reserves, and wilderness areas are often flooded by people from urban areas. They are the repositories of the aesthetic and spiritual values and are the inspirational source for creative human imagination.

2.2.1.6 Evolutionary Values

The loss of biodiversity and ecosystems through habitat destruction will lead to a reduction in the gene pool, which will severely affect the processes of evolution and speciation. A reduced gene pool will not only limit the potential for evolution but also possibly terminate certain evolutionary processes (Odum, 1969). When a population is reduced to a small size, it becomes vulnerable to natural variations in climate, disease outbreaks, and anthropogenic disturbances. Three biologically important processes known as inbreeding, genetic drift, and the bottleneck and founder effects would become operational as a result of the reduction in genetic variability in a small population. A brief discussion of these processes is presented here to show how they determine the evolutionary fate of such a population.

2.2.1.6.1 Inbreeding

Inbreeding refers to the mating between genetically closely related individuals. This happens within a population because when there are only a few individuals, the possibility of them being related is extremely high. Inbreeding does not cause changes in allelic frequencies but does change genotype frequencies. It increases the proportion of homozygotes to heterozygotes, thereby increasing the expression of the recessive variants in the population (Miller, 1979). Since genes for defective, nonadaptive, and lethal traits are usually recessive and hidden in heterozygotes, inbreeding leads to the exposure of these characters in the population. Apart from this phenomenon, natural selection will have to act on a gene pool with a reduced proportion of heterozygotes and may produce a reduction in variability (Dobzhansky, 1970). The net effect of inbreeding in a small population will be the manifestation of more weak, lethal, and nonadaptive recessive individuals and a reduction in the genetic survivability of the population. The deleterious effects of inbreeding in small populations of plants, animals, and humans have been documented (Myers, 1984; Miller, 1979; Dobzhansky, 1970).

2.2.1.6.2 Genetic Drift

Genetic drift can be defined as the change in gene frequency that results, in each generation, from the random sampling of alleles in a panmictic population (Miller, 1979). The random drift of gene frequency in a small population can lead, over generations, to the fixation or loss of the alleles within the population. When this happens, the population is said to have no genetic variability or plasticity. Such a population faces extinction under the strong pressure of natural selection and unfavorable environmental conditions.

In a panmictic (randomly interbreeding) population, the gametes for each generation are a random selection of all the possible gametes in the gene pool. When the gene pool is large, the chance variability that occurs in this random process is extremely small in relation to the size of the gene pool and the population remains stable unless affected by other factors. In a small population, however, the chance variability can be large in relation to the smaller gene pool and the frequencies in the population can change over time just due to chance. Again, just by chance, the changes over time can lead to fixation of certain alleles, loss of other alleles, a reduction of variability, and a higher risk of extinction due to the loss of genes needed to adapt to new environmental conditions.

2.2.1.6.3 Bottleneck and Founder Effects

Miller (1979) describes the bottleneck and founder effects as genetic changes that may arise when a population temporarily becomes reduced to a fraction of its usual size or when a few individuals establish a population in a new area. In both situations, the few individuals may possess and pass on only a fraction of the genetic endowment of the original population, which may result in a significant loss in genetic variability (Mayr, 1954). The degree of variability loss in the succeeding population is directly related to the reduction in population size. Both bottleneck and founder effects cause a decline in the average heterozygosity per locus. A low, effective population size during a bottleneck or the foundering of a new population may result in a significant, long-term reduction in average heterozygosity and a lowering of the fitness of the population.

There is a strong positive correlation between diversity and the effective population size of the organism or species. In general, the higher the population size, the greater is the gene pool and the genetic variability. Since genetic variability is the basic input upon which natural selection and evolutionary forces act, it has a direct effect on refining, processing, shaping, and determining the course of evolution of an organism or a species.

2.2.1.7 Educational Values

Protected habitats, parks, and Nature reserves provide a great educational service to humankind. They can be used to foster an understanding of environmental values and awareness and can be made available to carry out environmental studies. The use of these protected areas by educational communities, especially, will have a "multiplier" effect in communicating the value of Nature and the need for its conservation. The ultimate goal of environmental education is to achieve an understanding of man's intricate relationship with the natural environment and teach humankind how to live in harmony with Nature. National parks, nature reserves, and any protected natural habitat serve as living natural laboratories for this kind of nature and environmental study. The public in general, and young generations in particular, must have the opportunity to be able to enjoy, observe, and experiment in natural laboratories, which cannot be replicated or simulated by any man-made environment.

2.2.1.8 Ecosystem Service Values

Biological diversity must be regarded as a system of individual organisms and species interacting among themselves and with their environment. The importance of biodiversity as a system must be understood in terms its contribution to the generation of ecosystem services. Hence, the most compelling argument for the protection of biodiversity and ecosystems emerges from the facts of their involvement in the generation and delivery of ecological services and goods such as material input and energy flow, ecosystem stability, waste assimilation, water, soils, foods, energy, nutrients, and gasses (CO_2 and O_2). A brief discussion of the ecological services and material goods generated by biodiversity and ecosystems provides a better perspective on our understanding of why their protection is immensely important.

2.2.1.9 Material Input and Energy Flow

An ecosystem, like an economic system, depends on fixed stocks of material resources and components constantly being recycled throughout the system via food webs at the local level and biogeochemical cycles on the global scale (Rees, 1990). Ecosystems are inherently self-sustaining and self-organizing entities. They possess a fundamental property called autopoiesis, which, according to Rees, is the organizational property by which living systems continuously reproduce themselves. Autopoiesis arises from the complex interdependent relationships and flows linking the major components of the biosphere. The structural integrity of these relationships is essential for the production and maintenance of the participating components themselves.

Autopoiesis is related to the homeostatic behavior of the biosphere. Over geological time, life processes and environments of Earth have acted upon each other and modified and regulated each other and evolved accordingly. In this process, they have stochastically created and maintained conditions favorable to life processes by negative feedback mechanisms. As a self-regulating and self-producing system, the biosphere exhibits an important property of ecosystem dynamics: numerous positive and negative feedback mechanisms (Lovelock, 1992). As a constant supply of an external source of free energy, the Sun charges and drives ecosystems and makes autopoiesis an event of continuous occurrence. This steady stream of solar energy essentially sustains all biological activities and makes diversity of life on Earth possible through its conversion into chemical energy by photosynthesis.

Thermodynamically, photosynthesis is the most important materially productive process on the planet and is the ultimate source of all the renewable resources used by the human economy.

Since the flow of solar radiation is constant, steady, and reliable, resource production in the ecosystem is potentially sustainable over any timescale relevant to human beings; however, ecosystem productivity is limited by the availability of nutrients, photosynthetic efficiency, and rate of energy inputs. Unlike the economy, which expands through resource conversion and positive feedback, ecosystems are held in a steady-state or dynamic equilibrium by limiting factors and negative feedback. This distinction between ecosystem equilibrium and economic growth is important because human economic systems operate and grow within natural ecosystems, even though the consumption of ecological resources has exceeded the sustainable rate of biological regeneration (Costanza et al., 2014). Habitat destruction reduces energy flow in natural systems by simply eliminating a part of the system. Here, habitat destruction refers to the conversion of forests to croplands and pastures. Wright (1990) estimated the total reduced energy flow through natural systems due to habitat destruction to be 799 EJ. This amounts to a 30% reduction in the total photosynthetic energy flowing through natural ecosystems by anthropogenic intervention. The loss of biodiversity and degradation of ecosystems through habitat destruction causes not only the loss of energy flow in the natural ecosystem but also severely obstructs the functioning and the emergence of community-level properties such as trophic structures and successional stages. Brinck et al. (1988) argues that the restoration and redevelopment of degraded ecosystems must proceed to check the growing impoverishment and alienation of people. Human economic subsystems ought to be conducted in a manner that least disrupts the material and energy flow within an ecosystem. For this to happen, it is necessary that the diverse living systems that constitute the very fabric of an ecosystem must remain protected and functionally intact.

2.2.1.10 Stability and Resilience of Ecosystem Value

Both structural integrity and functioning of an ecosystem are sustained by synergistic feedback among the components (biotic and abiotic) of the system. The biological system and physical environments have acted upon and modified each other and have become interlocked and connected in a nonlinear web of interrelations with lags and discontinuities, thresholds, and limits (Perring et al., 1992). Solar energy drives the development, self-organization, and maintenance of the system through the cyclic use of materials and compounds. As Perring et al. (1992) state: "….it is the self-organizing ability of the system or more particularly the resilience of that self-organization, which determines its capacity to respond to the stresses imposed by predation or pollution from external sources. The importance of biodiversity lies in its role in preserving ecosystem resilience." From a short-term ecological perspective, some ecosystems may appear stable and resilient. Brinck et al. (1988) describe four transient phases, which are quite common in ecosystems, produced by natural or human-induced changes in the environment. These phases pass through four functions: exploitation, conservation, creative destruction, and renewal. Many ecosystems are deliberately kept in the exploitative phase until they are depleted of resources and destroyed beyond the limit of self-renewal. Controlled renewal and redevelopment must be initiated to bring these systems back to their functional phase.

We may argue about the inadequacy of empirical research to substantiate and account for the exact mechanism by which biological diversity confers productivity and resilience on an ecosystem, but there is no doubt that their elimination reduces a productive system to a degraded oligotrophic system and, ultimately, to a desert one. There must be a minimum level of biological diversity necessary to produce and sustain ecological services and goods by the continuous interactions between and among themselves and the environment. An ecosystem, like the population of an organism, requires biological variability to cope with changes in the physical environment. If a minimum level of variability is not maintained in the system, then succession cannot take place. Do we really need to conduct research to test this hypothesis? There are historical precedents that show this pattern.

2.2.1.11 Resource and Waste Assimilation Value

The interacting components of biologically diverse ecosystems (population and communities) provide all the necessary ecological services and resources that human societies require. Living and the nonliving systems are the two interacting complexes of an ecosystem. These systems have been acting, reacting, and responding or interacting with each other and evolving in an entangled relationship accordingly over millions of years. Interactions between living systems and nonliving systems have generated materials and ecological services that are vital for the maintenance and survival of the whole interacting biotic systems. These materials and services cannot be substituted by human ingenuity (innovative science and technology), and, even if they can be substituted, the cost of substitution would be enormous and unthinkable.

The material inputs that enter a human economic system from a natural ecosystem, after being processed and consumed, again enter the natural ecosystem as wastes. Hence, natural ecosystems not only serve as sources of resources for mankind but also serve as sinks for waste assimilation. The loss of biological diversity, thus, eventually leads to the loss of resources and ecological services (the assimilative capacity of a system). Ecologists believe that there is a critical threshold of biodiversity below which an ecosystem cannot function (Piementel et al., 1992; Ehrlich and Ehrlich 1992). The direct consequence of the loss of biological diversity can be measured by the corresponding loss of the ecological services of the natural ecosystems. Hence, the value of biological diversity and an ecosystem can be assessed by the valuation of the ecological services and goods generated by ecosystems (Gee, 1992).

2.2.1.12 Biogeochemical Cycle Value

Ecosystem destruction has been directly linked to the increase in carbon dioxide, which is a major contributor of the greenhouse effect and global warming. Carbon dioxide is released into the atmosphere from the combustion of deforested biomass, decreased photosynthesis by deforestation, and soil degradation. Carbon dioxide stabilization is primarily a biological process because the bulk of carbon dioxide flows through biotic processes (Goreau, 1990). Increased photosynthesis and carbon storage in vegetation, soils, and sediments is needed to stabilize carbon dioxide. Preventing the destruction of habitats and ecosystems and promoting the restoration and redevelopment of the degraded ecosystems will not only expedite the conservation of biological diversity but also provide the ecological basis for sustainable living.

The bulk of the biomass and photosynthesis occurs in the tropics, and, hence, balanced and sustainable development must be focused on the tropics to alleviate carbon dioxide buildup. Preservation of tropical forests is essential not only for the preservation of the biological diversity but also for the resilience and stability of the ecosystem's metabolism and maintenance of the constant energy flow within the biosphere. The effects of tropical deforestation and habitat destruction do not limit the release of carbon dioxide from burning forest biomass and decreased photosynthesis. Recycling of atmospheric carbon dioxide through photosynthesis, biota, and soil is the major method to control carbon dioxide levels in the atmosphere (Goreau, 1990; Houghton, 1990). A decreased photosynthetic capacity of vegetation causes the released carbon dioxide to remain longer, reach a higher level in the atmosphere, and absorb more heat, ultimately contributing to global warming and the greenhouse effect.

2.2.1.13 Regulating Hydrological Cycle Value

We know that natural vegetation cover on water catchments regulates and stabilizes water runoff. The root mass of vegetation cover makes soil more permeable to rainwater, thereby significantly lowering the runoff as compared to a cleared barren land. A large portion of rainwater is absorbed and retained by the soil mass under vegetation cover. As a result of this process, streams in forested regions continue to flow even in dry weather and floods are significantly minimized during rainy weather. The peak runoff per unit area of forested catchment in Malaysia has been found to be half of a rubber and oil palm plantation (Daniel & Kulasingham, 1974). McNeely et al. (1990) report that the cloud forest of La Tigra National Park in Honduras produces well-regulated, high-quality water flow throughout the year, supplying 40% water requirement to the capital city of Tegucigalpa. It is important to understand the roles played by biodiversity and ecosystems or vegetation cover in catchment areas in regulating and stabilizing the hydrological cycle that is as vitally important as drinking and irrigation water for the people living in the region.

Likewise, the role of the Amazon forest and tropical rainforests in regulating the hydrological cycle at the global level cannot be overemphasized. Ecologists (Houghton, 1990; Miller, 1991; Perring et al., 1992; Daniel & Kulasingham, 1974; Ehrlich & Mooney 1983; Costanza et al., 2014; Upreti, 1994) have warned that destruction of the Amazon forest and tropical rainforest will have unimaginable consequences on the regulation of the global hydrological cycle. Ecologists and climatologists (Houghton, 1990; Miller, 1991; Perring et al., 1992; Sage, 2020) have pointed out that the Amazon forest is a major player in determining global climate. Since it pulls the most important greenhouse gas from the air and stores it, transpires water, and creates clouds that carry moisture around the world, it plays a dominant role in the regulation of the global hydrological cycle. It provides ecological services and is home to much of the world's biodiversity. Rowan Sage (2020) points out that as the forest is degraded and destroyed, the power of the Amazon to mitigate global climate change is weakened, and it is adversely affected by the implications of climate change. Empirical studies have established that when deforestation increases, global warming is exacerbated, global temperature increases, and precipitation patterns are altered. According to Lovejoy and Nobre (2018), the Amazon rainforest is indeed highly vulnerable to deforestation, and they have warned that it could reach a tipping point if approximately 20–25% of it is deforested. The repercussions of this would be large-scale, transforming much of the forest into a savannah-like ecosystem, leading to drastic changes in regional climate, including increases in temperature and alterations to precipitation patterns. Amazonia is the region most affected by the effects of deforestation and would experience a temperature increase of up to 3.8 °C and a precipitation decrease of around 15% of annual rainfall if deforestation were to continue (Sage, 2020).

Because of the erosion and destruction of biodiversity as a system, the integrity and resilience of our planet's ecosystems are declining rapidly as witnessed by the collapse of biomes such as coral reefs, grasslands, and offshore fisheries. Loss of ecosystems and biodiversity will reach a stage that is great enough to threaten the biosphere itself, since these two are the elements that give the biosphere its structure and function. Humanity can avoid such inevitabilities of mass extinction with wise management policies and strategies; however, the piecemeal approach of saving a species here and there will not prevent such inevitabilities. It requires a landscape and ecoregion level of management strategies to protect and conserve biodiversity and ecosystems (Upreti, 1994, 1996).

2.2.1.14 Protecting Soil Value

Biodiversity as a system of interacting component of the ecosystem is actively involved in the generation and protection of soil. The interaction between biotic and edaphic factors over a long period of time (usually 300–500 years) results in the formation of surface soils of 1–2 inches. The protection provided by natural vegetation cover and litter can preserve the productive capacity of the land and prevent dangerous landslides and the destruction of riverbanks, freshwater, and coastal fisheries by siltation. The protection of watershed ecosystems can greatly reduce sediment loads and therefore contribute significantly to the longevity of reservoirs and irrigation systems downstream.

Roberts and Johnson (1985) provide a startling example of how protected parks and watershed areas provide soil conservation services. The example relates to the soil conservation service provided by Nepal's Royal Chitwan National Park, where villagers have cleared and grazed the north bank of the river Rapti (park boundary) so intensively that the land mass has been subjected to severe degradation and rapid erosion. On the other hand, the south bank, which lies within the park, has the well-protected vegetation line that binds the soil so that when monsoon rains cause flooding in the river Rapti, it is the northern bank that is washed away. As a result of this process, the course of the river has shifted north and taken more than 100 hectares of land from the villagers.

It is apparent that the importance of biodiversity as a system must be understood in terms of the benefits of ecological services that it provides to the humankind. It is in this broader perspective of maintaining the functional integrity, resilience, and stability of the components of the biospheric ecosystem that humankind should understand how significantly vegetation, forest ecosystems, and biodiversity contribute to its own ultimate well-being and survival. Should mankind not pursue an economic system that is more ethical, kindlier, and harmonious with the natural ecosystem?

2.2.2 *Intrinsic Values*

Intrinsic value is the value that an entity has in itself, for what it is, or as an end in itself. It is traditionally understood to be the value that a thing has by virtue of its own nature or its own intrinsic properties. Thus, a thing has intrinsic value "in itself," or "for its own sake," or "in its own right." This implies that intrinsic value is "nonderivative" or "nonrelational" since things that have an intrinsic value do not have it because of their relation to other things. For example, many writers argue that pleasure is intrinsically valuable because pleasure is good in itself and not because of its relation to something else. Things said to be intrinsically valuable are, for example, happiness, virtuous acts, knowledge, or experiences of beauty, friendship, or love. What is contested (Norton, 1995; Sarkar, 2005) is whether ecosystems and species have noninstrumental value, value as an end, or value in themselves as well (intrinsic value). Instrumental value is always derivative on the value of something else, and it is always conditional. Something's instrumental value fluctuates based on changes in the desirability of the end to which it is a means and whether alternative, more efficient, means are available.

Many people value species and ecosystems intrinsically for their organizational complexity, diversity, spiritual significance, wildness, beauty, or wondrousness. As a result, species and ecosystems are said to have subjective intrinsic value. How much subjective intrinsic value they have, in general or with respect to particular systems and species, depends upon the prevalence, strength, and stability of the valuation. Many people value some species and ecosystems (charismatic megafauna and old-growth forests) more than others (microorganisms and deserts). As a result, they possess more subjective intrinsic value. If species and ecosystems have objective intrinsic value, then their value is discovered by human valuers and is not created by them. Soule (1985) postulates that: "Species have value in themselves, a value neither conferred nor revocable, but springing from a species' long evolutionary heritage and potential." Some environmental ethicists (Sterba, 2001; Johnson, 1991) have argued that species and ecosystems also have a good of their own and that their good needs to be considered, i.e., that they have inherent worth. If nonhuman organisms, species, or ecosystems have (subjective or objective) intrinsic value, then their value is not dependent upon whether alternative means are available (economic or medicinal), and they cannot be traded or substituted for without loss. For this reason, proponents of intrinsic value argue that it is more stable and robust than instrumental value with respect to justifying conservation goals. They also believe that intrinsic value is relevant to developing particular conservation and management plans, strategies, and methods, since these need to reflect the values at stake.

When we observe and speculate about Nature, we cannot remain without being impressed by the properties of natural systems such as organizational complexity, order, harmony, interconnectedness, self-organization, and diversity. We find this impression well-articulated in Aldo Leopold's (1949) famous maxim "land ethic" where he says: "A thing (i.e., an action) is right when it tends to preserve the integrity, stability, and beauty of the biotic community. It is wrong when it tends otherwise...." All ethics rest upon a single premise: that the individual is a member of a community of interdependent parts. The biotic community is the sum total of interconnectedness, self-organization, diversity, and organizational complexity, and these attributes can be considered to have intrinsic values because these values originated from a long evolutionary process in Nature. This creative process of Nature should be considered as an end in itself and for itself because the man, the evaluator himself, is intrinsically the product of this creative process.

The intrinsic values of other life-forms are grounded in human feelings, and we find this in many cultures expressed in a range of intensities. Buddha appreciated this value and questioned whether we have the right to kill other life-forms (Ehrlich and Ehrlich 1992). He believed in the sacredness of all life-forms and founded a philosophy based on nonviolence, peace, coexistence, compassion, wisdom, and living in harmony with Nature. The American Indians also adopted a lifestyle that was tailored to live in natural harmony and balance, although, the dominant European culture, so grounded in egocentric Cartesian instrumentalism, ridiculed the Indian's lifestyle and value system and took pride in its destruction. Thoreau (1845) was one American philosopher who revolted against Cartesian instrumentalism and took refuge in the beautiful bounty of Nature. Walden, perhaps, presents some intrinsic values of Nature and deserves to be revisited in our mind's observance.

Chapter 3
Biodiversity and Ecosystem Destruction

Only when the last tree has been cut down; only when the last river has been poisoned; only when the last fish has been caught; only then you will find that money cannot be eaten.

An American Indian proverb

3.1 Introduction

This chapter provides some perspectives on the causes and challenges associated with the loss of biodiversity and ecosystem services and policy implication for their protection and conservation. This chapter focuses on modern ecologically hostile consumerism, increasing population pressures, inequitable development patterns, human poverty, and the current faulty development paradigm as the principal causes accelerating the destruction and loss of biodiversity, natural ecosystems, and ecosystem services all over the planet. If appropriate cultural, socioeconomic, and political measures are not taken in time with a holistic development paradigm to alter our current production and consumption patterns and to reduce population growth and human poverty, the global environmental changes (global warming, climate change, and environmental destruction) may become irreversible and any conservation strategies will inevitably fail, no matter how scientifically sound they may be. Humanity may simply be running out of time.

The destruction of biological diversity and natural ecosystems currently occurring, especially in tropical centers, is unprecedented and shockingly horrifying. The primary causes of this destruction are "population pressure," "poverty," "lack of equitable development" in developing countries and "ecologically hostile consumption" and "production patterns" in rich, industrialized countries. The loss of biodiversity and ecosystem destruction are taking place in areas marked by abject poverty, lack of food, and other necessities of life, on one hand, and by the unsustainable commercial profiteering activities on a large scale all over the world, on the other hand. Sustainable resource use policy measures must be implemented to meet the

G. Upreti, *Ecosociocentrism*, https://doi.org/10.1007/978-3-031-41754-2_3

basic human needs in such areas so that the remaining biodiversity and natural ecosystems can be saved and protected. The critical challenge is how to prevent the destruction of biodiversity, ecosystems, and ecosystem services emanating from the unfettered profiteering of the global corporate world of the developed countries. In developed, industrialized nations, resource extraction and consumption patterns have contributed the most to global environmental destruction (biodiversity, natural ecosystems, and ecosystem services) (Durning, 1992; Upreti, 1994, 1996).

Without comprehensive and ethically guided sustainable global policy measures that incorporate a multitude of action plans, strategies, and activities specifically designed to address the issues of changing anthropogenic and corporate behavior and attitudes toward resource extraction and consumption patterns, population growth, poverty, equity, and resource distribution, it is impossible to conserve and protect biodiversity, natural ecosystems, and ecosystem services, which will determine the well-being of mankind and other beings on planet Earth. It is unthinkable to preserve humanity without preserving and protecting the very environmental resource base that provides life support to it. We do not need empirical studies to prove or disprove the hypothesis that the human economic system operates within the boundary of the planetary ecosystem and is going to be constrained by the laws of "thermodynamics." The sooner we realize this, the better would be the fate of both humanity and living systems on planet Earth.

3.2 Global Trends in Destruction

The global trend in the destruction of biodiversity and natural ecosystems, especially through deforestation and forest degradation in the tropics, is both shocking and horrifying. The World Resources Institute (Reid & Miller, 1989) estimated a loss of 5–6 billion hectares of forest since preagricultural times. In the past, most of these losses were concentrated in the temperate forests of Europe, Asia, and North America. After the end of the last ice age about 10,000 years ago, 57% of the world's habitable land was covered by forest, which was 6 billion hectares. Today, only 4 billion hectares are left, and the world has lost one-third of its forest, an area twice the size of the United States, and half of this loss occurred in the last century alone (Hanan, 2021). At present, it is the tropical forests of developing countries of Latin America, Asia, and Africa that have been rapidly disappearing. Every year, 19–23 million hectares of tropical forest vanish as trees are cut for timber and land is used for agriculture and other developments (Schlesinger et al., 1990). Brazil, with the largest remaining tropical forest area, has been losing between 6 and 11 million hectares a year. India has been losing its forest area at a rate of 1.7 million hectares a year. Indonesia has been losing 1 million hectares a year. The annual loss of tropical forests in nine key countries is astonishing (Table 3.1).

The loss of tropical forests has been occurring at alarming rates with profound impacts on ecosystems, climate, and livelihoods, thus prompting renewed commitments to halt its continuation.

Table 3.1 The annual loss of tropical forests in some countries in the tropics

Countries	Estimated loss (thousands of acres)
Brazil	19,768
India	3707
Indonesia	2224
Myanmar	1673
Thailand	981
Vietnam	427
Philippines	353
Costa Rica	306
Cameroon	247

Source: New York Times (June 8, 1990)

Global tree cover loss reached a record of 29.7 million hectares (73.4 million acres) in 2016, according to new data from the University of Maryland released on Global Forest Watch (2016). Despite concerted efforts to reduce tropical deforestation, tree cover loss has been rising steadily in the tropics over the past 17 years. Natural disasters like fires and tropical storms are playing an increasing role, especially as climate change makes them more frequent and severe. However, clearing of forests for agriculture and other uses continues to drive large-scale deforestation.

Agricultural expansion has been reported to be the primary cause of tropical deforestation and a key driver of greenhouse gas (GHG) emissions, biodiversity loss, and degradation of ecosystem services, which are vital not only to the livelihoods of forest-dependent rural people but also to the stabilization of global climate change. Pendrill et al. (2022) report that most (90–99%) of the deforestation across the tropics from 2011 to 2015 was driven by agriculture and that only 45–65% of the deforested land became productive agricultural land within a few years. Among different land uses and commodities, pasture expansion was the most important driver accounting for half of the deforestation across the tropics, followed by oil palm and soy cultivation accounting for at least a fifth, and six other crops, namely, rubber, cocoa, coffee, rice, maize, and cassava, accounting for the rest. They argue that ending deforestation in the tropics requires combining measures to create deforestation-free supply chains with landscape governance interventions in deforestation risk areas that focus on strengthening sustainable rural development and territorial governance.

According to recent statistics from the University of Maryland and published on Global Forest Watch, the tropics have lost 11.1 million hectares of tree cover in 2021, as Weisse and Goldman (2018) noted. The 3.75 million hectares that were lost from tropical primary rainforests, which are crucial for carbon storage and ecological richness, are of special concern. The country with the greatest primary rainforest loss has consistently been Brazil. It lost 1.5 million hectares, or about 40%, of the world's tropical primary forests in 2021 (Weisse & Goldman, 2022). Achieving the global zero deforestation goals of the 2021 Glasgow Leader's Declaration on Forests and Land Use, where 141 countries committed to jointly "halt and reverse forest loss by 2030," appears to be an impossible task given the current situation. The current trend highlights just how much action it will take to achieve these goals (Fig. 3.1).

Deforestation is different from loss of tree cover. According to Weisse and Goldman (2018), "tree cover" includes both plantation-grown trees and trees found in wild forests, whereas "tree cover loss" refers to the disappearance of the tree canopy owing to anthropogenic or purely natural factors, such as fire. According to the authors, the continual decline in tropical tree cover is disturbing and the new data further underscore the inadequacy of ongoing attempts to stop deforestation. In addition to protecting biodiversity and supplying jobs for people, trees are essential for storing carbon. Only 2% of climate money goes to the forest sector, despite the fact that forest conservation can contribute roughly 30% of the solution to keeping global temperature rise at 2 °C, which scientists believe is necessary to avoid the worst effects of climate change. Because they operate as a carbon sink, soaking up carbon dioxide (CO_2), which would otherwise be free in the atmosphere and contribute to continuous changes in climatic patterns, forests are essential for mitigating climate change. Their crucial role as a carbon sink is compromised by deforestation. Deforestation is believed to be the cause of 15% of all greenhouse gas emissions. Because tropical rainforests are home to much of the world's biodiversity, deforestation in the tropics is particularly alarming. For instance, the Amazon has lost 17% of its forest in the past 50 years, primarily as a result of the conversion of the forest for cattle ranching (Weisse & Goldman, 2022) (Fig. 3.2).

At present, forests account for 28% of the world's land area or 3.6 billion hectares. Currently, the annual loss of forests has been estimated to be 17–20 million hectares, with the major loss occurring in tropical moist forests (Munashinhe, 1992). Tropical moist forests are also the major habitat centers for most of today's biodiversity, which has been estimated to be 5–10 million species (Myers & Sayensu, 1983; Carson, 1990). Piementel et al. (1992) have provided this estimate to be

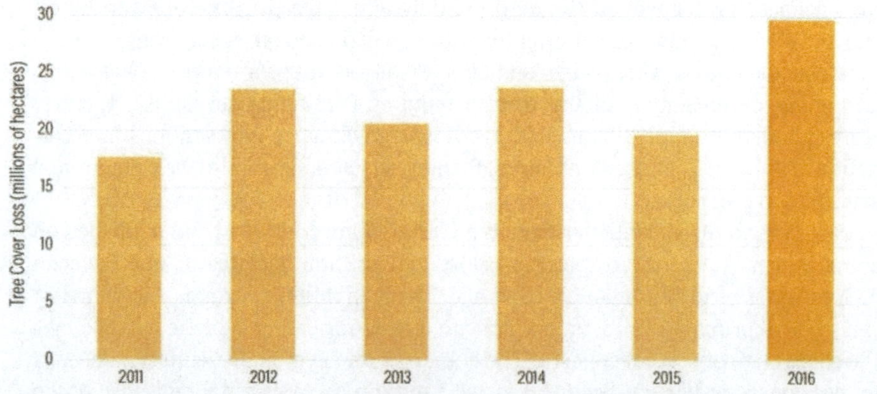

Fig. 3.1 Global tree cover loss reaches a new high in 2016. (Source: Adapted from global tree cover loss from the World Resources Institute 2016)

Fig. 3.2 Top 10 tropical countries for tree cover loss in 2017. (Source: Adapted from the World Resources Institute (2017); 2017 was the second-worst year on record for tropical tree cover loss)

10 million (Table 3.2). Since tropical moist forests account for more than half the world's total biodiversity, hundreds of species are eliminated each year (McNeely et al., 1990). It has been estimated that early in the next century, we could lose more than a million species of plants, animals, and other organisms from planet Earth (Myers & Sayensu, 1983; Wilson, 1988; Gee, 1992). This figure alone is more than the total mass extinction in the evolutionary history of the species, including the extinction of dinosaurs (Wilson, 1988; Gowdy, 1992) (Tables 3.3 and 3.4).

The president of the International Institute of Species Exploration (IISE), Quentin Wheeler's remarks, while celebrating the inclusion of the top 10 new species for 2017, as Christopher Dunagan (2017) point out, are worth pondering over: "This would be nothing, but good news were it not for the biodiversity crisis and the fact that we're losing species faster than we're discovering them. The rate of extinction is 1000 times faster than in prehistory. Unless we accelerate species exploration, we risk never knowing millions of species or learning the amazing and useful things they can teach us. Of all the devastating implications of climate change, none is more dangerous than accelerating species extinction. We can engineer our way through many impacts of climate change, but only hundreds of millions of years will repopulate the planet with biodiversity." It is argued here that if the world is serious about curbing climate change, all countries need to step up genuine efforts to reduce deforestation and increase tree cover over Earth's vast land surface.

The combined impacts of human-induced deforestation and climate change have accelerated the extinction of species at an unprecedented rate, and, if this trend is not slowed down or reversed, then there will be total annihilation of the diverse living systems that came into existence through billions of years of the evolutionary process. This will go down in history as the most horrific event of the "Anthropocene."

Table 3.2 Estimates of the diversity of plant and animal species on planet Earth

Organism groups	Number of species
Plants	
Vascular plants: dicots, monocots, and fern	260,000
Fungi	47,000
Liverworts and mosses	25,000
Diatoms	17,000
Lichens	16,000
Bacteria	14,000
Animals	
Arthropods	9,000,000
Mollusks	50,000
Protozoa	30,000
Fish	22,000
Helminths	14,000
Annelids	12,000
Nematodes	12,000
Coelenterates	10,000
Birds	9000
Flatworms	9000
Reptiles	6000
Echinoderms	6000
Sponges	5000
Mammals	4500
Moss animals (Bryozoa)	4000
Amphibians	2500
Other small animals	1000
Total plant and animal species	**10,000,000**

Source: Adapted from Piementel et al. (1992)

From the tropical rainforest facts complied by Rhett A. Butler (2020), it can be seen why these forests are fundamentally indispensable for the preservation of the remaining diverse species, generation of vital ecological services, and maintenance of the global hydrological cycle upon which depends the very survival of humanity.

Tropical Rainforest Facts, Characteristics, and Distribution (Compiled by Butler, Rhett A. (2020))

Rainforest Facts

- Tropical forests presently cover about 1.84 billion hectares or about 12% of Earth's land surface (3.6% of Earth's surface).
- The world's largest rainforest is the Amazon rainforest.

Table 3.3 Estimated numbers of described and undescribed extant species as of 2009 based on Chapman's (2009) report

Major component group		Described	Global estimate (Des + Undes.)
Chordates		64,788	~80,500
↳	Mammals	5487	~5500
↳	Birds	9990	>10,000
↳	Reptiles	8734	~10,000
↳	Amphibia	6515	~15,000
↳	Fishes	31,153	~40,000
↳	Agnatha	116	Unknown
↳	Cephalochordata	33	Unknown
↳	Tunicata	2760	Unknown
Invertebrates		~1,359,365	~6,755,830
↳	Hemichordata	108	~110
↳	Echinodermata	7003	~14,000
↳	Insecta	~1000,000	~5000,000
↳	Archaeognatha	470	
↳	Blattodea	3684–4000	
↳	Coleoptera	360,000–~400,000	1,100,000
↳	Dermaptera	1816	
↳	Diptera	152,956	240,000
↳	Embioptera	200–300	2000
↳	Ephemeroptera	2500–<3000	
↳	Grylloblattaria	24	
↳	Hemiptera	80,000–88,000	
↳	Hymenoptera	115,000	>300,000
↳	Isoptera	2600–2800	4000
↳	Lepidoptera	174,250	300,000–500,000
↳	Mantodea	2200	
↳	Mecoptera	481	
↳	Megaloptera	250–300	
↳	Neuroptera	~5000	
↳	Odonata	6500	
↳	Orthoptera	24,380	
↳	Phasmatodea (Phasmida)	2500–3300	
↳	Phthiraptera	>3000–~3200	
↳	Plecoptera	2274	
↳	Psocoptera	3200–~3500	
↳	Siphonaptera	2525	
↳	Strepsiptera	596	
↳	Thysanoptera	~6000	
↳	Trichoptera	12,627	
↳	Zoraptera	28	

(continued)

Table 3.3 (continued)

Major component group	Described	Global estimate (Des + Undes.)
↳ Zygentoma (Thysanura)	370	
↳ Arachnida	102,248	~600,000
↳ Pycnogonida	1340	Unknown
↳ Myriapoda	16,072	~90,000
↳ Crustacea	47,000	150,000
↳ Onychophora	165	~220
↳ Noninsect Hexapoda	9048	52,000
↳ Mollusca	~85,000	~200,000
↳ Annelida	16,763	~30,000
↳ Nematoda	<25,000	~500,000
↳ Acanthocephala	1150	~1500
↳ Platyhelminthes	20,000	~80,000
↳ Cnidaria	9795	Unknown
↳ Porifera	~6000	~18,000
↳ Other invertebrates	12,673	~20,000
↳ Placozoa	1	–
↳ Monoblastozoa	1	–
↳ Mesozoa (Rhombozoa, Orthonectida)	106	–
↳ Ctenophora	166	200
↳ Nemertea (Nemertina)	1200	5000–10,000
↳ Rotifera	2180	–
↳ Gastrotricha	400	–
↳ Kinorhyncha	130	–
↳ Nematomorpha	331	~2000
↳ Entoprocta (Kamptozoa)	170	170
↳ Gnathostomulida	97	–
↳ Priapulida	16	–
↳ Loricifera	28	>100
↳ Cycliophora	1	–
↳ Sipuncula	144	–
↳ Echiura	176	–
↳ Tardigrada	1045	–
↳ Phoronida	10	–
↳ Ectoprocta (Bryozoa)	5700	~5000
↳ Brachiopoda	550	–
↳ Pentastomida	100	–
↳ Chaetognatha	121	–
Plants sens. Lat.	~310,129	~390,800
↳ Bryophyta	16,236	~22,750
↳ Liverworts	~5000	~7500
↳ Hornworts	236	~250

(continued)

Table 3.3 (continued)

Major component group	Described	Global estimate (Des + Undes.)
↳ Mosses	~11,000	~15,000
↳ Algae (plants)	12,272	Unknown
↳ Charophyta	2125	–
↳ Chlorophyta	4045	–
↳ Glaucophyta	5	–
↳ Rhodophyta	6097	–
↳ Vascular plants	281,621	~368,050
↳ Ferns and allies	~12,000	~15,000
↳ Gymnosperms	~1021	~1050
↳ Magnoliophyta	~268,600	~352,000
Fungi	98,998 (Lichens 17,000)	1500,000 (Lichens ~25,000)
Others	~66,307	~2600,500
↳ Chromista (including brown algae, diatoms, and other groups)	25,044	~200,500
↳ Protoctista (i.e., residual protist groups)	~28,871	>1000,000
↳ Prokaryota (bacteria and archaea, excluding Cyanophyta)	7643	~1000,000
↳ Cyanophyta	2664	Unknown
↳ Viruses	2085	400,000
Total (2009 data)	**1,899,587**	**~11,327,630**

Source: Adapted from Chapman (2009). *Number of Living Species in Australia and the World, (2nd edn)*, Australian Biodiversity Information Service

- Brazil has the largest extent of rainforest cover, including nearly two-thirds of the Amazon.
- Rainforests also exist outside the tropics, including temperate North America, South America, Australia, and Russia.
- An estimated 50% of terrestrial biodiversity is found in rainforests.
- Rainforests are believed to store at least 250 billion tons of carbon.
- Deforestation and degradation of tropical forests account for roughly 10% of global greenhouse gas emissions from human activities.

Rainforest Characteristics

Rainforests are forest ecosystems characterized by high levels of rainfall, an enclosed canopy, and high species diversity. Although tropical rainforests are the best-known type of rainforest and are the focus of this section of the website, rainforests are actually found widely around the world, including temperate regions in Canada, the United States, and the former Soviet Union.

Tropical rainforests typically occur in the equatorial zone between the Tropic of Cancer and the Tropic of Capricorn, latitudes that have warm temperatures and relatively constant year-round sunlight. Tropical rainforests merge into other types of forests depending on the altitude, latitude, and various soil, flooding, and climatic

Table 3.4 The International Institute for Species Exploration (IISE, 2011) state of observed species

Group	No. of described species
Insecta	1,023,430
Plantae	285,885
Arachnids	105,055
Fungi	101,702
Crustacea	48,719
Mollusca	86,506
Prokaryotes	8838
Protoctista	29,470
Fishes	31,972
Nematoda	25,519
Platyhelminthes	20,442
Annelida	17,109
Amphibia	6792
Chromista	46,346
Cnidaria	10,050
Reptilia	8983
Porifera	6107
Mammalia	5569
Algae	12,340
Echinodermata	7065
Tunicata	2816
Aves	10,004
Hemichordata	111
Other invertebrates	41,045
Other chordates	149
Total	**1,941,939**

Source: Adapted from the International Institute for Species Exploration (IISE) (2011)

conditions. These forest types form a mosaic of vegetation types, which contribute to the incredible diversity of the tropics.

The bulk of the world's tropical rainforests occurs in the Amazon Basin in South America. The Congo Basin and Southeast Asia, respectively, have the second and third largest areas of tropical rainforests. Rainforests also exist on some of the Caribbean islands, in Central America, in India, on scattered islands in the South Pacific, in Madagascar, in West and East Africa outside the Congo Basin, in Central America and Mexico, and in parts of South America outside the Amazon. Brazil has the largest extent of rainforest of any country on Earth.

Rainforests provide important ecological services, including storing hundreds of billions of tons of carbon, buffering against floods and droughts, stabilizing soils, influencing rainfall patterns, and providing a home to wildlife and Indigenous people. Rainforests are also the source of many useful products upon which local communities depend.

Although rainforests are critically important to humanity, they are rapidly being destroyed by human activities. The biggest cause of deforestation is conversion of forest land for agriculture. In the past, subsistence agriculture was the primary driver of rainforest conversion, but, today, industrial agriculture, especially monoculture and livestock production, is the dominant driver of rainforest loss worldwide. Logging is the biggest cause of forest degradation and usually proceeds deforestation for agriculture.

Rainforest Distribution

The global distribution of tropical rainforests can be broken up into four biogeographical realms based roughly on four forested continental regions: the "Afrotropical," "Australasians," "Indomalayan/Asian," and the "Neotropical." Just over half the world's rainforests lie in the Neotropical realm, roughly a quarter are in Africa, and a fifth in Asia (Fig. 3.3).

These realms can be further divided into major tropical forest regions based on biodiversity hotspots, including:

1. Amazon: Includes parts of Bolivia, Brazil, Colombia, Ecuador, French Guiana, Guyana, Peru, Suriname, and Venezuela.
2. Congo: Includes parts of Cameroon, Central African Republic, Democratic Republic of the Congo, Equatorial Guinea, Gabon, and Republic of Congo.
3. Australasia: Includes parts of Australia, Indonesian half of New Guinea, and Papua New Guinea.
4. Sundaland: Includes parts of Brunei, Indonesia, Malaysia, and Singapore.

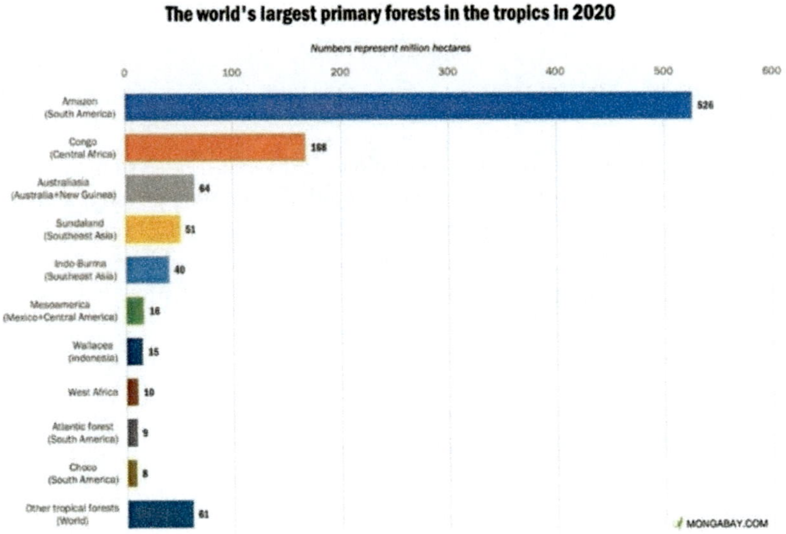

Fig. 3.3 Bar chart showing the world's largest rainforests as defined by the area of primary forest cover. (Source: Adapted from Butler (2020). *Rainforest Information*. https://rainforests.mongabay.com/)

5. Indo-Burma: Includes parts of Bangladesh, Cambodia, China, India, Laos, Myanmar, Thailand, and Vietnam.
6. Mesoamerica: Includes parts of Belize, Costa Rica, El Salvador, Guatemala, Honduras, Mexico, Nicaragua, and Panama.
7. Wallacea: Includes Sulawesi and the Maluku islands in Indonesia.
8. West Africa: Includes parts of Benin, Cameroon, Côte d'Ivoire, Ghana, Guinea, Liberia, Nigeria, Sierra Leone, and Togo.
9. Atlantic forest: Includes parts of Argentina, Brazil, and Paraguay.
10. Choco: Includes parts of Colombia, Ecuador, and Panama.

Why are They so Important?

Biodiversity

Rainforests have extraordinarily highs level of biological diversity or "biodiversity." Scientists estimate that about half of Earth's terrestrial species live in rainforests.

Ecosystem Services

Rainforests provide critical ecosystem services at local, regional, and global scales, including producing oxygen (tropical forests are responsible for 25–30% of the world's oxygen turnover) and storing carbon (tropical forests store an estimated 229–247 billion tons of carbon) through photosynthesis, influencing precipitation patterns and weather, moderating flood and drought cycles, and facilitating nutrient cycling, among others.

Source: Compiled by Butler (2020). *Rainforest Information and Facts*, last updated Aug 14, 2020. https://rainforests.mongabay.com/

3.3 Causes of Destruction

The causes of environmental destruction (biological diversity, ecosystem, and ecosystem services) can be broadly described as population growth, poverty, and lack of rural development in developing countries and ecologically hostile consumption and production patterns in rich, industrialized countries. Each of these is briefly discussed to provide a perspective on how seriously each could affect the destruction process.

3.3.1 Global Population Pressure

If we look into the population growth rate worldwide since 1830 to now, we cannot remain unimpressed at the procreative capacity of the *Homo sapiens*. It took 100 years to go from 1 to 2 billion, but only 44 years to go from 2 to 4 billion. Since 1980, although the annual growth rate has decreased slightly, the number of people added each year continues to increase so that now 1 billion new people are being added about every 11 years. The total world population had been projected to exceed 6 billion before 2000 and to reach 9 billion by 2025, unless the present level of increase is reduced (Myers, 1984). More than 90% of the projected increase in world population between now and 2025 is expected to take place in tropical developing countries in Asia, Africa, and South and Central America. At the present growth rate, these countries would be double in population in just 33 years. Such rapid population growth would almost inevitably produce low quality of life, high human suffering, and severe environmental degradation due to deforestation, erosion, and desertification (Upreti, 1987, 1994; Erickholm, 1975; Myers, 1984a, b).

3.3.1.1 Snapshot of the Global Population in 2017

According to the results of the 2017 Revision of World Population Prospects, the world's population numbered nearly 7.6 billion as of mid-2017 (Table 3.5), implying that the world has added approximately one billion inhabitants in the previous 12 years. In all, 60% of the world's people live in Asia (4.5 billion), 17% in Africa (1.3 billion), 10% in Europe (742 million), 9% in Latin America and the Caribbean (646 million), and the remaining 6% in Northern America (361 million) and Oceania (41 million). China (1.4 billion) and India (1.3 billion) remain the two most populous countries of the world, comprising 19% 18% of the global total population, respectively.

Much of the population increase predicted in tropical developing countries will be in rural areas, aggravating the pressure on agricultural and marginal and remaining forest lands. Very soon, all the low-grade and marginal lands will be in use, which means any further growth will be at the expense of forests and other natural environments. It is unlikely that such increasing rural populations will be able to restrict themselves to self-sustaining agricultural practices, leading to further degradation of natural ecosystems and ecosystem services. The governments in these countries will not be able to contain rural populations in self-sustaining agricultural

Table 3.5 Population of the world and regions in 2017, 2030, 2050, and 2100, according to the medium-variant projection

Region	Population (millions)			
Year	2017	2030	2050	2100
World	7550	8551	9772	11,184
Africa	1256	1704	2528	4468
Asia	4504	4947	5257	4780
Europe	742	739	716	653
Latin America and the Caribbean	646	718	780	712
Northern America	361	395	435	499
Oceania	41	48	57	72

Source: United Nations, Department of Economic and Social Affairs, Population Division (2017). World Population Prospects: The 2017 Revision. New York: United Nations practices because of the complexities of the problems inherent in sociopolitical and economic structures of the society.

In most developing countries, agriculture interacts almost inseparably with natural or forest environments. Forests provide villagers with fuelwoods, timber, and fodder and organic fertilizers to their animals and lands, respectively. There is a tremendous amount of one-way flow of energy and matter from the natural environment, such as a forest, to the farmland, which, in the form of fuelwood, timber, fodder, mulch, and litter, is primarily responsible for rendering forest ecosystems incapable of sustaining habitats for numerous other species. I cannot help quoting Paul and Anne Ehrlich (1993), who emphatically and correctly warned us about the consequences of uncontrolled human population: "We should not delude ourselves: the population explosion will come to an end before very long. The only remaining question is whether it will be halted through the humane methods of birth control, or by nature wiping out the surplus. We realize that religious and cultural opposition to birth control exists throughout the world; but we believe that people simply do not understand the choice that such opposition implies. Today, any one opposing birth control is unknowingly voting to have the human population size controlled by a massive increase in early deaths. Of course, the environmental crisis is not caused just by expanding human numbers. Burgeoning consumption among the rich and increasing dependence on ecologically unsound technologies to supply that consumption also play major parts. This allows some environmentalists to dodge the population issue by emphasizing the problem of malign technologies. And social commentators can avoid commenting on the problem too many people by focusing on the serious maldistribution of affluence."

The question is whether the nation states in the world, particularly in the developing world, can design an appropriate policy framework to address this problem. Both quality of life and improvement in human conditions in developing countries primarily depend upon how successfully human population growth is contained within the limits of the resource base. Unfortunately, it appears that human populations in most developing countries have already surpassed the sustainable threshold of the natural resource base. The delay in the realization of this fact and the failure to design rational national policy strategies to effectively curb human population growth rate in developing countries is to invite a catastrophe of human misery.

3.3.2 *Poverty, Inequity, and Wealth Transfer*

Poverty can be both the cause and the effect of increasing population growth. Both poverty and high population growth reduce the relative *per capita* share of the available resources of a nation, thereby creating a social situation that severely limits people's accessibility to economic, social, and educational opportunities. When the socioeconomic and the educational quality standards of a society are low because of limited access to resources, people have to desperately struggle for their immediate survival, no matter how grave the consequences of their immediate survival acts may be for their own survival in the long run. In simple terms, it can be stated that a hungry farmer and his family have a moral right to use natural resources for their very survival and that they cannot and should not be blamed for the detrimental effects of their actions if governments and development agencies cannot provide them with better alternatives. It has been estimated that there are more than 600 million "poorest of the poor," and this figure had been projected to reach at least 1 billion by the start of the year 2000 and 2 billion by the time the developing world's population would come close to leveling out at around 10 billion in the year 2100 (World Bank, 1983).

The Brundtland Commission (1987) argues that a five-fold to ten-fold increase in economic activity would be required in the next 4–5 decades to meet the needs and aspirations of the burgeoning world population and to reduce mass poverty. If the existing mass poverty in developing countries is not eliminated, then there is no way to stop the decline in our planet's stock of basic natural capital: its forests, soils, species, fisheries, waters, and atmosphere. The developing countries have been caught in a downward spiral of debt and financial flows to the developed countries. The cumulative debt of developing countries to developed nations is growing and has now reached more than $1 trillion, the interest of which alone amounts to $60 billion a year. According to the World Bank's account, there was a net flow of more than $43 billion (Table 3.6) from developing to developed countries (MacNeill, 1989). There exists such a glaring inequity between the developing and industrialized worlds in terms of the control of resources, wealth, and consumer goods that if developed countries continue to maintain their control, unwilling to share resources and goods essential for basic human needs, then our planet would experience a greater stress and insults from anthropogenic activities.

Debt and financial flows reflect the worsening financial situation of developing countries. In 1988, more than $43 billion a year was transferred from developing countries to developed countries. In 2013, more than $309 billion was transferred, but, in 2016, the net amount transferred was more than $81 billion (World Bank, 2018; Table 3.7).

Most of the developing countries have natural resource-based economies. Environmental resources such as soils, forests, fisheries, waters, biodiversity, and natural parks constitute their main natural capitals, and it is quite clear that their

Table 3.6 Long-term debt and financial flows in developing countries from 1983 to 1988. (billions of dollars)

Year	1982	1983	1984	1985	1986	1987	1988
Debt disbursed	562.5	6449	686.7	793.7	893.8	996.3	1020
Debt service	98.7	92.6	101.8	112.2	116.5	124.9	131
Principle payments	49.7	45.4	48.6	56.4	61.5	70.9	72
Interest payments	48.9	47.3	53.2	55.8	54.9	54	59
Net flows	67.2	51.8	43	32.9	26.2	15.8	16
Net transfer	**18.2**	**4.6**	**−10.2**	**−22.9**	**−28.7**	**−38.1**	**−43**

Source: Adapted from MacNeill (1989)

Table 3.7 Long-term debt and financial flows in developing countries from 2008 to 2016

Year	2008	2009	2010	2011	2012	2013	2014	2015	2016
Debt disbursed	621	528	657.5	776.3	866.5	934.5	977.4	794.9	953.9
Debt service	506.9	510.5	490.6	522.5	558.5	621.3	719.7	753.6	865.4
Principle payments	391.4	400.3	384.2	398.3	406.7	479.4	565.7	595.2	687.5
Interest payments	115.4	110.2	106.4	124.1	151.8	142.1	154.1	158.5	177.9
Net flows	215.8		256.8	373.6	456.5	451.9	401.9	188.6	259.1
Net transfer	**100.4**	**−10.8**	**150.4**	**249.5**	**304.7**	**309.8**	**247.8**	**30.1**	**81.2**

Source: The World Bank 2018 International Debt Statistics (all developing countries). The figures are in dollars

long-term economic development depends on maintaining these stocks of natural capitals and on enhancing their ability to support agriculture, forestry, fishing, mining, and tourism for local use and export. The greatest problem that the developing countries are facing now is that their environmental and renewable resources (the basic natural capitals) are being consumed faster than can be restored or replaced. The consequence of such catastrophe is not only increased hunger and death but also increased social instability and conflict because resource depletion and degradation drive millions of environmental refugees across national borders. The existing inequity between the poor southern and the rich northern countries in terms of the distribution and control of goods and natural resources and wealth cannot be morally justified. People in developing countries are struggling for the most basic needs that are necessary to sustain their lives let alone talk about the needs necessary for the self-actualization of a human being, whereas, in rich, industrialized nations, the creation and cultivation of environmentally destructive needs have been recklessly progressing. This is quite obvious from the fact that developed, industrialized countries, consisting of one-quarter of the world's population, consume about 80% of the world's resources and goods and that developing countries with three-quarters of the world's population have access to less than 25% of the world's resources and wealth (MacNeill, 1989).

People in industrialized nations spend billions of (Americans spend $ 8 billion) dollars each year to lower their calorie consumption, whereas the world's poorest 880 million people suffer from severe undernourishment, stunted growth, mental retardation, and death and 5.15 billion people are forced to live below the poverty line. In the developing world, poverty is intricately linked with environmental problems and has become an environmental phenomenon. The distribution of cost and benefit of environmental externalities is extremely disproportionate. The rich harvest the benefits, whereas the poor suffer from environmental destruction caused by the rich. The poor have been pushed to ecologically marginal lands by increasing population pressures and inequitable development patterns. Hence, the economic deprivation of the poor, inequitable development patterns, and environmental degradation have all reinforced each other to form a "downward spiral," a whirlpool of misery, which is pulling more and more people into its grip.

Most of the world's environmental problems, such as global warming, air pollution, water pollution, and climate change, are the creation of affluence not poverty; nevertheless, poverty has been driving the ecological deterioration in developing countries where desperate people have been overexploiting their resource base in an attempt to keep themselves alive. This immediate need for survival has forced landless families to plow fragile mountain slopes and the rainforest floors. This has manifested in severe environmental decline, which, in turn, further increases poverty, as a degraded ecosystem loses its productivity, thereby diminishing the crop yield to its poor inhabitants. Hence, a catastrophic "downward spiral" of economic deprivation and ecological degradation becomes inevitably operational, seriously undermining the very survival and security of the people in developing countries. Poverty and inequity dilemma have created three distinct but interrelated phenomena, all of which eventually manifest their effects in the negative spiral of environmental degradation in all developing countries. These are summarized as follows:

Locally, poor people have to compete for environmentally marginal land and natural resources. The rapid population growth of the poor further makes them even poorer, and the pressure on the marginal resources increases accordingly, thus exacerbating the degradation process.

Nationally, policies favor urban over rural because the ruling elites have their interests and power base in the urban centers. There is a massive flow of wealth and resources from rural to urban areas.

Internationally, wealth and resources are transformed from the poor, developing world (Global South) to rich, industrialized world (Global North) by debt patterns, trade patterns, and capital flight.

If these negative trends in ecological degradation and inequitable development in developing countries are not reversed and the transfer of wealth from developing countries to the developed, industrialized countries in the form of debt servicing, capital flight, and trade patterns are not stopped, the global environmental degradation, ecosystem destruction, and social unrest will accelerate at an unprecedented rate.

3.3.3 *Ecologically Hostile Consumerism*

The singular cause that degraded natural ecosystems far more than any others can be attributed to the "egocentric consumerism" (faulty paradigm, misconception of development, greed, and self-aggrandizement) that patronized the reckless mechanization (industrialization) of the extraction and production of material goods and services. The essential elements of this paradigm can be characterized by the notion of man as being the ultimate conqueror of Nature and Nature as being the adversary that ought to be conquered or subdued for the benefits of mankind and its self-aggrandizement (Jones, 1990). Anthropogenic activities, followed by the Industrial Revolution in particular, have been directed to achieve the ultimate control of

Nature, seemingly with good intentions for the benefits of humanity but ironically with wrong attitudes toward Nature. Ecologically hostile consumerism can best be described as material growth without regard to the effects on the environment and is characterized by the following inherent anomalies:

It assumes infinite economic growth and regards Nature just as a resource to be exploited.

The growth is driven by greed and self-aggrandizement rather than need satisfaction.

Growth is not for improvement in human conditions of masses of people but for the satisfaction of insatiable desires of powerful political and corporate elites to accumulate wealth.

The current industrialism and neoliberal market economy does not recognize the material balance perspective, the ecological limit to economic growth, and the entropic effects of its activities. Contrary to this, it presupposes a universal and unequivocal acceptance of industrial growth and expansion and the pursuit of unlimited material growth, which has become its very *raison d'etre*. Such conception of material growth forces one to regard the environment or nature as something external to humans and just as a resource to be exploited for human needs. This very attitude has become the unifying principle of the neoliberal market's economic paradigm, which reflects the dominance of material growth over improvement in human conditions, quality of life, aesthetics, and natural, spiritual, and other nonhuman life values. The ultimate implications of such economic views can be seen in the creation of a false impression in the minds of people to associate good life with an ever-increasing supply of goods and services produced by the industrial consumer society. Under such circumstances, as Jones (1990) argues, "No ethical imperatives are attached to consumer behavior other than the exhortation to consume. Consumption thus becomes an end, rather than a means and ties consumers not just to their possessions, but also to the virtually unconscious ideology of consumerism upon which the very existence of advanced industrial society depends."

Modern consumerism and free market economy are casually interlinked and are indispensably interdependent upon each other. The criticisms of free market economy and modern consumerism can be summarized in the following points:

Free market economy requires a constant material growth and profit.

It does not internalize the environmental cost of production and consumption.

It has created inequities in developing countries by making them the suppliers of raw materials and wealth in the form of debt servicing to the developed economy, creating elite consumer class and urban, biased, development patterns in developing countries.

Benefits from the free market economy did not and will not trickle down to the poor. Instead, only the elites in both developing and developed countries and in the corporate world reaped all benefits from the free market economy. Consequently, rich areas and the global corporate elites are getting richer, whereas poor areas and the poor are getting poorer. This has manifested even in the socioeconomic life of developed countries like the United States where income disparity has widened so much that there is already a third world class of people in USA.

Consumption, upon which free market economy thrives, is directly proportional to environmental destruction. Hence, the more that a society consumes, the more successful the free market economy becomes, the more negative environmental impacts it creates, and the more throughputs it puts in, thus accelerating the degradation of the planetary ecosystem.

Durning (1992) emphatically points out that the so-called free market economy of Western consumerism has produced a "negative trickle-down effect" and has proved to be a disappointing source of economic incentives for the poor in developing countries. The free market economy has been successful only in creating an "elite consumer" class in poor, developing countries and world class consumers in every nation. Developing countries' corporate elites profit massively from the exports of natural resources from the Global South to the Global North, and the poor in developing countries have gained nothing except degraded homelands. The environmental problems that are occurring all over the world today have an intricate connection with Western consumerism and neoliberal free market economy. One can easily see the connection between the massive deforestation occurring in Amazonia and the flourishing hamburger industries in Brazil, the United States, and Japan (Table 3.8 and Fig. 3.4).

Based on data from 2017, China consumes more than 6.3 trillion kilowatts of energy per hour annually. The United States is the second largest consumer of electric energy in the world with more than 3.9 trillion kilowatts per hour used each year. Other nations that use at least 1 trillion kilowatts per hour per year include Russia and India (Fig. 3.5).

Table 3.8 Per capita consumption of energy in selected countries in 1989

Country	Energy (kilogram coal equivalent)	Consumption ratio
United States	10, 127	146.7
Soviet Union	6, 546	94.8
West Germany	5, 377	77.9
Japan	4, 032	58.8
Mexico	1, 689	24.4
Turkey	958	13.8
China	810	11.7
Brazil	798	11.5
India	307	4.4
Indonesia	274	3.9
Nigeria	192	2.8
Bangladesh	69	1

Source: Adapted from Durning (1992). *How Much is Enough? The Consumer Society and the Future of the Earth*

Consumption ratio is the ratio of energy consumption of the countries to that of Bangladesh

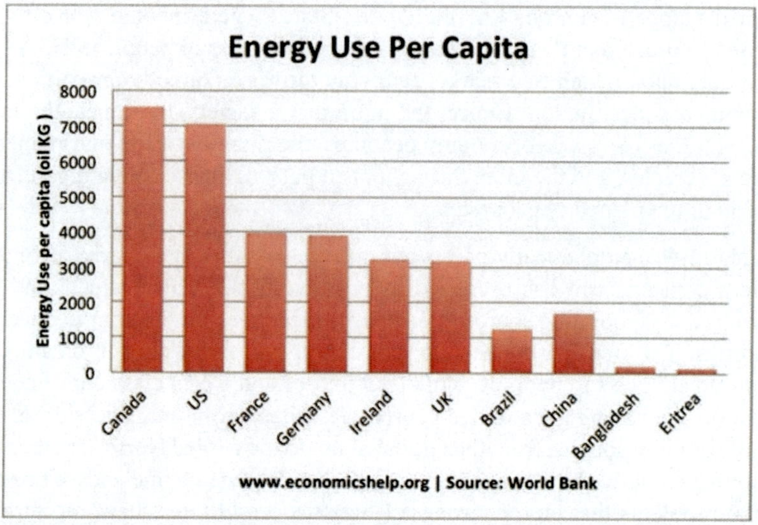

Fig. 3.4 Per capita energy consumption by the major countries in the world. (Source: Adapted from Pettinger (2017). List of Countries Energy Use per Capita)

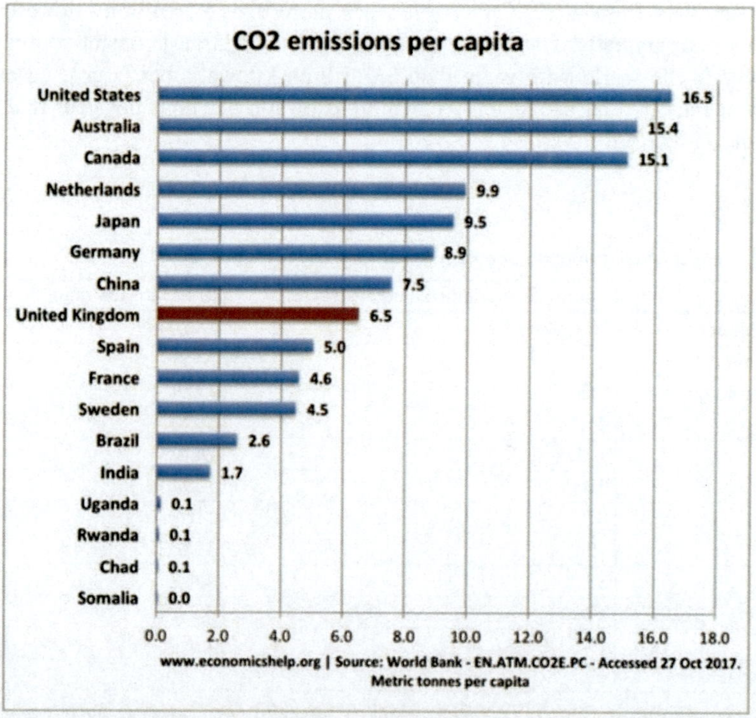

Fig. 3.5 Correlation between energy use and CO_2 emissions per capita. (Source: Adapted from Pettinger (2017). List of Countries Energy Use per Capita)

Global energy consumption has been growing significantly since 2000 due to rapid economic growth and rising demands in China, the world's largest energy consumer since 2009. The variation in the per capita contribution of different countries to greenhouse gas emissions is enormous. The level of economic growth is directly associated with the amount of greenhouse gas emissions into the atmosphere, and this is reflected by the scale of consumption. The per capita energy consumption of rich, developed countries is not at all comparable to that of developing countries (Table 3.8 and Fig. 3.4). This also means that the richest countries have benefited the most because their process of economic growth has contributed the most to greenhouse gas emissions. The per capita material consumption in developed, industrialized nations is about 30 times the per capita consumption in developing nations. The magnitude of the environmental impact is directly proportional to that of the consumption itself, which means that the environmental impact of an individual's consumption in developed countries is about 30 times more than that of an individual's consumption in poor, developing countries. The simple fact is that the rich in the developed countries are many times more responsible for environmental destruction (biodiversity, natural ecosystems, and ecosystem services) and ecological catastrophes than are the poor in developing countries. There is, however, one important difference between the ways how both the rich and poor cause environmental destruction: the environmental destruction caused by the poor is basically out of desperation to keep themselves alive, whereas that caused by the rich is basically out of wealth accumulation through profiteering and ever-increasing consumption of ecologically harmful and culturally created needs cultivated by modern Western consumerism. Industrialized countries are solely responsible for the creation of today's Updatedserious environmental crisis such as air pollution, greenhouse effects, acid rains, ozone depletion, global warming, and climate change.

Kirk Smith (1991) has developed an index of accumulated carbon dioxide that each country has put into the atmosphere (Table 3.9). He calls this accumulated carbon dioxide, a country's "natural debt" and strongly argues "As with the national debt, the present natural debt reflects our efforts to maintain a strong economic growth rates by burrowing from the future, in this case burrowing from the capacity of the environment to assimilate carbon dioxide. ......there comes a time when the size of this natural debt threatens to have serious repercussion."

Until 1991, the United States (US) had the highest per capita carbon dioxide emissions and China was way below many other countries in this regard. Over the course of three decades (30 years), China has climbed the ladder and has become the largest emitter of carbon dioxide, not only replacing the United States but also doubling its emissions (China's emissions are nearly twice those of the United States). There is not only significant variability in how much CO_2 countries emit across the world today but also large differences in how much each has emitted in the past. Who has contributed the most to global CO_2 since 1750? As Hannah Ritchie (2019) reports, the distribution of cumulative emissions around the world is used to compare countries or regions in relation to others and relative to the total. Let me present the key summary points from Hannah Ritchie's 2019 report:

Table 3.9 Per capita carbon dioxide emitted since 1950

Countries	Tons/capita
United States	186.00
United Germany	140.00
Canada	134.00
Czechoslovakia	132.00
United Kingdom	125.00
Australia	102.00
Former USSR	94.00
Japan	69.00
South Korea	39.00
North Korea	17.00
Malaysia	13.00
Iraq	10.00
China	10.00
Philippines	4.40
Thailand	4.20
India	3.40
Indonesia	2.90
Vietnam	2.60
Bangladesh	0.51
Nepal	0.23
World average	**30.00**

Source: Adapted from Smith (1991). The Natural Debt. East-West Center Views Vol. 1 and 3

- The United States has emitted more CO_2 than any other country to date: at around 400 billion tons since 1751, it is responsible for 25% of historical emission.
- This is twice more than China, the world's second largest national contributor.
- The 28 countries of the European Union (EU-28), which are grouped together here as they typically negotiate and set targets on a collaborative basis, are also a large historical contributor at 22%.
- Many of the large annual emitters today such as India and Brazil are not large contributors in a historical context.
- Africa's regional contribution relative to its population size has been extremely small. This is the result of its extremely low per capita emissions both historically and currently.

Carbon dioxide (CO_2) at an optimum concentration is an essential gas for the life process (photosynthesis) and is vital for sustaining life. It is called a greenhouse gas (GHG) because it absorbs and emits thermal radiation and creates the "greenhouse effect." Along with other greenhouse gases (nitrous oxide and methane), CO_2 is important for sustaining a habitable temperature for the planet. Had there been absolutely no GHGs, Earth would simply be too cold to support the living system as we know. It has been estimated that without these gases, the average surface temperature of Earth would be about −18 °C. Since the Industrial Revolution, however, energy-driven consumption of fossil fuels has led to a rapid increase in CO_2 emissions, disrupting the global carbon cycle and leading to a planetary warming impact. Global warming and a changing climate have a range of potential ecological, physical, and health impacts, including extreme weather events (such as floods, droughts, storms, and heat waves), sea-level rise, altered crop growth, and disrupted water systems. The Intergovernmental Panel on Climate Change (IPCC) report (2014a, b), after extensive analysis, presented the full coverage of all impacts. In the light of the evidence of its impacts, the United Nations (UN) member parties have set a target of limiting average warming to 2 °C above preindustrial temperatures, but unfortunately, if the current trend continues, it is impossible to limit it to 2 °C.

3.4 Tropical Rainforests and Greenhouse Gases

Tropical rainforests function as carbon sinks by absorbing carbon dioxide from the atmosphere and storing it as carbon in the forest biomass. When forest biomass is burned, it releases the stored carbon back into the atmosphere, thus increasing the concentration of carbon dioxide, which is a major greenhouse gas that contributes to global warming. The role of tropical forests in regulating global climate is becoming increasingly clear; however, their destruction is proceeding at an unprecedented rate. The destruction of tropical forests can be regarded as one of the most serious global environmental problems. Forests, through photosynthesis, assimilate huge quantities of carbon dioxide, the most pollutant of the gases accumulating in the atmosphere. If this gas is not kept in check, then it will cause a significant increase in the warming of Earth through a process known as the greenhouse effect (Weisse & Goldman, 2018; Sage, 2019). It is estimated that forests hold about 1200 BMT (billion metric tons) of carbon, which is 60% more than the atmosphere, and deforestation has contributed 125 BMT carbon to the atmosphere, compared to 200 BMT carbon contributed by the combustion of fossil fuels by the industrialized nations over the last 100 years (Munashinghe, 1992). Tropical forests contain the highest density of biomass per unit area and account for 55% of the world's organic carbon, and it is quite clear that their clearance causes a significant corresponding reduction in the absorption of carbon dioxide from the atmosphere, on one hand, and the emission of more carbon dioxide into the atmosphere through burning and decay, on the other hand. This will accelerate the onset and the magnitude of the global warming phenomenon, accelerating global climate change (World Bank, 1991). Although forest deforestation certainly increases the buildup of greenhouse gases in the atmosphere, its relative contribution is much less compared to fossil-fuel burning and the total amount of greenhouse gases in the atmosphere. Apart from functioning as "carbon sinks," tropical forests have other important functions such as being a reservoir of diverse life-forms, regulation of the hydrological cycle, floods, erosion control, essential environmental services, etc. (Sage, 2019).

Tropical rainforests play a crucial role in regulating the rainfall in and around the region of their prevalence. As Ehrlich and Wilson (1991) point out, the rainforests of Amazonia create moist conditions required for their survival by recycling rainfall. Deforestation could produce serious regional effects in Brazil outside of Amazonia by reducing rainfall in the agriculturally important areas in the south. Devegetation of watersheds also severely affects the regional hydrological cycles, leading to a higher incidence of floods, drought, and erosion and modification of the natural streams and aquifers (Upreti, 1987; Munashinghe, 1992). The Amazon rainforest is indeed highly vulnerable to deforestation, and Lovejoy and Nobre (2018) have warned that it could reach a tipping point if approximately 20–25% of it is deforested. According to Costa and Pires (2010), deforestation can lead to significant increases in temperature in the Amazon region as the average temperature over deforested areas has increased by about 1.0 °C, at most, relative to the neighboring forested areas. Regarding precipitation, Spracklen et al. (2012) found a negative correlation between deforestation and rainfall. According to their study, large-scale deforestation would result in a drastic reduction of rainfall across the Amazon basin, particularly in the dry season. The tropical rainforests may already have reached a critical threshold beyond which they may not be able to maintain resilience for regeneration. Tropical forests are the richest sources of biodiversity. They contain the largest and the most diverse populations of plant and animal species of any habitat in the world (Wilson, 1988). As the forests in the tropics disappear, these valuable diverse forms of life, the marvelous products of millions of years of evolutionary processes, also disappear once and for all, before they have even been discovered, named, and analyzed for possible use by human beings.

3.5 Current Trends of CO_2 Emissions

The cumulative contribution of the United States began to rise in the second half of the nineteenth century into the twentieth. By 1950, its contribution peaked at 40%; since then, it has declined to approximately 26% but remains the largest in the world. By 2015, China accounted for 12% of total cumulative emissions and India 3%. Emissions from a number of growing economies have been increasing rapidly over the last few decades. Fast-forwarding to annual emissions in 2014, we can see that a number of low-to-middle-income nations are now within the top global emitters. In fact, China is now the largest emitter, followed by (in order) the United States, EU-28, India, Russia, Indonesia, Brazil, Japan, Canada, and Mexico. It should be noted that a number of nations that are already top emitters are likely to continue to increase their emissions as they undergo development. In contrast to CO_2 emission growth in low-to-middle-income economies, trends across many high-income nations have stabilized and, in several cases, has decreased in recent decades. Despite this downward trend across some nations, emission growth in transitioning economies dominates the global trend and, as such, global annual emissions have continued to increase over this period.

According to the Global Carbon Project, global carbon dioxide emissions from fossil fuels and industry grew by 0.1% in 2019, reaching a record high of 36.8 gigatons (Gt) of CO_2 (Le Quéré et al., 2020). The increase in emissions in 2019 had been much lower than that in recent years, owing to a sharp decline in emissions from the power sector in advanced economies. However, the year 2020 saw an unprecedented drop in CO_2 emissions due to the COVID-19 pandemic and associated restrictions on movement and economic activity. A study published in *Nature Climate Change* estimated that daily global CO_2 emissions decreased by 17% by early April 2020 compared with the mean 2019 levels and also that emissions for the year could decrease by about 7% compared with the 2019 levels, depending on the trajectory of the pandemic and measures to control it (Le Quéré et al., 2020).

As Table 3.10 clearly shows, 15 countries generate more than two-thirds of global CO_2 emissions (72%) and the rest of the world's 180 countries produce nearly 28% of the global total, which is close to the amount that China alone produces. Interestingly, two countries, China and the United States, are responsible for more than 40% of the world's CO_2 emissions. As Sean Fleming (2019) reports, before the Industrial Revolution, levels of atmospheric CO_2 was about 280 parts per million (ppm). However, by 2013, these levels had already breached the 400-ppm mark for the first time (Fig. 3.6).

Table 3.10 Global CO_2 emissions by the top 15 countries

Rank	Country	Emissions in 2017 ($MtCO_2$)	% of Global Emissions
#1	China	9,839	27.2%
#2	United States	5,269	14.6%
#3	India	2,467	6.8%
#4	Russia	1,693	4.7%
#5	Japan	1,205	3.3%
#6	Germany	799	2.2%
#7	Iran	672	1.9%
#8	Saudi Arabia	635	1.8%
#9	South Korea	616	1.7%
#10	Canada	573	1.6%
#11	Mexico	490	1.4%
#12	Indonesia	487	1.3%
#13	Brazil	476	1.3%
#14	South Africa	456	1.3%
#15	Turkey	448	1.2%
	Top 15	26,125	72.2%
	Rest of World	**10,028**	**27.7%**

Source: Adapted from Visual Capitalist (2019)

Fig. 3.6 Major carbon dioxide (CO_2)-emitting countries in 2014. (Source: Adapted from Boden et al. (2017), Carbon Dioxide Information Analysis Center)

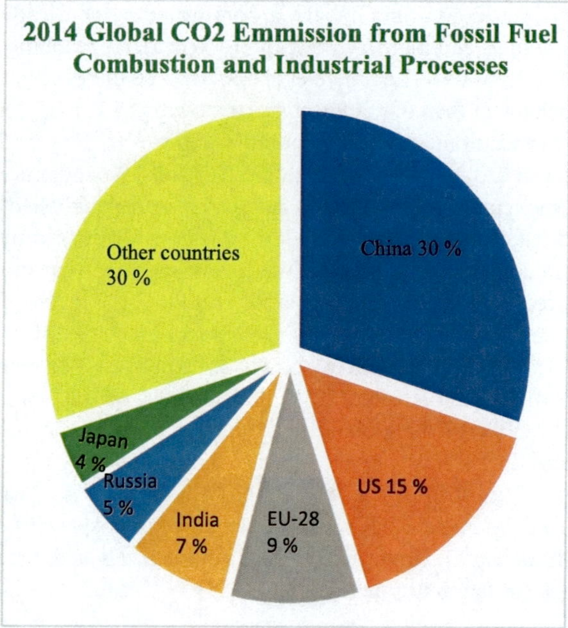

In 2014, the top carbon dioxide (CO_2) emitters were China, the United States, the European Union, India, the Russian Federation, and Japan. These data include CO_2 emissions from fossil fuel combustion as well as cement manufacturing and gas flaring. Together, these sources represent a large proportion of the total global CO_2 emission (Boden et al., 2017). Emissions and sinks related to changes in land use are not included in these estimates. However, changes in land use can be important: estimates indicate that the net global greenhouse gas emission from agriculture, forestry, and other land use were more than 8 billion metric tons of CO_2 equivalent or about 24% of the total global greenhouse gas emissions. In areas such as the United States and Europe, changes in land use associated with human activities have the net effect of absorbing CO_2, partially offsetting the emissions from deforestation in other regions. One disturbing fact is that coal, besides fossil fuels, is another main culprit when it comes down to CO_2 emission. The CO_2 emission report released by the International Energy Agency (IEA) in 2022 informs us that CO_2 emissions rose by 6% in 2021 to 36.3 billion tones, which is the highest level, after the world economy rebounded from the COVID-19 crisis, and the main factor driving this rise was the use of coals.

The United States, China, and India are responsible for about half of the global natural debt. As Smith (1991) points out "Since World War II, the average American's share of this natural carbon debt is more than eight times that of the average person in the world." The magnitude of the environmental problems, whether it be the production of greenhouse gasses, water pollution, land degradation, or loss of biodiversity, habitats, and ecosystem destruction, is directly proportional to that of the consumption, and excessive consumption is the "engine and the heart and soul of the neoliberal market economy," an economy that has operated within the framework of human cunning and environmental destruction. It is not a free economy other than that it freely destroys the environment without any reckoning. Thomas Berry (1990a, b), a visionary environmentalist warns us about the consequences of this egocentric, reckless consumerism, the most powerful cultural paradox created by the current dominant development paradigm: "In our time human cunning has mastered the deep mysteries of the Earth at a level far beyond the capacities of earlier peoples. We can break the mountains apart; we can drain the rivers and flood the valleys. We can turn the most luxuriant forests into throwaway paper products. We can tear apart the grass cover of the plains and pour toxic chemicals into the soil and pesticides onto the fields until the soil is dead and blows away in the wind. We can pollute the air with the acids, the river with the sewage, the sea with oil, all this in a kind of intoxication with our power for devastation on an order of magnitude beyond all reckoning…. The issue goes beyond economics or commerce or politics…. We are … losing ourselves." Modern consumerism has given rise to the assumptions, expectations, and values that do not acknowledge the ecological and social sustainability of resource use and consumption. This kind of consumerism, in effect, not only undermines the very ecological base of the resources but also seriously undermines the survival of humankind and the rest of the living systems on planet Earth.

What is extremely worrisome is that the major CO_2-emitting countries such as the United States, China, Russia, India, EU, and other developing countries in South America and Africa have not been able to come up with a strong political commitment to reduce greenhouse gas emissions, particularly CO_2, despite clear evidence that global warming due to increased greenhouse gas emissions has been casually implicated in climate change, the symptoms of which is felt by every country in the world, including the United States. The most frustrating and alarming irony is that the political leadership of the largest emitter of greenhouse gases (the United States) simply does not believe in climate change due to increased greenhouse gas emissions, even though both national and international scientific communities have clearly stated that to deny this scientific evidence and not take the necessary measures to reduce the emissions is to invite devastating catastrophes and human misery. The sooner the political establishment realize this undeniable truth and takes necessary measures to deal with climate changes, the lesser would be human suffering and pains. The question is, are not we running out of time?

3.6 Global Governance and Strategies

It is important to correctly identify and analyze the socioeconomic, political, demographic, and cultural factors that are responsible for driving the destruction of ecosystems and loss of biodiversity and natural capitals all over the world and thereby accelerating the global environmental and climate crisis. It is also important to note that the environmental problems that have emerged in developing countries are not exactly the same as those that have emerged in developed countries, but their cumulative effects will certainly create tremendous pressure dynamics that will increase and accelerate the rate of environmental destruction all over the world. Most of the environmental problems in developing countries are deeply rooted in poverty, debt servicing, inequitable development pattern, and population growth, whereas those in developed countries are rooted in the unsustainable extractive nature of market economy and excessive consumerism. Developing countries, on one hand, cannot resist the temptation of market-driven rapid economic growth and Western consumerism. This is the principal reason why they have adopted Western development models, which are based on the very same philosophy that has already done enough damage to the environment and humanity as well (free market economy and ever-increasing consumerism). This has, in effect, compounded the problems of developing countries rather than solved them. The inequitable development patterns, increasing debt burden and massive capital flights from the Global South to Global North, and rapid depletion of natural resources in developing countries, to a large extent, are the outcomes of the current neoliberal development paradigm.

The prospect for environmental conservation both in developed and in developing countries appears bleak if not totally hopeless in the context of continuing the existing development paradigm without any alteration in its basic assumption, values, and approaches. The development strategies and governance policy instruments to curb the social and environmental problems of developing countries differ from those of developed countries because of the different level of socioeconomic development and environmental problems. The following governance strategies need to be seriously considered in the development agenda of developed countries if the pace of negative environmental change is to be minimized and the goal of social and environmental sustainability is to be pursued.

3.6.1 Minimizing the Scale of Economy

The scale of economy is the function of consumption. The more a society consumes, the greater is the scale of its economy. A negative environmental impact is directly proportional to the level of the consumption of material goods and services. Modern consumerism, which includes production, distribution, consumption, and disposal of material goods and services, is the most destructive force of nature and is ecologically hostile (environmentally destructive). "Ecologically unsustainable consumerism" is the one that produces, distributes, consumes, and disposes goods and services without any regard to the very environment that provides the biophysical foundation (material basis) of all the resources required for the production of those goods and services. Most of modern consumerism is based on culturally created needs and the false perception of those needs as necessary. For example, animal skin, fur, bones, and tusks have a lucrative and unregulated market in the West, and this market is directly or indirectly responsible for the killing and poaching of wild and rare animal species such as elephants, rhinos, tigers, crocodiles, and many others. Likewise, the cosmetics industry is a multibillion dollar industry and is similarly responsible for the loss of many valuable species of plants and animals in addition to the enormous negative environmental impacts that the industrial complex creates in the process of the production, distribution, and consumption of cosmetic goods and disposal of waste materials. Most of the cosmetic goods and services are culturally created needs that have nothing to do with essential human functioning.

The marketing industry has been extremely successful in creating a false impression of the needs of these goods in the minds of modern consumers. Human beings unconsciously attach a great deal of importance to these needs and become an active participant in the culture that promotes and adores the consumption of false needs as an end. People in Western countries spend billions of dollars (Americans alone spend $ 5 billion) to cut down their excess dietary calories intake, whereas millions of people in poor, developing countries die from hunger, starvation, malnutrition, and malnourishment. It is true that Westerners cannot be blamed for all the problems of developing countries, but if we look into the historical perspective of political, economic, and cultural colonialism of the developing world by the West, the concomitant environmental destruction that followed, and the present international economic arrangement in which most of the developing countries have an enormous debt burden, it appears that most of the problems in developing countries can be eliminated with appropriate development policy and international cooperation. Western countries must take some responsibility for the problems that developing countries are facing today. It is impossible for developing countries to engage in environmental protection and conservation by ignoring the very basic needs of their peoples while Western countries continue the ecologically hostile production and consumption patterns of material goods and services.

It is interesting to note that most of the policy analysts, development experts, and scientists and conservationists genuinely seem to be concerned about environmental protection and conservation in developing countries and their problems but do not seem to be concerned about industrialized nations' production and consumption patterns of material goods and services, the values and assumptions of the very development paradigm that has created most of today's environmental and human problems all over the world. Since Western countries and some emerging large economies (China, India, Brazil, etc.) are responsible for the creation of most of today's environmental crisis, the burden lies more with them to pursue effective environmental protection and conservation actions through appropriate social, cultural, behavioral changes, and international development strategies; nevertheless, it is not to suggest that the developing countries have the moral rights to plunder and destroy the environment that provide their own resource base.

3.6.2 Equitable Development Patterns

Environmental and resource management strategies that fail to address the linkage between rural poverty and institutionalized socioeconomic inequities in developing countries will further aggravate environmental degradation. Deforestation in tropical developing countries is intricately linked to ever-increasing numbers of landless peasants, and, therefore, it is important that environmental conservation and development programs not only include appropriate alternatives to these people but also require that such development programs be equitable to overcome institutionalized and systemic inequitable socioeconomic development patterns. It is important to realize that most people living in tropical forest regions have been pushed to entrenched poverty, and it is essential that the development initiatives and innovations move toward freeing these people from such an impasse.

3.6.3 Biomass-based Resource Development

One of the main causes of forest degradation and, consequently, the loss of biodiversity in developing countries is fuelwood, fodder, forest products, and timber extraction from the forest ecosystem. About 50% rural people in most of the developing countries depend on natural forest products for their needs. Fuelwood and fodder extraction and livestock grazing in forests have been ongoing activities since time immemorial and are responsible for the destruction of undergrowth vegetation and severe degradation of forests. Rural communities must have access to an alternative energy source before the outright "protection" of forest policy is implemented. Establishment of agroforestry or community forestry projects in village communal or waste lands has a great potential to meet rural energy needs, thereby reducing the pressures and encroachment of rural people on forest areas. Agroforestry, as a system of growing ecologically suitable trees and shrubs together with that of food crops in the same land unit, holds great promise not only in meeting the fuelwood, fodder, and other needs of the villagers but also in securing an environmentally more sustainable pattern of agricultural development in developing countries.

There are some promising developments in some countries like Nepal where the Community Forestry Program managed by the local communities has been hailed as a global innovation in participatory environmental governance (Ojha et al., 2009). Around 34% of Nepal's forests are managed by more than 22,000 community forest user groups. Under the community forestry principles pioneered by Nepal in the 1970s, each community manages its forest for its own use and benefits based on an operational plan approved by the divisional forest officer, a representative of the federal government. Under the program, which began with the goal to reverse deforestation and protect existing forests, community members can collect wood, fodder, and other forest products up to a limit prescribed by the government based on the availability of wood, fodder, and timber and the condition of the forest. Nepal's community forestry approach has received worldwide acclaim for helping the country increase its forest cover from 25% to 45% of its total area (Joshi, 2022).

Forests in Nepal cover about 40% of the country, and a great majority of the population, which lives in rural areas, depends on forest resources for their livelihood. Today, forests under community management represent more than one-third of the total forest area. When forests came under the management ownership of the local communities, the quality and coverage of the forests improved, forest user groups, mainly farmers, benefited from the sustainable uses of the forest products, and forests became a source of diversified investment capital and raw materials for new market-oriented livelihoods. Experience of Nepal's Community Forestry Program suggests that participation of the local communities in the management of forests have improved the environmental sustainability of forests as manifested by the lower incidence of fires and illegal harvesting of forest products, better controlled grazing, higher tree density in formerly degraded forests, increased species diversity, and regeneration of important species (Dougill et al., 2001; Acharya, 2002; Dev et al., 2003; Yadav et al., 2003). Ojha et al. (2009) report that there have been improvements in forest conditions, forest land uses, and biodiversity following community management and that the strong interest of local communities in forest governance and their adoption of a sustainable approach to forest management are the key foundations of sustainability of community forestry in Nepal. Such a management success story can be replicated in other parts of developing countries where a large number of rural people, particularly farmers, have to depend upon forest products such as fuelwood, fodder, timber, grass, and organic manure for their agriculture-based livelihoods.

3.6.4 Natural Resource Governance Policies

Natural resources are the major contributors to the overall economy of many developing countries. The development policies that sanction the reckless exploitation of natural resources are primarily responsible for deforestation in several countries with large tropical forests. Development policies that subsidize natural resource extraction must be replaced by more innovative and environmentally sound policies that encourage the preservation of valued forests and natural ecosystems, promote regenerative capacity and sustainable use of natural resources, and ensure equitable distribution of development benefits. The maintenance of "natural capital" stock (natural resource base), especially for the biomass-based economy of developing countries, is essential not because it is the best option for them but because it is the only option available to them for their continued survival. The development programs and strategies of developing countries must entail sustainable management of natural resources and socially and environmentally responsible governance policies.

The policy instruments and strategies that can ensure protection and conservation of biological diversity, natural ecosystems, and sustainable development must emerge from a sociopolitical and ecological understanding of the problems that human beings are facing today. Politicians and policymakers must attempt to understand how "social ecology" is interlinked with the "ecology" of a natural system. Understanding the nature of interaction between social and natural systems ("sociosphere" and "ecosphere", respectively) is impossible without integrating information and knowledge systems from "system ecology" and human "sociopolitical" and "economic" systems. Planning, as an institution to generate appropriate governance policy instruments and strategies, has often failed in its mission in both developed and developing countries to address environmental and socioeconomic problems, primarily because of the "trickle-down" and "growth-oriented" approach of economic growth policies. This approach had, in effect, not only destroyed the natural base of resources in the pursuit of its growth agenda but also brought tremendous social disharmony via inequitable income distribution through inequitable development patterns. The economic growth that has taken place within the framework of this approach has seriously undermined both "ecological" and "social sustainability." It has pushed the larger section of humanity to the brink of social disaster and dehumanization, on one hand, and, on the other hand, created an "ecologically hostile" consumer society that adores and values consumption as an end in itself.

The biggest challenge we face today is to come up with a development approach that provides a balanced and comprehensive perspective on both social and ecological dimensions of issues with respect to why natural systems (biological diversity and ecosystems) are important and valuable, what are the social and economic causes responsible for the destruction of natural systems (loss of biodiversity, habitat, and ecosystem destruction), and, finally, what measures can be used to reduce or reverse the destruction of Earth's natural systems. This requires, more than anything else, a reassessment of our development paradigm, particularly its basic assumptions, values, and belief structures. What we need today is a new concept and definition of development itself and the policy framework that stems from such conceptualization. Only from such a policy framework can one see the multifunctionality of environmental systems, especially their roles in generating life support systems and services that can be recognized in the development planning of the national governments precisely for the following reasons as Pearce et al. (1990) point out: "Our imperfect understanding of the life support function of natural environments; our lack of capability to substitute for those functions and the irreversibility of the losses of the environmental resources."

Only from the willingness to recognize the values of "life support system" and "services" generated by planetary ecosystems, the understanding of how the social and economic forces (poverty, population growth, and the scale of economy) are driving the destruction of planetary ecosystems and biodiversity, and the understanding of the impacts of such destruction on the well-being of human and biotic communities can emerge appropriate policy instruments and measures. Only from such understanding can planners design a policy framework for natural resource management in the world. The forces that are responsible for the loss of biological diversity, global warming, climate change, degradation of planetary ecosystems, and social misery, as we discussed above, include "human population growth," "poverty and inequitable development patterns," "ecologically hostile production and consumption patterns," "faulty development paradigm (ego-satisfying rather than need-satisfying)," and "lack of internalization of environmental externalities in the economic analysis."

Furthermore, human ignorance, lack of understanding, and unwillingness to recognize the life support systems and ecological services of Earth's systems may very well become the basis for their destruction and irreversible loss, which may inevitably reinforce our own destruction and misery. Environmental thinkers and professionals (Upreti, 1994; Gowdy, 1992; Cohen & Polunin, 1990) have long argued that the policies that promote continuation of all life-forms, preservation of human dignity, and social capital can lead to a greater social benefit (human happiness and security) and environmental harmony. "What we need today" as Upreti (1994) points out "are the development policies of the nation states that must maintain ecological processes and ecosystem health, protect biological diversity, and ensure sustainable use of the ecosystem and promote human dignity, equity, mutualistic, and pluralistic cooperation in social system rather than unrestrained competition, control, and domination." This is possible only when there is greater cooperation and willingness among nation states with the sincere intention of not only sharing Earth's resources and wealth but also designing a comprehensive global governance policy framework with firm commitments to address the causes and challenges associated with environmental destruction, degradation of planetary ecosystems, and climate crisis. Science and education policies should be directed toward greater understanding of "the complex dynamics involving human interactions with global planetary ecosystem, governance policy options for promoting the goal of sustainable development, and human cultural adaptation through behavioral changes." We know enough about the problems associated with environmental destruction and what needs to be done to protect the environment and Earth's systems. The question is one of choice and determination to deal with these problems. The problem is that market and political institutions avoid making hard choices and decisions and always opt for popular and profitable short-term solutions, all of which, over time, accumulate as negative cumulative effects on our planet Earth.

How can the environment of planet Earth be separated from humanity, when, in fact, it constitutes the very basis of humanity's existence? Market and social and political institutions must be restructured around the ethics of environmentally sustainable development. An investment in environmental protection and conservation is, in fact, an investment in the preservation and protection of humanity itself, but willingness for such an investment must stem from a new social and "ecological wisdom consciousness," which seeks a comprehensive governance policy instrument and framework to restructure society socially, culturally, and materially and reestablish its lost harmony with nature, the planetary ecosystem. This is not an impossible proposition in view of the end of decades' long superpower rivalry and cold war, but, it requires, more than anything else, a change in the values, perception, and understanding of political decision-makers, corporate elites, and planners with regard to their treatment of nature (planet Earth) in particular and of humanity's relation with nature in general. The evolution of world political leadership guided by "ecological vision" and social justice is necessary to establish a harmonious and synergistic relationship not only among people in a society but also between humanity and nature and is inevitable given the logic of evolution, but the only question is when? If it comes in time before nature exerts its own selection pressure and the planetary ecosystem loses its resilience and integrity, then the cost of human adaptation can be greatly minimized; if not, then the cost would be horrendous and immeasurable. Humanity, with its knowledge system, technological capability, vision, and consciousness, is capable of repairing its ruptured relationship with nature and, perhaps, possesses the potential to direct its own evolution and that of other beings within the context and limit of ecological laws. The question is one of freedom of choice, freedom to choose our own destiny on Earth, our only home.

The powerful nation states of the world have wasted tremendous amount of world resources and energy in the past nine decades and are still continuing to create weapons of mass destruction, drawing images of enemies in the minds of people in the name of "national security," "defense," and "development ideology." This, in turn, has effectively evaded and put aside the issue of environmental problems and crisis for decades, which, in fact, poses a greater threat to humanity today. Even if only 10% of the world's total military budget is allocated for global environmental security (eliminating the causes and problems associated with environmental destruction), both humanity and the biotic community would have a far greater chance to flourish together and save the planetary ecosystem from breakdown. To save humanity and other living systems, we need to save Earth first. That is why we need a paradigm that can be called the Earth-first paradigm, which we will discuss in the last chapter of this book.

Chapter 4
Understanding Ecosystem Evolution and Behavior

> *Every living thing survives by numerous and subtle relationships with all living things and the inanimate environment. When all living things are considered together, these relationships appear as complex, interdependent, and self-regulating structures or ecosystems, in which any one form of life depends on the rest of the system to provide the conditions necessary for its existence.*
>
> Jim Corbett (1981)

4.1 Introduction

This chapter explores the fundamental principles of ecosystem evolution and the dynamic processes shaping biodiversity and ecosystems over time. Understanding these principles and mechanisms is crucial for interpreting ecosystem responses to disturbances and predicting future changes. Ecosystem evolution concerns long-term changes driven by biotic interactions, abiotic factors, and stochastic events (Matthews, 2016). It is a complex interplay between biotic interactions, adaptation, and environmental changes. Over long timescales, ecosystems evolve due to gradual shifts in climate, plate tectonics, and species interactions (Willis & Whittaker, 2002). A recent work has integrated traditional perspectives with insights from evolutionary biology, recognizing the role of adaptation and speciation in shaping ecosystems (Matthews, 2016).

4.2 Ecological Principles

Ecosystem change can be observed in two different dimensions. One is the modification of the physical environment by the processes of life, as in plant succession, formation of marine islands by corals, or formation of limestone by the deposition of calcareous skeletons of mollusks. These processes modify the physical environments in which the species itself lives as well as those of other species. The physical

G. Upreti, *Ecosociocentrism*, https://doi.org/10.1007/978-3-031-41754-2_4

modification of the environment can take place without any genetic change. The other change is the one brought about by biological evolution itself. This takes place in all the species of an ecosystem, with each species affecting some of the others directly or indirectly. Ecosystem change brought about by biological evolution appears to be not only the main cause of species extinction but also the origin of new species as well.

Historically, natural systems arose after passing through a series of traumas and shocks created by droughts, floods, and geo-ecological changes. The systems that thus came into existence, absorbed and adapted to these traumas, and, such systems, as Holing (1986) points out, are not "fragile" but are the creation of change. They are not, however, infinitely resilient. A forest can be turned into a desert or a river into an open sewer. However, this can only happen as a result of human overexploitation of such ecosystems. Ecological systems are affected by destructions/disturbances, both human-made and naturally occurring. Such destructions/disturbances can push humanity toward disasters, but, with careful design and management, they can benefit humanity tremendously. Is variability, not constancy, a feature of ecological systems that contributes to their persistence and to their self-monitoring and self-correcting capacities? Holing (1986) argues that ecosystems display a pattern of connections resulting in sub-assemblies that are tightly connected within themselves but loosely connected to others. It is generally presumed that once a disturbance is removed, the system will ultimately return to its original condition but only if such disturbances or destructions have not exceeded the threshold point of the resilience of the system. If the disturbance exceeds the threshold of resilience, then systemic breakdown of the ecosystem becomes inevitable. Perspectives and principles from biological, evolutionary, and ecological sciences help us understand how ecosystems develop, interact, and change over time. Some of these concepts include evolution by natural selection, ecosystem succession, coevolution, diversity and stability, interconnectedness and interdependence, mutualism, environmental carrying capacity, and system complexity.

4.2.1 Evolution by Natural Selection

If humanity was completely under the control of Nature, then natural laws and natural selection would have kept the anthropogenic activities in a state compatible with the maintenance of ecological equilibrium. The *Homo sapiens* has been liberated from the immediate bondage of natural selection by its intelligence and cultural and technological innovations that could modify Nature, and use these to its advantages, though without understanding the impact effects of "entropy law" and the thermodynamics of ecosystem equilibrium. One important evolutionary principle of relevance is that a species has a genetically programmed rate of potential adaptation to environmental change. In the process of adaptation, some species could readily adapt, whereas others could not. Certain higher animals, including humans, have the capability to manipulate environments to counterbalance or circumvent the negative selection pressure. If environmental changes exceed the genetic or cultural capacity of living organisms, including human beings, to change, evolve, or manipulate the environment, then the result may be total extinction.

It is important to view evolution in the context of "entropy law." We know that energy flows through every organism. It enters the living system at a high level and leaves the system in a more degraded state. Organisms survive by being able to accumulate "negative entropy" from their environment. Natural selection selects those organisms that are better equipped to capture the available energy (Odum, 1968). It is interesting to discuss how this energy flowthrough operates in ecosystem development through ecological succession. It has been acknowledged that energy flowthrough is maximized in the early "ecological succession" when there is still an excess of available energy present. However, when various species begin to fill up a given ecological habitat, they are forced to adapt to the ultimate "carrying capacity" of that habitat using less energy flow more efficiently (Rifkin, 1989). The early stages of maximum energy flowthrough can be identified by the "colonizing phase" of primary succession and the later stages of minimum flowthrough by the "climactic phase" of mature ecosystems (Odum, 1968; Allen et al., 1982). Mature ecosystems are complex, energy-efficient, have more specialized species occupying varied "niches," and are interconnected through a network of positive and negative feedback. Hence, from the point of view of bioenergy, ecosystem development proceeds from the maximum-energy flowthrough in the early stage to the minimum-energy flowthrough in the later stage of the climactic/climax phase, thus giving rise to organizational complexity.

4.2.2 Diversity and Stability

Stability is the propensity of a system to attain either a steady state or stable oscillation, and "resilience" is the capacity of a system to reestablish itself after a perturbation (Holling, 1973). Ecologists seem to intuitively believe that increased diversity offers some pathways through a system for stabilizing signals. This was the earliest idea relating diversity to stability. May (2019) and Levins (1974) suggest that increased diversity increases system lag, which has been viewed as a destabilizing force. There is no consensus among ecologists over the view that diversity always leads to stability. There are ecologists who argue that the nature of "connectedness" is important for rendering stability or instability to a system. They argue that a system may lose either "stability" or "resilience" by virtue of being overconnected because over connectedness introduces a lag in significant signals. Some signals may not be transmitted because they are lagged out of significance, whereas some may become destabilizing because they elicit the wrong response by the time they are received. The ecosystem components become unable to act as a cohesive whole and thus undergoes uncontrolled change (Allen et al., 1988). A system may also become unstable or lose resilience by being underconnected, i.e., it lacks a pathway that allows the passage of sufficient signals between its significant components.

Hence, diversity *per se* does not necessarily contribute to the stability of a system. It is the nature of the relationship of the newly added form to the system that determines the stability or loss of stability of the system. As Starr and Allen (1988) make it clear in the following statements: "Thus, it is that an increase in diversity might have opposite effect on the stability depending upon whether the added individuals become a part of an already overconnected or an already underconnected system. A further complication is that a new individual will often increase diversity but would be either a stabilizing or destabilizing force, depending upon whether the individual is itself connected to or disconnected from the main line of communication of the system to which it is added. Adding a strongly connected individual to an already overconnected system will tend to destabilize the system by further increasing the total system connectedness." The gamut of the argument presented by these ecologists is that there has to be an "optimum" level of connectedness (interaction) of individuals with a system for that system to be stable and resilient. Overconnectedness or underconnectedness of the individuals added to the system will act as a destabilizing force and cause the system to lose its stability or "resilience." As MacArthur et al. (1972) also point out: "If species interact weakly, their communities are vulnerable to invasion by additional species, thereby increasing the interaction: if they interact strongly, they are vulnerable to almost all hazards of existence and some will extinct, thereby reducing the interaction. The in-between degree of interaction is surprisingly robust."

One of the most well-articulated ecological laws is the "law of the diversity and stability." This has been refined by ecologists to hold that "stability leads to diversity." According to this law, a stable environment tends to develop diverse ecological communities, whereas unstable environments (desert, tundra, and alpine) tend to develop less diverse communities over geological time (Baldwin, 1985). Less diverse communities would tend to be more vulnerable to environmental disruptions. This concept can be extrapolated to the system of food production that relies on a monoculture, which is vulnerable to crop failure, similar to what happened with potato blight in Ireland in 1848. What relevant conclusion can we draw from this ecological principle in the context of human beings in terms of their relationship with natural environments? From the short-term ecological perspective, some ecosystems may appear stable. Brinck et al. (1988) describe four transient phases, which are quite common in ecosystems formed by natural or human-induced changes in the environment. These phases pass through four functions: "exploitation," "conservation," "creative destruction," and "renewal." Many ecosystems are deliberately kept in their exploitative phase until they are depleted of their resources and destroyed beyond the limit of self-renewal. Controlled renewal and redevelopment must be initiated to bring back the systems into their functional phase. Ecosystem (habitats) destruction reduces energy flow in natural systems by simply eliminating a part of the system. Here, habitat destruction refers to the conversion of forests to croplands and pastures. Wright (1990) estimates the total reduced energy flow through natural systems due to habitat destruction to be 799 EJ. This amounts to a 30% reduction in the total photosynthetic energy flowing through natural ecosystems by anthropogenic intervention. Vitousek et al. (1986) estimate this loss to be up to 39%. Species elimination through habitat destruction is directly related to this loss of energy flow in natural ecosystems.

Biological diversity has long been recognized as a crucial determinant of ecosystem functionality and resilience (Isbell et al., 2017). Yet, the specific mechanisms and nuances of this relationship remain a topic of ongoing debate and investigation (McCann, 2000). Ecological stability encompasses, as Donohue et al. (2016) point out, several dimensions: resilience (the speed of return to the equilibrium state following perturbation), resistance (the capacity to withstand disturbance), variability (fluctuations around a mean state), and persistence (the capacity to maintain structure and function over time). Biodiversity can potentially enhance ecological stability through several mechanisms, such as the "portfolio effect," where independent responses of species to environmental change average out at the community level, thus reducing variability (Tilman, 1999). Additionally, high species diversity may increase the likelihood of including species that can confer resistance or resilience on specific disturbances, a concept known as the "insurance hypothesis" (Yachi & Loreau, 1999). Recent studies have further underscored that biodiversity's impact on ecological stability can be context-dependent. For instance, Gross et al. (2022) found that the stabilizing effect of biodiversity was more pronounced in stressful environmental conditions. The relationship between biodiversity and ecological stability, while complex, remains fundamentally important in our rapidly changing world. Conservation efforts should strive to maintain high biodiversity, which could potentially buffer ecosystems against a wide range of disturbances.

Ecosystem destruction has been directly linked to the increase in carbon dioxide (CO_2), which is a major contributor of the greenhouse effect and global warming. CO_2 is released into the atmosphere from the combustion of deforested biomes, decreased photosynthesis by deforestation, and soil degradation. CO_2 stabilization is primarily a biological process because the bulk of CO_2 flows through biotic processes. Increased photosynthesis and carbon storage in vegetation, soils, seas, and sediments is strongly needed to stabilize CO_2. This can be done only by preventing the destruction of habitats and ecosystems and redeveloping or restoring the degraded ecosystems. The bulk of the biomass and photosynthesis occurs in the tropics; hence, balanced and sustainable development must be focused on the tropics to alleviate CO_2 buildup. "Thus, preservation of tropical and Amazon forests is essential not only for the preservation of the biological diversity but also for the stability of the ecosystem processes, metabolism and maintaining of the constant energy flow within the biosphere."

The effects of tropical and Amazon deforestation and habitat destruction do not limit the release of CO_2 from burning forest biomass and decreased photosynthesis. Recycling of atmospheric CO_2 through photosynthesis, biota, and soil is the major method to control CO_2 in the atmosphere. A decreased photosynthetic capacity of vegetation causes the released CO_2 to remain longer, reach a higher level in the atmosphere, and absorb more heat, ultimately contributing to global warming and the greenhouse effect.

4.2.3 *Carrying Capacity*

The concept of "carrying capacity" has been derived from biological science. In simple terms, it refers to the maximum population of a species that can be sustained by an ecosystem over time. According to Rees (1990), "Carrying capacity is the maximum population that can be supported indefinitely in a given habitat without permanently damaging the ecosystem." He extended this concept to human society and defined human carrying capacity as "The maximum rate of resource consumption and waste discharge that can be sustained indefinitely in a defined planning region without progressively impairing ecological productivity and integrity."

The idea of carrying capacity is critically important to understand the limit to how much growth an area can accommodate without seriously undermining the environmental quality. The analysis of carrying capacity leads one to the realization that there are limits to the amount of growth that a piece of landscape can accommodate. Ortolano (1984) argues that such analysis ultimately leads to the determination of the maximum level of growth consistent with maintaining socially acceptable levels of environmental quality and public welfare. The concept of carrying capacity becomes meaningful only in the context of two other interrelated concepts: the concept of "growth variable" and the "limiting factor." A growth variable may represent either a population or a measure of human activity, such as the number of housing units per year or the number of individuals consuming resources such as water or forest products. Limiting factors are natural resources, physical infrastructures, and other material elements that are not available in infinite quantity and may restrain growth (Fig. 4.1).

We all know that our society cannot move forward without developmental activities to satisfy the basic and emerging needs of human beings. In the remote past, society did not have to think about the consequences of its activities on Nature, but,

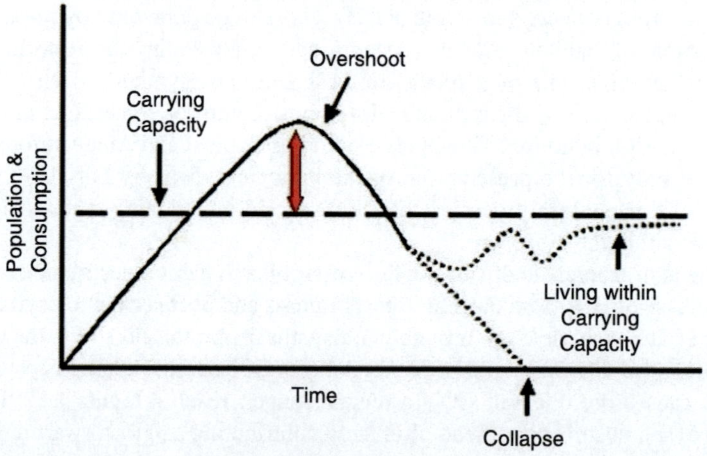

Fig. 4.1 Consequences of carrying capacity overshoot. (Source: Adapted from Foss (2020). COVID-19 Curves and Carrying Capacity. As the consumer population increases beyond the carrying capacity of the resource base, the system collapses)

recently, the human cultural and technological capacity has reached a level that not only affects Nature but can also force Nature to change itself in a way that could be detrimental and disastrous for the survival of humankind. It is in this context that we have to ask certain questions such as what is the carrying capacity of planet Earth for humans? What is the maximum population that can be achieved at all costs versus sustained population without permanently disrupting the life-sustaining ecosystem services (ESs) of planet Earth? Given the current level of production, consumption, and waste throughputs, ecologists argue that the carrying capacity of Earth for humans has already exceeded its limit because the total throughput is more than the regenerative biocapacity of planet Earth.

Scientists have expressed their concerns about the quality of life of billions of people and the integrity of natural ecosystems that sustain human and nonhuman life. Earth may support a much bigger population than the current one at a minimum subsistence level, but the issue here is one of quality and sustainability of the planetary system and humanity. The optimum quantity for quality is always less than the maximum quantity that can be sustained. Earth can support more human bodies in less dignified and more dehumanized conditions for a short period of time. The concept of carrying capacity is extremely simple. Organisms that overshoot the carrying capacity of their environmental resource base that sustain them will face a terrible consequence of mass extermination and possibly extinction. Biologists long ago recognized this when they were studying the population ecology of certain species with respect to their specific habitat. The implication of this concept in the sphere of human socioeconomic and cultural life has just been recently realized, though Malathus long ago pointed this out and Darwin was quick to realize its direct implication in the context of individuals' competition for the limited resources in Nature. The concept of the survival of the fittest was directly derived from Malathus's theory of population increase on the planet of limited resources. Increased consumption, waste production, and exploitation of resources have significant implications for Earth's carrying capacity. Living sustainably necessitates operating within Earth's regenerative biocapacity. According to Rockström et al. (2009), we have already transgressed several planetary boundaries, including climate change, biodiversity loss, and biogeochemical flows, raising serious concerns about sustainability. If the present rate of resource consumption continues for some time in future and if the current developmental activities continue to be dictated by the "egocentric consumerism" of neoliberal corporate capitalism and not by the biocapacity of Earth and laws of ecology, then humanity will inevitably enter the community of endangered species.

4.2.4 *The Principle of Connectivity*

This principle states that everything is connected to everything else. This connection becomes highly transparent and obvious if we look into how the biotic and abiotic components of an ecosystem are functionally and structurally connected to each other and how living organisms within ecosystems are interrelated to each other and finally dependent on autotrophs by way of food chains. Corbett's (1981) observation provides testimonial for this fact: "Every living thing survives by numerous and subtle relationships with all living things and the inanimate environment. When all living things are considered together, these relationships appear as complex, interdependent, and self-regulating structures or ecosystems, in which any one form of life depends on the rest of the system to provide the conditions necessary for its existence. Human beings are as much a part of this ecosystem as any other form of life and depend on the rest of the ecosystem for a food, a breathable atmosphere, drinkable water, and a survivable climate. The Earth has not always provided a suitable environment for humans but was made hospitable over the millennia by functioning ecosystems."

It is obvious that all things, no matter how great and small they may be, are intertwined and entangled in a complex web of apparently subtle relationships that are often beyond the capacity of human comprehension. Since our cultural (economic and social) and physical life depend on this intricate interconnected web for nourishment and sustenance, it is highly imperative that human beings must exercise a great deal of caution in their actions and not upset the delicate balance of the planetary ecosystem for their own existential well-being. Today, humanity is on the verge of either a complete systemic collapse or an evolutionary breakthrough in consciousness that can increase our capacity for connectivity and pull ourselves together to live under the stewardship of planet Earth for our survival and sustainable living.

4.2.5 The Principle of Interdependence

The principle of interdependence is the key to the existence of nature's entire system within which also lies the human cultural system (socioeconomic system). Just like how the cells, tissues, and organelles in the human body are interrelated and connected with one another and function in a coordinated manner for the benefit of the entire body, the physical, ecological, and biological components and processes are interrelated and connected with one another and function in a coordinated manner for the benefit of the entire planetary ecosystem.

One of the key concepts of sustainable development is the interdependence of society, economy, and the planetary ecosystem. From the concept of the carrying capacity of the environment, it is clear that our own existence is dependent upon the sustainable utilization of the resources in the environment and respecting the law that there is a limit to how much Nature can provide without being damaged and degraded. Scientific advances in physical sciences, biology, and ecology have revealed that all living things, including humans, depend upon each other, and are interconnected through natural cycles, ecological systems, and processes. Such cycles, systems, and processes are naturally and continually subjected to change that can harm or enhance the ability of different species to survive and flourish. Unfortunately, mankind, through unsustainable patterns of resource consumption, has now accelerated this rate of change and consequently damaged and ruptured the planet's ecological systems and processes irreparably.

4.2.6 The Brontosaurus Principle

The Brontosaurus principle, as Miller (1979) stated, is "up to a point, bigger the better; beyond that bigger the worse." The Brontosaurus dinosaur was the biggest land animal ever known. It was 65 feet long and weighed between 35 and 40 tons. It became extinct at the end of the Cretaceous period (75 million years ago). Nobody is certain exactly why the Brontosaurus and other species of dinosaurs that dominated the Mesozoic era became extinct. It has been speculated that these creatures were not able to adapt to the changing physical (climate change, radiation or cosmic events, and destruction of low-land vegetation habitats) or ecological conditions (epidemics, eating of dinosaur eggs by early mammals). Their body size was so huge that they needed an enormous amount of food materials to stay alive, and the habitats in which they lived simply could not provide the same. Irrespective of the cause of their extinction, the Brontosaurus could not adapt to the changing environment and paid the price for it by going extinct most probably due to the constraints placed on them by the ecological law of the carrying capacity.

Baldwin (1985) argues that the Brontosaurus principle also applies to the size of socioeconomic systems and units such as communities, unions, corporations, and governments created by the human social system. When these structures grow beyond a certain point, the marginal benefit of growth diminishes and reverses the benefit–cost ratio. Thus, bigger is better only to a certain point beyond which continued growth becomes a liability rather than an asset. This principle is closely related to the economist's "law of diminishing returns." The Brontosaurus principle has an important implication for the current rate of resource consumption by the human population, particularly in developed Western countries. The *per capita* amount of resource consumption in Western countries is so huge that it makes no sense to talk about sustainable development because the rate of resource consumption has far exceeded that of resource regeneration by the planetary ecosystem and there is nothing that can sustain such consumption patterns. The current ecological footprint clearly demonstrates that humanity requires 1.5 planetary ecosystems to sustain the current rate and amount of resource consumption. Is humanity moving toward its own doom's day to meet the same fate of the Brontosaurus? This has become the logical question to ask today.

4.2.7 Popular Ecology

Ecosystems are universally fragile, where everything in Nature is linked to everything else. It is necessary to study all the components of an environment before one can evaluate the impact of a development project. Martin Holdgate (1984) believes that these issues are the exaggerated tenets of popular ecology. He stresses that there is a need for a fundamental understanding of the structure and dynamics of ecosystems as entities. Ecosystems are complexes of plants and animals interacting with one another and with their immediate habitat. Although a link exists between ecosystems, it is by no means always necessary or possible to trace these to their ultimate termination in order to understand the functioning of systems. Ecosystems, like species, exhibit "resilience" and are in a state of dynamic equilibrium, the "balance of Nature," which is the result of continuing change. They have evolved in such a manner as to be able to withstand considerable stress before their structure and integrity are damaged. Indeed, controlled stress can enhance the useful productivity of a system. It is true that most of the ecosystems that have evolved from a long evolutionary process can withstand considerable stress, but such stress must not exceed the capacity for the resilience of ecosystems beyond which systemic collapse or ecological catastrophes become inevitable. Where exactly are we standing today in terms of exerting pressure and stresses on planetary ecosystems? Have not humanity's throughputs exceeded or begin to exceed the biocapacity of planet Earth?

4.3 Ecosystem Evolution and Its Implication

An ecosystem, as defined by Tansley (1935), represents a holistic, functional unit in ecology that includes both biotic components, such as plants, animals, and microorganisms, and abiotic components, such as minerals, climate, and soil, all interacting as a system. The concept places emphasis on the interactions and reciprocal influences between living organisms and their physical environment, underscoring the intrinsic interconnectedness that characterizes natural systems (Odum, 1971). These interacting entities (system) serve as a functional unit and show the general properties of a living system with a tendency to achieve stability. Ecosystems consist of subsystems within larger systems. The biosphere can be considered as the most inclusive ecosystem composed of all living organisms (plants and animals) inhabiting it. However large it may seem, this inclusive system acts as a unit with respect to the exchange of energy. In summary, ecosystem evolution involves a complex interplay of succession, disturbances, feedback mechanisms, trophic interactions, and the actions of ecosystem engineers, each of which contributes to the ever-changing nature of ecosystems.

We believe that Nature is independent of human existence simply because it existed long before humans came into being. It means that the human existence is a recent phenomenon in the evolutionary history of the biological world. It is believed that life arose spontaneously from a complex interaction between matter and the energy furnished by the Sun. From this earliest form of life, arose other forms of life through the process of natural selection. The fuel that drives the engine of natural selection is the variability inherent in the life-forms. What is natural selection that we so often use in our description of the evolutionary phenomenon? Natural selection implies the combined effect of all those physical, chemical, and biological forces operating in environments at a particular point of time. When we talk about natural selection, it makes sense only in the context of a particular spatial and temporal scale because forces of natural selection or their effects do not remain the same on all spatial and temporal scales. It means that the selection forces in nature are also subject to change. Hence, natural selection is a relative concept in terms of its spatial and temporal relationship.

There is no doubt that the inherent biological variability or, more specifically, genetic variability is the very basis or foundation of biological evolution, but where does this variability come from in living organisms? It comes from mutations in the genetic constitutions of organisms, but the question of what causes mutations to occur in living organisms is crucial for understanding the whole mechanisms of the evolutionary phenomenon. It is the environment or the forces of nature or natural selection that induces genetic mutations in a living organism. In other words, natural selection not only operates on the available genetic variability and selects those organisms that are best fit or adaptive to the environment at a particular time but also induces genetic variability through mutations in organisms when the organisms reach a point of having no genetic variability and thus act on this newly created variability. It is generally believed that mutation and natural selection are two opposing forces, where one is involved in the creation of biological variation and the other is involved in reducing such variation, respectively. As we know, a mutation involves the creation of new variation and natural selection works toward creating uniformity, thus eliminating the variation that does not contribute to the organism's survival. Again, when there is change occurring in the environment, the organisms again have to change themselves or change the environment to their own advantage. Hence, there is a coevolution of organisms and their environment, which has been taking place throughout the history of the evolutionary phenomenon.

Does the abiotic physical environment remain static or is it a dynamic system, always in the flux of change? From an evolutionary perspective, it appears that when life first arose on planet Earth, it did not exert influence of any consequence on the environment, but when progressively more complex living organisms evolved from the interactions between early life-forms and their immediate environment, the living organisms, in turn, began to influence (modify) their environments. Hence, the biotic and abiotic factors in the environment constantly acted upon each other and changed each other. The process of evolution can be better understood if we view it as a process of interaction. This interaction always leads to new conditions for life, and life-forms modify or create new conditions. The living organisms that adapted to the previous environmental conditions may no longer be able to adapt or fit into the new environmental conditions. The nature of this interaction is complex because it is not one factor (biotic or abiotic) that is influencing the other factor, but it is both factors that are dialectically influencing and changing each other, thus constantly giving rise to new relational properties of the system.

There are two fundamental aspects of this interactive evolutionary process: these are temporal-spatial dimension and the trajectory of the evolutionary change. If we view the evolutionary process on a short temporal scale, then we may not see any tangible change taking place in the system and, naturally, we may assume that the system is in a steady-state condition, but quite a different picture of the system emerges when we view it from a longer temporal perspective. It took almost three billion years for life to evolve and develop from a primitive unicellular form into a highly complex multicellular organism and, eventually, to the evolution of the *Homo sapiens*. Apart from the scale, the direction of the evolutionary process is another aspect that has important bearing on the understanding of the evolution. It appears that life arose from the interaction of matter with energy about four billion years ago on this planet. The earliest life-form has been believed to be the simplest unicellular organism. From this unicellular organism arose multicellular organisms with higher levels of organizational and functional complexity from which emerged more organisms with even more higher levels of organizational and functional complexity, eventually giving rise to the *Homo sapiens*, the most complex organism, being in a continuous interactive interplay between biotic and abiotic forces over billions of years. In other words, natural selection selected organisms that were more adaptive or organisms with higher levels of organizational and functional complexity. Is there a correlation between organizational and functional complexity and adaptiveness? Or are organisms with higher levels of organizational and functional complexity more adaptive than are those that do not possess these traits? This is a highly disturbing question because this implies that evolution is always directional, which means that it always proceeds from simpler beings to progressively more complex beings. This suggests another disturbing possibility, which is the teleological design for the evolutionary process. In other words, is the evolutionary process teleological? Or is there a special scheme for the evolution of life that always proceeds from the simplest to the most complex? If we accept the evolutionary process as teleological, then we must also accept the fact that there is a designer for this whole scheme of things. If that is the case, then, the biblical theory of special creation should become the logical explanation of the whole business of the evolutionary process.

Modern evolutionary theory certainly does not prescribe the teleological design for the evolutionary process. First, we must be clear that evolution is a process that comes into being from the interaction between biotic and abiotic forces. In this process, both abiotic and the biotic forces influence each other and change each other first in a quantitative way, which, then later, paves the way for a qualitative change. The process is not teleological but stochastically dialectical. The nature of interactions between the biotic and abiotic components of an ecosystem is not teleological but stochastic or probabilistic. This means that the evolution that takes place at a particular time in a given environment (system) is the function of the stochastic process operating on the biotic and abiotic communities in that environment at that particular point of time.

4.3.1 Ecosystem Succession and Adaptation

One of the primary mechanisms driving ecosystem evolution is succession, a process first formalized by Clements (1916). In this process, an ecosystem undergoes sequential stages of development following a disturbance event, such as a wildfire or hurricane. The succession process, which includes primary and secondary succession stages, moves the ecosystem toward a "climax" state, a dynamic equilibrium that represents the maximum biodiversity and stability attainable in a given set of environmental conditions (Clements, 1916; Odum, 1969). Succession is defined as the phenomenon of orderly change of an ecosystem. It is based on the physiological and interactive processes that give rise to a habitat, becoming less favorable for the survival of some species and more favorable to other species adapted in different ways. This is particularly true of the succession of plant communities. Animal species may also show a similar tendency through the accumulation of waste products and destruction of plant foods, but the effects are less obvious because of the mobility of the animals to new habitat localities to find food. The physiological change processes that drive succession involve differing metabolic requirements, patterns of growth, and means of dispersal. Such processes are limited and guided by climatic factors and the availability of seeds from the various species involved (Scott, 1989). Succession becomes a predictable phenomenon in a place with a given set of climatic and soil conditions and knowing the species for recolonization. For example, in the north temperate zone of North America, each case of succession recapitulates what happened after the last ice age when great stretches of land were covered by barren sand and gravel by glaciers (Scott, 1989). It seems that succession eventually results in stability, which resists further ecological change. On a regular basis, ecosystems cannot be disorganized or reorganized, but phenomena such as fire, floods, volcanic eruption, and climate change can cause the necessary breakdown of the system.

Odum (1969) defines an ecosystem as a "Unit of biological organization made up of all organisms in a given area.... interacting with the physical environment so that a flow of energy leads to characteristic trophic structure and material cycles within the system." The biotic entities, plants and animals, interact with each other as a community. He suggested that the processes of ecological succession can become the basis for the broader concept of development. Succession or ecosystem development is an orderly and predictable process, resulting from the modification of the physical environment and culminating in a stabilized ecosystem with maximum biomass per unit of energy flow. This is an organizational process, but, unlike the organization of a single plant or animal, which persist for only one generation, the process of ecosystem development extends over many generations, and the final ecosystem may persist thousands of years. Biological evolution continues during the development of ecosystems, with a reciprocal interaction between the two processes.

The important change processes in succession are physiological and may be affected by a number of factors, which may be collectively called the forces of natural selection. All physical, chemical, climatic, and bioclimatic factors operating in a particular place constitute the forces of natural selection. The effects of such forces may bring about certain changes in the physiological processes that drive the development of ecosystems. Changes in the organization of an ecosystem are reflected by changes in the absolute and relative numbers of an organism in each species. As Scott (1989) correctly points out: "If a species that is intermediate in a food chain becomes extinct, all species dependent upon it must either change their food habits or become extinct themselves. Or if a primary food source does not prosper, the whole chain may be threatened." Apart from succession, the most significant causes of changes in the numbers of a species population are the climate change affecting the entire ecosystem and the genetic evolutionary changes within each of the component species. Drastic climate change is a long-term phenomenon and usually consists of short-term fluctuations that may produce corresponding fluctuations in numbers over several years. Evolutionary genetic changes are usually slow and gradual, and they bring about changes in the organization of the system on a different level. Genetic changes result in more successful adaptations that ensure increases in the number of the species; if the genetic changes result in less successful adaptation, this causes a reduction in the number of the species and, in more extreme case, extinction. Successful adaptation of a species requires that other species directly dependent upon it in the ecosystem adapt to the ecosystem through their own genetic changes. Failure to successfully adapt may result in extinction, whereas successful adaptation introduces a new balance of organization between species (Scott, 1989).

Adaptation is a process by which the population of a species organize themselves to a changing environment and this is a continuous never-ending process. As soon as an organism adapts to a particular environment, it has to prepare itself to adapt to the changes taking place in that environment. In this process, the organism and the environment interact with each other, thereby influencing each other and entering into a new organizational relationship that may possess its own properties and characteristics. From such an interactive and adaptative process emerges an "ecosystem" with higher levels of organizational complexity. One of the characteristics of such a complex ecosystem is that each species tends to occupy an "ecological niche" in the ecosystem with little direct competition with other similar species.

4.3.2 Evolution of the Biota

Biotas evolve. They have structures and dynamics and are the units of the evolutionary natural process. Energy flow is central to biota evolution. Biotas, like organisms, evolve at many spatial-temporal scales and with equilibrium and nonequilibrium processes. The evolution of the biota requires us to broaden our concept of natural selection. Van Vallen (1991) argues that ecological systems do evolve but not in the same way as do organisms. It appears that the center of system ecology is energy flow. Free energy is what powers life, and most of the content of physiology and biochemistry consists of the manifold effects and control of the flow of this energy through a single individual. Biota, population, or species exist only to the extent that it can obtain and process free energy. It is this amount used, of free energy or available energy, which determines the scope of the biota. It appears that energy is casual. All biotas change over time at various timescales. As Van Vallen concludes (1991): "The ecological interactions among organisms evolve though not entirely in the same manner as the organisms themselves. Which of this sort of evolution is primary is a matter of our perspective, not of the biological processes. Some properties of biotas are just the conjunctions of properties of their constituents, while others are emergent (the same is true of species and the higher taxa). Whether emergent or not, they are valid properties at the biotal level because they are parts of the structure of the biota." The biotic world is, among other things, a system of energy flow. Biotas have diverse manifestations; a natural group from one casual perspective may be unnatural from another. For any biota, we can consider the energy flow itself. We can ask such questions as how the flow is partitioned, what causes and regulates the flow and its partitions, how these change over time at various scales, and what processes cause these changes. One does not need to be a system ecologist in order to appreciate and study such questions.

Terrestrial ecosystems or biotas have been believed to play an important role in determining regional and global climate; the most spectacular example of this is in Amazonia, where destruction of tropical rainforests have led to warmer and drier conditions (Shukla et al., 1990; Nobre et al., 1991). Bonan et al. (1992) report that the location of boreal forest and of the correlated climate indices were the outcome of the coupled dynamical interactions in which the geographical distribution of boreal forest affects climate, and vice versa. They conclude that boreal deforestation may initiate long-term irreversible feedback in which the forest does not recover, and the tree line moves progressively farther south. Biotas are not superorganisms, but they are organized and have diverse processes and regularities. The phenomena are real, and they deserve real attention from an evolutionary perspective.

4.3.3 Coevolution

Animals and plant species are the entities within an ecosystem. Part of the ecosystem change consists of the biological evolution of each component species. These changes modify the relationship between the species within the ecosystem. Hence, the processes of evolution include all ecologically related species. This genetic aspect of ecosystem change comes under the domain of coevolution. Coevolution consists of two important aspects of the evolutionary processes: genetic changes taking place in plants and animals and the how they affect each other. The adaptation in plants is entirely limited to physiological processes based on structural organization; however, through coevolution, plants can make use of animal behavior for the purpose of fertilization and dispersal (Scott, 1989). The best example of coevolution is that of flowering plants and insects. For example, the coevolution of the "Yucca moth" and "Yucca flower" is strikingly convincing. The Yucca moth cross-fertilizes the Yucca flower and leaves its larvae on Yucca seeds. The Yucca plants benefit by providing food to the Yucca moth so that there will be a new generation of moths to carry out the cross-fertilization, which maintains the continuity of the Yucca plant, but, at the same time, Yucca larvae must not eat all the seeds or there will not be any Yucca plants on which they are so dependent. "The coevolution of this remarkable mutualism suggests that both plant and the insect species are intricately interdependent upon each other for their survival." Tackey and Gray (2017) argue that the relationship between Yucca moths and Yucca plants is an example of obligate mutualism not coevolution. They argue that the traits found in Yucca moths that enable them to cooperate appear to have been present before Yuccas became the host plant, suggesting that natural selection's pressure-driven acquired adaptation of Yucca moths may have been the main driving force behind the evolution of a mutualistic relationship between the two species. The coevolution phenomenon seems to indicate how the evolution of one species affects that of another species in Nature and how they are interrelated and interconnected. As Scott (1989) eloquently pointed out: "Evolutionary changes in one species must inevitably affect those species with which it has ecological relationship, and vice versa. If the Yucca plant became extinct, the Yucca moth would have to evolve in other directions or to become extinct also."

Another good example of coevolution is that of seed dispersal of the pine species by nutcracking birds. Pines are cross-pollinated species and are fertilized by windblown pollen. They develop heavy seeds and are eaten, carried about, and occasionally dropped elsewhere by nutcracker birds. It works to the advantage of the pine seeds to provide good nutrition to nutcracker birds and to the advantage of nutcracker birds to carry the seeds about and occasionally drop them. The process of coevolution involves a coupled phenomenon characterized by the mutual modification in both agents and systems. Hence, biological and/or physical systems enter into well-connected webs and networks of interrelations that are essentially nonlinear, with lags and discontinuities, thresholds, and limits. The coevolutionary ecological systems are in a dynamic process of self-organization and self-maintenance (Berkes & Folke, 1992).

4.3.4 Ecosystem Behavior

Ecosystems can be considered as systems that are defined as complex units in space and time systematically cooperating in a way that the integral configurations of structure and behavior can be restored after nondestructive disturbances (Weiss, 1970). Muller (1992) points out that a system consist of elements, which can be grouped into subsystems, and relations, which are the interactions between the system's elements. The arrangement of elements and their relations (interactions) in space and time determine the structure of the system, whereas the function of a system is defined as the order inherent in the interactions (Bahg, 1990). Laszlo (1978) and Muller (1992) have pointed out four fundamental behavioral features of ecosystems:

Order and irreducibility: Systems are entities. The properties and behavior of higher levels cannot be described as the sum of the properties and behaviors of its components if the components are studied in isolation. Higher levels can be comprehended only through the knowledge of the ensemble of their parts and the relations between them (Bertalanffy, 1972; Muller, 1992). Emergent properties arise at a higher level, which display additional qualities rather than being the simple sum of lower levels.

Self-regulation: Systems react invariantly on certain transformations. They can return to their specific referential states after receiving limited inputs (Muller, 1992).

Self-organization: Certain physical and chemical processes help the system develop toward an ordered state or an ordered sequence of states from initial conditions in nonlinear dynamic systems. The evolution of this state of higher complexity is not forced from the exterior but is reached spontaneously, caused by symmetry-breaking interactions and the properties of the variables that take part in irreversible processes (Ebeling, 1989; Muller, 1992).

Hierarchical structure: All living systems are organized hierarchically (Bertalanffy, 1968; Miller, 1978). Observers are allowed to define subsystems according to the aim of their investigation. As Muller (1992) stated: "Organic systems are open systems exchanging energy, matter, and information with their environment. These fluxes come into existence because open systems are able to maintain gradients of electrical, chemical, hydro-dynamical, and thermal potentials. This maintenance depends on a continuous input of energy, therefore organic systems are dissipative structures far from thermodynamic equilibrium. Their integrative structures are based on processes of self-organization."

Solar energy is the gradient that drives development in any system, allowing the cyclic use of matter required for self-organization, which results in the evolution of complex, interdependent hierarchical structures (Odum, 1969; Berkes & Folke, 1992). It is the "self-organizing" ability or "resilience" of a system that determines the capacity of the system to respond to stresses and perturbations. It is the diversity within a system that confers ability on the system for self-organization or resilience in the face of outside perturbations. Hence, this is where the significance and importance of biodiversity lies, in the context of their role in preserving ecosystem resilience. Holing (1986) describes ecosystem behavior in the sequential interaction between four phases of system functions:

Exploitation: This phase consists of those ecosystem processes that are characterized by rapid colonization of disturbed ecosystems.

Conservation: This phase consists of resource accumulation characterized by energy and material buildup and storage in the system.

Creative destruction: This phase consists of the abrupt changes in the system caused by external disturbances. Such abrupt changes result from the release of energy and materials accumulated during the conservation phase.

Reorganization: The energy and materials released by the abrupt changes in the conservation phase offer a new opportunity for reorganization of the system. This reorganization takes place through the mobilization of released energy and materials and the reorganized system becomes available for the next exploitative phase.

Perring et al. (1992) point out that the last two phases of system functions are related to the concept of "resilience." The ability of a system to maintain its "self-organization" in the face of external stress or disturbances is crucial. Studies conducted on the relation between the complexity and resilience of ecosystems so far have not yielded consensus on the question, though the reported results on the link between stability and complexity of the ecosystems are contradictory (Orians & Kunin, 1990). One can argue that the seemingly contradictory relation between stability and complexity may have resulted from our inadequate understanding of system complexity and the concept of stability itself.

The fundamentally important assumption of hierarchy theory is that smaller subsystems undergo a more rapid dynamic change than do larger systems (Allen & Starr, 1982; O'Neil, 1988; Norton & Ulanowicz, 1992). Stability and complexity may make sense only in relation to which hierarchical level they are being viewed. Lower-level subsystems may be less stable and less complex compared to higher-level systems in the hierarchy. Hence, scale becomes a subject of central concern not only from the perspective of understanding the stability–complexity phenomena in nature but also from the standpoint of pursuing an appropriate environmental policy goal. Today, the scale on which we should address environmental policy constitutes, perhaps, the most important policy question pertaining to conservation and preservation of natural ecosystems. Determination of the appropriate scale and perspective from which to address environmental problems involves a complex interaction of value definition, concept formation, and scientific description (Norton & Ulanowicz, 1992).

Ecosystems are considered as the fundamental units of biological organization, and their structure and function are sustained by synergistic feedback between the biotic community and the physical environment. Usually, the growth and development of a biological subsystem is constrained by the physical system, which is, in turn, modified by the biological subsystem. Biological systems (populations and communities of species) do not merely adapt to the physical system; they also constantly modify their physical systems in the process of adaptation (Lovelock, 1991). The process involves a coupled phenomenon characterized by modification in both the biological subsystem and physical system. Evolution of an ecosystem involves a coupled phenomenon that is characterized by the mutual modification of both biological and physical systems (Schlesinger et al., 1990). Hence, biological and physical systems enter into well-connected webs and networks of interrelations that are essentially nonlinear, with lags and discontinuities, thresholds, and limits. Ecological systems are in a dynamic process of self-organization and self-maintenance (Berkes & Folke, 1992).

4.3.5 *Complex Systems and Their Characteristics*

Complex systems are defined by their dynamism, nonlinearity, adaptability, and emergence. They comprise interacting components, or agents, that exhibit collective behavior, including patterns of self-organization, robustness, and adaptability, among other nontrivial phenomena (Simon, 1962). Complexity emerges in natural (e.g., ecosystems), social (e.g., economic systems), and engineered (e.g., electrical power grids) systems (Mitchell, 2009). Parrott and Kok (2000) define a complex system as "A network of many components whose aggregate behavior is both due to, and gives rise to, multiple-scale structural and dynamical patterns which are not inferable from a system description that spans only a narrow window of resolution." Typical examples of complex systems include ecosystems, economies, transportation networks, human neural systems, etc. Research studies on complex systems have shown that many complex systems, whether ecosystems, brain neural systems, or social systems, share common structural and dynamic properties. Ecosystems are nonlinear, self-organizing, seemingly chaotic structures in which individuals interact both with each other and with the myriad of biotic and abiotic components of their surroundings across geographies as well as spatial and temporal scales. Ecosystems are complex systems, and complex systems are typically described by the age-old metaphor "the whole is more than the sum of the parts." Generally speaking, complex systems are those in which the sum of the parts is insufficient to describe the macroscopic properties of the systems' behavior and evolution. Interactions among parts of the system, at different scales, in a nonhierarchical, nonlinear, and self-organizing manner lead to emerging properties that fail to have a single, certain unfolding in the future (Furtado et al., 2019). Complex systems exhibit the following characteristics:

Nonlinearity: Complex systems exhibit nonlinear dynamics, meaning that the behavior of a system is not directly proportional to the input. Minor changes can result in major effects (the so-called butterfly effect) or, conversely, major changes can have only a minor impact (Bar-Yam, 1997).

Emergence: The concept of emergence in complex systems refers to the appearance of unexpected, novel patterns or properties in a system, which are not present in the individual components of the system but emerge from their interactions (Holland, 1998).

Adaptability: Complex systems often have the capacity to adapt to their environment, making them robust yet flexible. They can evolve and change over time (Gell-Mann, 1994).

Self-organization: Self-organization refers to the spontaneous formation of structures or functions in a system without external control. The system's structure emerges from interactions among its components (Camazine et al., 2001).

Interconnectivity and interactions: In complex systems, components are interconnected, often in networks, with interactions and relationships that may not be immediately evident. The interactions can be physical or involve the exchange of information (Strogatz, 2001).

Feedback loops: Feedback loops are common in complex systems and may involve positive (reinforcing changes) or negative feedback (counteracting changes), leading to dynamic behavior over time (Sterman, 2000).

Ecologists have long sought tools to streamline and aggregate information about ecosystem complexity. Starr and Allen (1988) propose a methodology called hierarchy that reflects the assimilation of hierarchy theory into ecological research and has been successfully applied to the understanding of complex systems. Hierarchically informed data analysis has enabled a revolution in the ecological understanding of complex ecosystems. The essential characteristics of a complex system include local interactions between individual components, feedback between processes occurring at different scales, amplification of minor variations in initial conditions, and the emergence of patterns in the absence of a global controller. There is widespread consensus that ecological complexity emerges from the interactions between organisms and their biotic and abiotic environments, and attempts to clarify these basic notions has produced several useful conceptual definitions of ecological complexity (Madhur et al., 2010). For example, Levin (1998) considers ecological systems to be prototypical complex adaptive systems (CASs) in which macroscopic system properties such as trophic structure, diversity–productivity relationships, and patterns of nutrient flux emerge from interactions among components and may provide feedback to influence the subsequent development of those interactions.

Maturana and Varela (1980a, b) formulated the theory of "autopoiesis" as the organizing network of a living system. The product of an autopoietic system is its own self-organization, and, over time, this process leads to the development of a complex self-organizational network. The spontaneous origin of life-forms and their progressive evolution into more complex forms of life and living systems and the emergence of complex ecosystems can be seen as products of the self-organizing property (autopoiesis) of a living system. Ecosystems are living systems, and, like all living systems, they form multilevel structures of systems nesting within other systems. An ecosystem at the largest level consists of communities of plants and animals in the landscape extending over millions of square miles known as the biome. As can be seen in nature, every living entity continually renews itself, as its cells break down and build up structures and its tissues and organs replace their cells in continual cycles, thus maintaining its identity or pattern of organization. This self-renewal or self-organization can also be seen in ecosystems at all levels.

Most definitions about a complex system stem from information theory and reflect the belief that a system is complex when it comprises many different parts whose combined state or behavior are difficult to predict. Ecologists' perception of complexity invariably includes diversity, interactions that cross many spatial, temporal, and organizational scales, ecological memory (historical effects), and heterogeneous and fluctuating environments (Field & Parrot, 2017). Madhur et al. (2010) emphasize the importance of memory and environmental variability because these are the key features of ecological systems that create and maintain diversity. They further explain that diversity is not just a passive effect of ecological interactions but an important determinant of a system's persistence and adaptability in the face of environmental change. Ecological complexity might be generated by the interaction between a variable environment and internal self-organized dynamics, an aspect notably lacking from standard definitions of complex systems that typically exclude environmental effects. The combination of local interactions and feedback loops between different hierarchical levels in a complex system gives rise to self-organized structural, spatial, and temporal signatures that are neither completely ordered nor disordered.

4.3.6 Ecological Systems and Chaos

Chaotic systems can be defined as those systems that do not settle into equilibrium or a simple cycle but rather exhibit irregular and nonrepetitive dynamics (Berryman et al., 1989). Ecological systems contain the seeds of chaos because they all possess positive feedback processes (reproduction, cooperative behavior, interspecific competition) and because time lags frequently occur in their negative feedback components (predator numerical responses, food depletion and recovery, pollutant buildup). An ecological system, a type of complex system, can exhibit characteristics of chaotic behavior, such as sensitivity to initial conditions, unpredictability, and complex dynamics. Notably, chaotic dynamics are deterministic, meaning that they can be described by precise rules, but they may seem random because of their sensitivity to initial conditions (Lorenz, 1963). One important issue in ecology is that ecological systems often behave unpredictably. The issue is whether this unpredictability is due to deterministic chaos, stochastic environmental disturbances, or human-induced environmental disturbances. It is generally believed that chaotic dynamics emerge when positive/negative feedback systems are dominated by positive feedback growth processes (Schlesinger et al., 1990). Empirical evidence and evolutionary and ecological reasoning do not support the view that ecosystems behave chaotically, but they can be driven to chaos by human actions that increase growth rate or induce delays in the regulatory (negative feedback) processes.

4.3.7 Natural Systems, Ecological Processes, and Services

The sustainability of natural systems is dependent on ecosystem services (ESs), the delivery of which would cease without a strong connectivity among the ecological processes underlying their production. However, research on the connectivity between multiple ecosystem services is in its early phase, and preliminary studies have focused on the flows of individual ecosystem service; but, as ecologists point out, research is yet to expand in terms of incorporating the interactions and feedback that occur between several different types of ecosystem services and how such complex relationships might influence landscape sustainability (Field & Parrot, 2017). The use of multi-ES networks can help evolve our understanding of landscape connectivity and resilience and incorporate complex ecosystem service relationships into applied planning and management of natural resources.

Parrott (2002) argues that ecological engineers must incorporate concepts arising from complex system studies such as emergence, scaling, self-organization, and unpredictability into their conceptual model of an ecosystem in order to effectively design, manage, or restore such systems. These concepts are introduced with reference to complex systems in general and then with specific reference to ecosystems. He provided the guidelines on how ecological engineering should be approached in the context of complex system studies that are applicable to anyone working in ecosystem restoration and natural resource management. "Ecological engineers" are those professionals who are involved in "the design, operation, management, and repair of sustainable living systems, in a manner consistent with ecological principles, for the benefit of both human society and the environment" as defined by the American Society of Association Executives (ASAE) Ecological Engineering Technical Committee.

With the ever-growing pressures on the environment caused by a rising human population and the associated consumption of natural resources, ecological engineers (ecologists, agriculturalists, forest scientists, etc.) are increasingly required to deal with problems related to environmental management, ecosystem restoration, or mitigation of human impact on wilderness areas. Ecological engineering differs from other branches of engineering in two key ways: "It is founded on an underlying ethics in which the preservation of the global ecosystem is acknowledged to be of key importance, and it has ecology as its fundamental science base."

An ecological engineer, therefore, works within the constraints of a code of ethics that requires designs that improve human welfare while, at the same time, protecting and sustaining the natural environment in which we live. It is argued that for ecological engineering to be successful as a viable profession, its practitioners must develop a new set of principles and practices that accommodate the variability and unpredictability of living systems. Parrot and Meyer (2012) assert that complexity science provides a valuable conceptual framework and quantitative tools for dealing with cross-scale interactions and nonlinear dynamics in social–ecological systems. They identified concepts and actions arising from complexity science that can applied by ecosystem managers to achieve sustainable future landscapes, the management of which requires an understanding of the myriad interacting human and natural processes operating on the landscape over a continuum of spatial and temporal scales.

The concept of "resilience" has been increasingly applied to ecological approaches for assessing complex, interconnected systems (Moberg & Simonsen, 2014). Field and Parrot (2017) define "resilience" as the long-term capacity of a system (e.g., a landscape) to deal with change while maintaining essentially the same structure, function, and identity, without crossing critical thresholds. Several properties contribute to system resilience, including adaptive capacity, diversity, redundancy, slow-changing variables, feedback, and connectivity (Biggs et al., 2012; Parrott & Meyer, 2012). In general, low resilience results in low-biodiversity landscapes that produce fewer ecosystem services and are more vulnerable to disturbances (Simonsen, 2014). Maintaining landscape-scale resilience is increasingly cited as a goal of environmental management and planning strategies; however, there exists no accepted method for measuring and assessing the ecological resilience of landscapes (Field & Parrot, 2017). It is generally accepted that an ecologically connected landscape that facilitates spatial movement of species and environmental flows should be more resilient than a fragmented landscape (Mitchell et al., 2013).

The goal of conservation and restoration activities is to maintain biological diversity and the ecosystem services provided by this diversity. These activities traditionally focused on the measures of species diversity that included only information on the presence and abundance of species. Cadotte et al. (2011) point out that how diversity influences ecosystem function and depends on the traits and niches filled by species. There is a correlation between functional diversity and species richness, and this aspect of diversity potentially affects community assembly and function. Given this explanatory power, functional diversity should be incorporated into conservation and restoration decision-making, especially for those efforts attempting to reconstruct or preserve healthy, functioning ecosystems. Ecological experiments, observations, and theoretical developments show that ecosystem properties greatly depend on biodiversity in terms of the functional characteristics of organisms present in the ecosystem and the distribution and abundance of those organisms over space and time (Hooper et al., 2005). The scientific community needs to come to a consensus on the many aspects of the relationship between biodiversity and ecosystem functioning, including the management of ecosystems. It is not only important to integrate the knowledge about biotic and abiotic controls into ecosystem properties, how ecological communities are structured, and the forces driving species extinctions but also necessary for management of Earth's ecosystems, ecosystem processes, and the diverse biota and the ecosystem services they provide.

4.4 Implications for Human Civilization and Living Systems

In Nature, the evolution of complex planetary ecosystems in general and the phenomenon of coevolution and mutualism in particular have important implications for human civilization and living systems on planet Earth. Humanity must learn to live in a mutually beneficial relationship with Nature without destroying the ecological processes that produce vital goods and ecological services for its own existential survival. Environmental or ecological changes in Nature inevitably affect the *Homo sapiens* and human civilization that cannot survive without keeping the ecologically interconnected relationship intact with Nature. If the ecological systems on Earth become degraded and dysfunctional, then the *Homo sapiens* possibly has to evolve in another direction or become extinct. Humanity cannot escape from the consequences of the breakdown and degradation of natural ecological processes unless it demonstrates its own capability to create an entirely new ecological system that can replace the existing Earth's systems that have evolved over billions of years of the evolutionary process. The cultural evolution of the *Homo sapiens* has taken a wrong direction that is characterized by the development of its destructive capacity and power manifested in the ever-increasing egocentric consumerism, domination, and destructive technologies. The trajectory of this cultural evolution ought to be and must be changed and redirected toward a better understanding of the ecological laws and processes, development of science and technologies to restore the ruptured and dysfunctional natural processes, and, more importantly, cultivating an ethical culture that guides human behavior to live sustainably within the "carrying capacity" of planet Earth's systems.

No human and nonhuman living beings could exist without ecosystem services; nevertheless, humans have neglected the responsibility of maintaining the ability of Earth's ecosystems to provide these services. It is important to recognize that both continuity and change in planetary ecosystems have fundamental and consequential influences on the cultural lives of human beings. Understanding the concept of interconnected interdependence is not only necessary for changing the direction of human cultural evolution but also for recognizing our responsibilities for the future generation of the *Homo sapiens* and other living systems. Humanity must rebuild and reconnect with natural systems and help sustain ecological systems to support the well-being and continuity of humanity and other living systems. There is a paradigm shift that recognizes that complex Earth systems are built upon a myriad of interactions, which result in emergent dynamics equating to more than the sum of individual components (Parrott & Meyer, 2012). Our epistemological approach and scientific research must move away from conducting isolated ecological assessments to realign and focus on multidisciplinary study of how interconnectedness and interdependence are the keys to the maintenance of natural ecosystems (Field & Parrot, 2017). What is needed is a novel scientific interdisciplinary approach to the assessment of multi-ecosystem service (ES) connectivity and resilience across planetary ecosystems, which sustain humanity and all living systems. The scientific community, scholars, thinkers, philosophers, and spiritual leaders have warned us that humanity has been approaching a stage of planetary systemic collapse in the "Anthropocene." There is no easy way out of this predicament unless there is a revolutionary breakthrough in human consciousness that can liberate humanity from the prison of the delusion of separation from Nature. What we need is an ecological consciousness that enables humanity to see this relational interdependence and interconnectedness with Nature (Earth's systems) and help develop a moral compass to guide its behavior. The view expressed by Stephen Long (2017) is worth pondering: "The Connectivity Principle can also contribute to the necessary shift in the collective consciousness that could prevent a worldwide societal breakdown and see us all on the high road to peace, international cooperation, economic security, and advances in our social development that are beyond our current imagination."

Biological systems (populations and communities of species) do not merely adapt to the physical system, but they also constantly modify their physical systems in the process of adaptation (Lovelock, 1991). Ecosystems are considered as the fundamental units of biological organization, and their structure and function are sustained by synergistic feedback between the biotic community and the physical environment. Usually, the growth and development of a biological subsystem is constrained by the physical system, which is, in turn, modified by the biological subsystem. The theory (Gaia) that Earth's biosphere (planetary ecosystem) evolved from the interaction of the self-organizing and the self-regulating process of a living system with its environment is becoming apparent, and the scientific community is finding it hard to dismiss this theory. A large number of scientists from a wide disciplinary spectrum, including evolutionary biologists, environmentalists, ecologists, microbiologists, system scientists, geologists, Earth scientists, etc., have raised their voices and concern over how the current hyper-anthropocentric neoliberal development model has pushed the planetary ecosystem (the biosphere) to the point of its systemic breakdown, i.e., the breakdown of the biosphere's self-organizing and self-regulating feedback process mechanism that evolved over two billion years. It is unfortunate to see how the evolution of the *Homo sapiens* as the most powerful species on planet Earth is moving toward the destruction of this interconnected web of living systems in which it has become the most virulent pathogen. The relationship between the *Homo sapiens* and planet Earth can be conceptualized as the relationship between "Yucca moths" and "Yucca plants." If a Yucca plant as a host cannot sustain the Yucca moth's greedy behavior, then the moth has to evolve in other direction or become extinct. So is the case with the *Homo sapiens*' s relationship with planet Earth. How can it survive with the destruction of Earth's systems that sustain its own existence and the existence of all other living systems? It is in this broader perspective of maintaining the functional integrity and stability of the components of the biospheric ecosystem that mankind should try to understand how significantly vegetation, forests, and biodiversity contribute to its own ultimate well-being and survival. "Should not mankind pursue a cultural (socioeconomic and technological) system that is ethical, kindlier, and compatible with planet Earth's systems?"

Chapter 5
Autopoiesis, Organizational Complexity, and Ecosystem Health

An ecological system is healthy only when its creative processes, represented by the free flow of energy and active competition to utilize it, remain intact. Unhealthy ecological systems will be characterized by a tendency to undergo rapid change, change such as the rapid disintegration of complexity and integrity.

Bryan Norton (1991)

5.1 Introduction

In this chapter, I have attempted to elucidate perspectives regarding how the autopoietic property (self-organization) intrinsic to life systems undergirds the emergence of complex ecosystem structures. Autopoiesis provides the fundamental basis for a system view of life. It pervades the biophysical realm of Nature as a ubiquitous phenomenon. Ecosystems can be comprehensively conceived as autopoietic systems, entities that engender and sustain themselves via homeostatic and homeorhetic responses to shifting environmental conditions. The notion of ecosystem health encapsulates its capacity for resilience, self-organization, and preservation of its functional integrity throughout time. I believe that the health of ecosystems should occupy the focal point of any policymaking and managerial strategy that aims to safeguard Nature, promote conservation, and guide ecologically informed management and societal values such as human health and well-being, which are intrinsically interconnected to the health of ecosystems.

5.2 Autopoiesis and the Evolution of Complex Systems

The term "autopoiesis" is a Greek term, which means "self-making" (capable of making itself). Self-organization has been described as a spontaneous and autonomous movement of a system from a more or less uniform state to a form a pattern

G. Upreti, *Ecosociocentrism*, https://doi.org/10.1007/978-3-031-41754-2_5

(Merino, 1992). Merino provided this description while explaining the emergence of "stylolites," a special kind of physical structure in geological systems. This spontaneous and autonomous organization of a system occurs as a result of the internal dynamics of the system and not from any outside force. Merino argues that all self-organization requires feedback in which the product of the feedback mechanism controls the rate of the mechanism. Merino explains this with an example of the stock market in which the amount of money that one makes (the product) may control how much one plays in the market. If one makes a lot of money, then one will be able to invest even more money and control the market. This is an example of positive feedback.

The simplest of chemical reactions to the most complex biological activities all occur on the surface of Earth in a nonequilibrium state. Madore and Freedman (1987) argue that the pervasive cosmic imbalance is the driving force behind producing an environment conducive to the formation of both structure and complexity. The Belousov–Zhabotinsky (BZ) reaction is known to be the most impressive demonstration of a system showing the power of self-organization in a chemical experiment undertaken 50 years ago. It is one of the most spectacular examples of nonequilibrium thermodynamics, indicating how simple chemical reactions can lead to complex behaviors, and has been studied as a type of chemical "clock" (Zhabotinsky, 1964). This reaction also indicates a possible link between chemically based organization of matter and biological activity (Field, 1985). The speculation that self-organizing chemical reactions were a step toward the origin of life seems highly likely. Morphogenesis, as Madore and Freedman (1987) indicate, is the creation of patterns and forms out of a previously random or uniform environment. This implies that biological "reactions" are analogous and therefore reducible to physics and chemistry: for example, the sequencing of amino acids into the self-replicating structure of deoxyribonucleic acid (DNA), slime mold organization, and the origin of the lens structure of the firefly. Biological systems are considered especially good at producing, maintaining, and reproducing extremely complicated forms.

The Belousov–Zhabotinsky reaction has been characterized by the spontaneous origin of forms, the subsequent growth and stability of patterns, and the increasing complexity of structures, and all of these constitute the characteristics of living creatures. Madore and Freedman (1987) present an example of slime mold amoeba (*Dictyostelium discoideum*) that could exactly replicate the same spatial structures as seen in the Belousov–Zhabotinsky reaction. As they describe, slime molds reproduce by means of spores, each of which is an independent, single-celled organism. However, the cells originating from spores divide repeatedly, and, eventually, the offspring swarm together in a heap to form a common amoeboid mass. The original cell boundaries sometimes disappear, and the once independent cells take on specialized functions, somewhat like organisms in a larger animal. The creeping mass of protoplasm, in some cases as large as 30 cm in diameter, is difficult to classify as a plant or an animal or, indeed, as either a collection of individuals or a larger, single organism. At any rate, it is one of the most curious examples of "self-organization" in biology.

Maturana and Varela (1980a, b) formulated a theory of "autopoiesis" as the organizing network of a living system. Their theory explains that the product of an autopoietic system is its own self-organization and that, over time, this process leads to the development of a complex self-organizational network. The interactions of living organisms (plants, animals, or humans) with the environment are cognitive. Maturana and Varela (1980a, b) postulate that an organism interacts with its environment through a cognitive process, whereby it modifies or creates its own environment and the environment, in turn, permits its actualization. When this complexity of a living organism increased over a long history of evolution, their cognitive processes also increased over time. This self-organization is a pervasive phenomenon in the biological world. The spontaneous origin of life-forms and its progressive evolution into more complex forms of life and living systems and complex ecosystems can be seen as the product of the self-organizing property (autopoiesis) of a living system. The role of feedback mechanism in the organization and stability of a biological system is so much that, feedback, especially negative feedback, seems to be the dominant regulatory mechanism in all biological entities, right from a cell to an organism to a community to an ecosystem (Margulis & Sagan, 1997; Schlesinger et al., 1990). As we can observe at the cellular level, organisms maintain their organization and stability through the process of cell division (mitosis and meiosis). The division of a cell into a pair of identical daughter cells depends on the accurate replication and segregation of chromosomes. The fidelity of this process depends on the completion of certain events before others start. For example, "eukaryotic cells" must finish DNA replication before they enter mitosis and align all chromosomes on a spindle before they segregate. Schrodinger (1994) attempted to solve the dilemma of living systems, which create and maintain an exquisite order from disordered elements dominated by carbon. He explained this puzzle by introducing the idea of "nonequilibrium thermodynamics" through which organisms create the state of order by importing free energy from outside and processing it to generate a lower entropy state within. This is called negentropy, and the organisms thrive on "negentropy." All living organisms are operationally closed systems, but, thermodynamically, they are open systems (Odum, 1988). Open systems such as cells, organisms, ecosystems, and civilizations all evolve through the process of the utilization of free energy imported from the outside environment. Schrödinger (1956) explains that living systems evade the decay to thermodynamical equilibrium by homeostatically maintaining "negative entropy" in an open system.

The spontaneous origin of life and its subsequent evolution into increasingly complex life-forms and living systems, including the rise of intricate ecosystems, can be regarded as a manifestation of the self-organizing characteristic (autopoiesis) of living systems. All ecosystems are living systems, characterized by a multilevel nested structure of systems within systems. At its largest scale, an ecosystem comprises communities of flora and fauna sprawling across millions of square miles, termed as the "biome." As highlighted by Capra and Luisi (2014), eight biomes have been delineated across Earth's terrestrial landscape: tropical, temperate, and coniferous forests; tropical savannas; temperate grasslands; chaparrals (shrublands);

tundra; and deserts. The culmination of millions of years of evolution, the biosphere, represents the grandest planetary ecosystem, encapsulating all ecosystems on Earth.

Autopoiesis, or self-organization, propels the spontaneous generation of new orders within complex systems, ruled by nonlinear dynamics. In recent times, systems ecologists have begun to perceive ecosystems as self-organizing, dissipative structures in operation significantly distant from equilibrium. The defining attribute of an autopoietic system is its constant structural alteration while maintaining its network-like organizational pattern (Capra & Luisi, 2014). Such changes may encompass self-renewal, visible in all living entities, from the microscopic to the macroscopic scale, wherein cells constantly break down and regenerate structures, replacing themselves in a cycle that sustains the overall identity or organizational pattern. The Gaia theory postulates that the biosphere is intimately linked to Earth's lithosphere (rock), hydrosphere (oceans), and atmosphere, forming a collective, self-regulating planetary system. The atmosphere is inextricably intertwined with Earth's life-forms, with the continual removal and replenishment of atmospheric gases by living organisms. The biosphere's metabolic processes continually create, maintain, and transform Earth's atmosphere, converting inorganic substances into organic matter and returning them to the soil, ocean, and air.

The biospheric system's defining feature (Gaia) is the intricate interplay between living and nonliving systems within a singular interconnected network, regulated and sustained by feedback loops known as bio-geochemical cycles. Capra and Luisi (2014) fittingly underscores this point, stating, "When we shift our perception from ecosystems to the planet as a whole, we encounter a global network process of production and transformation as described in the Gaia theory of James Lovelock and Lynn Margulis." Margulis's concept of a planetary autopoietic web system is a product of her extensive microbiological research spanning three decades. Margulis and Dorion (1997) assert that understanding the microcosm, metabolism, and evolution of microorganisms' roles in the early evolutionary history of living systems is crucial for comprehending the evolution of diversity, complexity, and self-organizing capability of the biospheric planetary ecosystem or the Gaian network.

Life on Earth began around 3.5 billion years ago, and the living systems consisted entirely of unicellular microorganisms for the first 2 billion years (Margulis & Dorion, 1997). In the first 1 billion years of evolution, bacteria (the most basic form of life) covered the planet with an intricate web of metabolic processes and began to regulate the temperature and chemical compositions of the atmosphere, which became conducive to the evolution of higher forms of life. The arrival of eukaryotic multicellular organisms with their photosynthetic capacity subsequently paved the way for the evolution of much complex higher life-forms of plants, animals, primates, and, ultimately, humans. Plants, animals, and humans (the visible world of living systems) as well as macrocosms emerged from the microcosm (the invisible world of bacteria), but, even today, the visible living systems function only because of the well-developed network of connection with the bacterial web of life. Margulis advanced the concept of a planetary autopoietic network because all life is embedded in a self-organizing network of bacteria involving a network of control systems. As Margulis and Dorion (1997) state: "It is the growth, metabolism, and gas-exchanging properties of microbes … that form the complex physical and chemical feedback systems which modulate the biosphere in which we live." As we can observe, for example, myriads of bacteria living in soil, rocks, and oceans as well as inside all plants, animals, and humans continually regulate life on Earth. It is inconceivable to imagine the Earth system without the network of complex physical, biological, and chemical systems.

The scientific understanding of the evolution of ecosystem complexity will ultimately improve our ability to manage and restore natural systems. New methods of ecosystem management and ecological engineering must accept the dynamic nature of ecosystems and incorporate concepts such as self-organization, emergence, and adaptation into intervention practices (Parrott & Kok, 2000). Systems theory considers natural complexity as a hierarchy of embedded systems represented on different scales. It has been applied to the understanding of complex biological processes and phenomena from the molecular level to the level of ecosystems. Complex systems are characterized by many levels of organizations that can be considered as forming a hierarchy of systems and subsystems, as proposed by Herbert Simon (1962) in his famous paper *The Architecture of Complexity*. A complex system is an entity that is organized into different levels of structures and properties that dynamically interact with the levels above and below and may exhibit causal regularities, feedback, and order. The best example of such a system is an ecosystem or the whole Earth system. The human brain, individual organisms, cells, and tissue systems of complex organisms are the living examples of a complex system. The cosmos, with its complex structure atoms, molecules, gases, liquids, and, ultimately, stars and galaxies, and clusters and superclusters, can be considered as a nonliving example of a super-complex system (Verhoeff et al., 2018). Levin (2000) was deeply concerned with the pressing issue of our time, which is the looming ecological crisis, and suggested how a new science of complexity can help solve this crisis. Levin argued that our biosphere is the classic embodiment of what scientists call complex adaptive systems. He believes that Earth is under severe environmental assault. His basic premise for coping with this problem is perfectly clear: "to have any hope of dealing with such a complex combination of threats to our survival, we must study the Earth as an integrated physical and biological system." He provided an ecological thought on how ecosystems and ecological communities are structured, with an emphasis on biodiversity. By answering a series of questions – "how did biodiversity arise; what maintains it; and how fragile is it and the services it provides?" – he discussed how ecosystems achieve stability and how resistant they may or may not be to human interference; he pointed out how the evolving science of complexity can shed light on Earth's ecology and help resolve the looming ecological crisis.

O'Neill (1986) point out that ecological complexity is a multidisciplinary field of research that borrows tools and concepts from the core disciplines of complex systems science (physics, mathematics, computer science) as a means of studying the relationships between patterns and processes in natural systems. Complexity theory differs from other analytical approaches, in that it is based upon a conceptual model in which entities exist in a hierarchy of interrelated organizational levels. Whereas in conventional approaches, systems are described at only one level of organization (e.g., a community or nation but not both simultaneously), complexity theory provides a framework in which the relationships between constructs at different hierarchical levels can be accommodated as shown in the Fig. 5.1 below (Parrott, 2002). In recent years, there has been an increased interest in the study of complex systems, particularly ecological complexity. Systems ecologists have defined a complex system as "a network of many components whose aggregate behavior is both due to and gives rise to multiple-scale structural and dynamical patterns which are not inferable from a system description that spans only a narrow window of resolution" (Parrot & Kok, 2000). The most defining characteristic of a complex system consists of the interactions among components and feedback processes occurring at different levels, giving rise to higher-level emergent entities as illustrated by the conceptual model proposed by Parrott (2002) (Fig. 5.1).

Network analysis has become an effective technique to the study of the complexity of the interconnectedness at multiscale levels from molecular, subcellular, cellular, organism levels to community to ecosystem and biome levels. As Zhang et al.

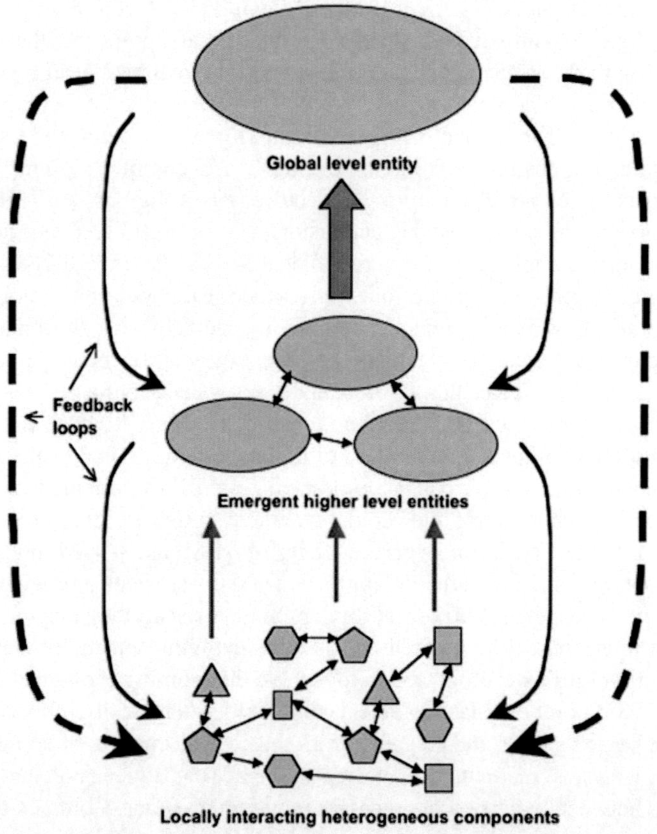

Fig. 5.1 Typical conceptual model of a complex system. (Source: Adapted from Parrott (2002). Complexity and the Limits of Ecological Engineering, 2002)

(2019) explain, in multicellular and even single-celled organisms, individual components are interconnected at multiscale levels to produce enormously complex biological networks that help these systems maintain homeostasis for development and environmental adaptation. Systems biology studies and explores how relationships between individual components give rise to complex biological processes. Network analysis has been applied to dissect the complex connectivity of mammalian brains across different scales in time and space in the Human Brain Project.

5.2.1 The Emergence of Ecosystem Complexity

I want to emphasize that the emergence of ecosystem complexity through the autopoietic (self-organizing) process is the most important question in modern ecological discourse and certainly deserves careful consideration. Complex ecosystems have evolved over a long evolutionary time, and the mechanism involved is the "autopoietic process" triggered by natural selection. At the heart of ecosystem evolution is the idea of ecosystem succession, the process by which the species composition of a community changes over time, typically following a disturbance (Connell & Slatyer, 1977). This theory was first systematically studied by Frederick Clements in the early twentieth century, who proposed that succession followed a predictable path, progressing through a series of serial stages to culminate in a stable "climax" community (Clements, 1916). Clements' view, often called the monoclimax theory, assumes that the climax state is solely determined by climatic factors. Our empirical observation on ecosystem succession reflects that early successional stages are characterized by intense competition among the short-lived plant communities to colonize. All opportunist species compete for resources in an effort to establish their colonies. Their competition among themselves and their interaction with the environment brings about a structural change in the species and also in the environment that becomes more favorable to certain specific organisms (plant and animal communities). Hence, the community that colonized in the beginning disappears, and a new community that can better adapt to changes in the environment emerges. The early community or the colonizer species cannot remain physiologically functional in this newly created environment. As a result, they must pave the way for those species or plant communities whose physiological processes can cope with the new environment. Hence, different kinds of plant communities gradually evolve and come into existence. The plant community that replaces the early colonizer species is characterized by having more specialized physiological processes and functions. It tends to be more complex and stable than its predecessors. This community tends to push the system toward equilibrium. Hence, such species are also called equilibrium species (Odum, 1988). Compared to their predecessors, they persist much longer in the system and create a system of much higher level of organization (a more complex system). The constant interaction between the biotic community and the environment over a long period of time forces both the community and the environment to change. This change, however slow and gradual it may be, in the long run, gives rise to a qualitatively different system characterized by an even higher level of organizational complexity and stability with the maximum amount of biomass per unit energy flow. Thus, it can be hypothesized that ecosystem development proceeds from a simpler and unstable system to a more complex and stable system through the process of autopoiesis. However, stability does not mean that ecosystems are static. Ecosystem stability is relative to spatial and temporal scales. We can say with certainty that ecosystems do not remain static over an indefinite period of time. There is no system that remains static (stable) without undergoing certain changes. From an evolutionary perspective, ecosystems, however complex they may be, are always in a constant flux of change. They change, organize, and reorganize themselves and remain resilient, but this does not induce instability. An ecosystem collapses when its structural components cannot function and maintain matter and energy recycling within it upon outside pressure of perturbation beyond a certain threshold.

Ecosystems are complex systems, and complex systems are typically described by the age-old metaphor "the whole is more than the sum of the parts" (Parrott & Kok, 2000). Generally speaking, complex systems are those in which the sum of the parts is insufficient to describe the macroscopic properties of the systems' behavior and evolution. Research in complex systems studies has shown that many complex systems, whether they are ecosystems, brain neural systems, or social systems, share common structural and dynamical properties (Furtado et al., 2019; Parrott & Kok, 2000). Ecosystems are nonlinear, self-organizing, seemingly chaotic structures in which individuals interact both with each other and with the myriad of biotic and abiotic components of their surroundings across geographies as well as spatial and temporal scales (Parrott & Kok, 2000; Odum, 1968). Interactions among the components of a system at different scales in a nonlinear, self-organizing manner have led to the emergence of the properties that characterize a complex system (Furtado et al., 2019).

From an interactive and adaptative process, it can be seen that a more complex ecosystem emerges with a higher level of organizational networks. For example, in a complex ecosystem, each species tends to occupy an ecological niche with little direct competition with other similar species in the ecosystem, and, over time, such an ecosystem maintains its integrity with a higher level of resilience.

5.3 Ecosystem Health and Its Implication

The interconnectedness between autopoiesis, diversity, organizational complexity, and ecosystem health merits careful consideration. In this context, autopoiesis has been employed to articulate the innovative aspects of ecosystems (Rees, 1990). This concept positions an ecosystem as a self-organizing entity that generates and preserves itself through homeostatic and homeorhetic responses to fluctuating environmental changes (Norton & Ulanowicz, 1992). A key attribute of self-organizing systems is their inherent capacity for creative adaptation to changes in the environment. This intrinsic creativity bolsters the ecosystem's resilience to economic exploitation and its capacity to absorb human-produced waste and pollution. As human activities encroach more heavily upon natural systems, this adaptive creativity becomes increasingly essential. Ulanowicz (1986) forcefully advocates that this capacity for resilience forms the cornerstone of ecosystem health.

As the scale of human activity on Earth expands and anthropogenic landscapes increasingly dominate, the pace of ecological change will unavoidably hasten. This accelerated rate of change will induce a corresponding deterioration of ecosystem health. The notion of ecosystem health encompasses its ability to preserve its self-organization and autonomy over time. As succinctly defined by Haskell et al. (1992): "An ecological system is free from any 'distress syndrome' if it is stable and sustainable—i.e., if it remains active and retains its organization and autonomy over time." This concept of ecosystem health has been recognized by Aldo Leopold (1949), who eloquently expressed it in his renowned land ethics principle: "A thing (i.e., an action) is right when it tends to preserve the integrity, stability, and beauty of the biotic community. It is wrong when it tends otherwise." As Leopold emphatically asserted, the maintenance of the integrity, stability, and beauty of the biotic community can be realized solely through the preservation of ecosystem health. In a similar vein, Rapport et al. (1998) defines ecosystem health as "The state of an ecosystem and its associated structure and processes in relation to its ability to function normally, particularly with regard to its ability to deliver ecosystem services." A system can be delineated as an ensemble of interconnected subunits, bound by quantifiable processes, for instance, an ecosystem, composed of species or aggregates of organisms linked by the transfer of material or energy. This exchange transpires in an orderly and connected manner, evident in the pattern of trophic connections and temporal sequences. This order manifests the components of the overall diversity of material or energy flows within the system. A system's capacity for growth and development is inextricably tied to its biodiversity (Norton & Ulanowicz, 1992). Therefore, biodiversity is directly related to creativity and ecosystem health. Biodiversity's ecological services become more easily quantifiable when it is treated as a system. Developing methodologies to accomplish this is an urgent research endeavor necessitating an interdisciplinary approach involving economists, ecologists, systems biologists, agriculturalists, and systems engineers.

5.3.1 Ecosystem Health and Ecosystem Services

An ecosystem can be defined as a dynamic complex of living organisms consisting of plant and animal (including human) communities located within a given environment. The demarcation of an ecosystem is often guided by the parameters established by humans to suit scientific, managerial, or policy objectives, rather than physical dimensions. Notably, the human community is a crucial component and a primary driving force within ecosystems. The interaction of all ecosystem elements and components is essential for the continued existence of all living organisms, including human communities. Human existence within an ecosystem is significantly shaped by these interactions, from which sustenance (livelihood) is derived. An ecosystem encompasses nurturing elements like clean air, pure and safe water, ecological processes (including hydrological, material, and nutrient cycling), fertile agricultural soils, and diverse plant and animal communities (biodiversity), alongside productive social behavior and a caring attitude. Collectively, these elements provide the material basis for human communities' existence and well-being within an ecosystem. Conversely, elements such as toxins, pathogens, pollutants, non-cooperative and non-caring social behaviors pose potential threats to human health and well-being.

Maintaining the functional integrity of all nurturing elements and ecosystem components is a prerequisite for preserving ecosystem health. As per Haskell et al. (1992), an ecosystem can be deemed healthy only if it exhibits resilience and can sustain its organization and autonomy over time. Resilience, which is the tendency of an ecosystem to return to its initial state after being perturbed, is an important property of a healthy ecosystem. Highly degraded ecosystems lose their resilience and cannot return to their earlier states. Therefore, a highly degraded or unhealthy ecosystem will have negative effects on the health and well-being of the people who live and depend on such an ecosystem. Norton and Ulanowicz (1992) propose that the central concern of any policy and management strategy should be ecosystem health, with all public values (like human health, economics, aesthetics, and morals) hinging on the protection of processes that bolster the health of ecological systems. The ecosystem approach presents substantial potential for enhancing human health through judicious ecosystem (natural resources) management. This approach is grounded in the understanding that the well-being of human communities is intertwined with their ecosystems, which provide all living entities with a home, along with both life-sustaining elements and health hazards. Although not all health risks originating from ecosystems are the product of human activities, many inherent hazards can be minimized or eliminated through astute ecosystem management.

The central tenet of the ecosystem approach to human health focuses on ecosystem management interventions leading to improved human health and well-being while simultaneously preserving or enhancing the health and productivity of ecosystems. Morand and Lajaunie (2018) discuss how alterations in biodiversity, coupled with ecosystem functional degradation, impact the epidemiological environment, triggering the emergence or reemergence of new infectious diseases. Ecosystem health and ecosystem services (ES) are intimately interconnected concepts, with the latter contingent on the former. This implies that healthy ecosystems can provide a comprehensive range of ecosystem services, whereas degraded ecosystems could decrease the quality and quantity of ecological services (Costanza et al., 1997; Norton & Ulanowicz, 1992; Morand & Lajaunie, 2018). Furthermore, the concept of ecosystem health is linked to the notion of natural or ecological capital – defined as the structure and functions of ecosystems that support the creation and flow of natural goods and services valuable to humans. De Groot et al. (2002) have identified more than 32 biological, physical, aesthetic, recreational, and cultural ecosystem services. These services, derived from the proper functioning of ecosystems, are pivotal in maintaining human health, underscoring the immense importance of ecosystem protection and management to mitigate adverse impacts on human health resulting from ecosystem degradation and dysfunction.

5.3.2 Human Health and the Environment

The World Health Organization (WHO) defines human health as "a state of complete physical, mental and social well-being, and not merely the absence of disease or infirmity." The scientific advance in medical science has been spectacular in that significant progress has been achieved in sanitation, health education, nutrition, immunization, and antibiotics, all of which have contributed to the reduction of infectious diseases, the leading cause of mortality in nineteenth century. The progress was based on advances in the specialized aspects of medical science that often focused on diagnosis, prognosis, and prescription in a clinical setting that separated human health from the environmental context in which people live and diseases occur. Despite this progress, the benefits of improved health care have not been shared equally among the people of the developing world and infectious and communicable diseases are still the most common causes of global mortality (Peden, 1998).

It is being increasingly recognized that the health-care system should be based on a more holistic understanding of human health in the context of the environment in which people live (WHO). There is a greater emphasis on understanding the links between human health and a range of biotic and abiotic pollutants that originate from different sources in the environment such as air, water, soil, and food. A review of the history of medicine shows that our greatest achievements in health have come from improving or altering our relationship with our environment. The Sanitary Reform of the eighteenth century serves as a prime example. As Stephen (1998) points out, through improvement in water quality, food quality, housing, and social support, more gains in human well-being and greater strides in reducing human disease were made than have occurred since. Eighteenth century's medical practitioners implicitly applied the teaching of Hippocrates, who instructed new physicians that to know a patient's health, one must know how the patients relate to the environment around them.

5.3.3 Application of an Ecological Model to Human Health

Ecology, as a science, attempts to understand how an organism relates to and is affected by its environment. The components of an ecosystem interact among themselves and interrelate with each other. Although the components of an ecosystem may not interact to produce an optimal state of health, it cannot be denied that all components of an ecosystem are linked together at some level and hence their health is interrelated. This is similar to how the health of a cow affects the health of the herd, which, in turn, affects the health of the farm, which, in turn, affects the well-being of the people working on the farm and people consuming the farm products as well as the general community. Similarly, human activities may degrade agroecosystems, which may accelerate the growth and accumulation of pollutants in the air, water, soil, and food, which, in turn, affect human and animal health, and this, in turn, further degrades the agroecosystems, thus, ultimately, threatening the very survival of animals and human beings. This is an irreversible cycle of marginalization of both human beings and agroecosystems. Only healthy people can maintain a healthy agroecosystem or the environment. As Stephen (1998) points out, ecology may reveal how farm management can affect human health by studying the interactions with agriculture that determine the distribution and abundance of health and disease in a community.

An ecological model of health contends that health is a product of the interactions between individuals and subsystems of ecosystems (Green et al., 1996). Paying attention to such interactions is imperative while investigating and managing health. Stephen (1998) maintains that ecological perspectives have been a part of public health and herd health programs since their inception. The early efforts at population medicine were almost exclusively concerned with finding the key environmental factors that could be controlled to reduce diseases. Health promotion programs for people and herd health programs for animals focused on providing environmental, social, and economic conditions that allowed the populations to meet their expectation for health and productivity. Knowing how a community relates to its biotic and abiotic environment is not only the basis for population medicine and public health but is also becoming a foundation for environmental resource management. The role of ecology in linking agriculture and human health lies in identifying management practices that could allow agroecosystems to meet the social demands and expectations for individual, community, and environmental health. If sound principles of agroecology and epidemiology are applied to the studies of interactions between agriculture production systems and human health, it would become much easier to identify the key components of the system that can be manipulated to reduce or prevent adverse health effects and reduce the likelihood that resource management plans and programs result in unanticipated or undesired effects. Kay and Schneider (1994) argue that we do not manage ecosystems but we do manage our interactions with them.

Stephen (1998) defines a healthy agroecosystem as one that produces safe and good quality food while insuring ecological sustainability and human health. Early efforts to adapt the ecosystem approach to resource management have often focused on the goal of identifying the characteristics of healthy ecosystems but have often forgotten to consider how the health of ecosystem components, including people, are affected by ecological interactions. People are typically seen as risk factors, catalysts of environmental degradation rather than victims or benefactors of ecosystem change. If we consider farming as ecosystem management, then we must acknowledge that people are the key drivers of the ecosystem that can affect or be affected by management practices. In order to improve our ability to predict the human health effects of agricultural practices, we should strive toward finding and evaluating the current links between people and the environment they share with agricultural production systems by trying to solve agriculture-related health problems in an integrated manner. This stems from a simple fact that humans, animals, and the environment make up a farming community and that altering the health of one component can affect the health of others. Agriculture, managed to promote human health, must strive to understand these relationships.

5.3.4 Agroecosystems and Human Health

An agroecosystem is a conceptual construct used to describe a geographically and functionally coherent domain of agricultural activities. It includes all living and nonliving components and the interactions among them. Peden (1998) points out that the boundary or the geographic scale of an agroecosystem can vary from a system consisting of a single farm to communities and watersheds composed of many farms and even beyond that to a large ecoregion. Agroecosystems are not closed systems and are characterized by driving variables or inputs that include immigration and inflows of capital, information, energy, fertilizers, chemicals, and human infrastructures and knowledge. Natural driving variables or inputs include solar radiation, rain, wind and water, soil conditions, biodiversity, etc.

Three key ecological processes regulate the sustainability of an agroecosystem or the ecosystem under natural disturbance. These processes are "soil surface hydrology," "soil organic matter dynamics," "nutrient recycling," and the "synchrony between soil processes and plant nutrient demands." An understanding of how these processes interact and regulate the sustainability of an agroecosystem is a prerequisite for not only understanding the problems of food production and the sustainability of the food production system but also devising appropriate management strategies for the conservation of the resource base upon which the long-term sustainability and the productivity of the system can be maintained (Upreti, 1987). Degraded agroecosystems are less productive and are less resilient to stresses caused by outside perturbations. Ecological processes regulate the health of agroecosystems, thus ensuring their productivity. Agricultural research and management interventions become imperative to find out the optimum level of operation of these processes in order for agroecosystems to remain in a healthy and productive state. How can one know whether an ecosystem is healthy or not? One of the major objectives of the research on ecosystem health is to identify and establish indicators or parameters that can adequately characterize and describe ecosystem health. Once these indicators or parameters are identified and established, we can devise an appropriate action research strategy to bring or maintain the ecosystem in a healthy state. Therefore, monitoring the changes in the indicators or parameters induced by management interventions is important in ecosystem management research. Considering the role of healthy agroecosystems in sustaining a healthy human life, learning how to manage the health of an agroecosystem presents an immediate challenge for agriculture all over the world. A basic working hypothesis, as Peden (1998) has reflected, is that better management of an agroecosystem is a cost-effective strategy for improving human health. This implies that agriculture must be viewed as ecosystem management and that the principles of natural resource management are applied to it (Nielsen, 1998; Paden, 1998).

5.4 Ecosystem Services (ES) Framework

Nature's ecological services are the fundamental foundation for human existence and well-being, but these services have historically been ignored and not integrated into the economic system. In terms of economic valuation, a seminal study by Costanza et al. (1997) estimated the global value of ecosystem services to be approximately $33 trillion per year, nearly double the global gross national product at that time. More recent studies have attempted to update this figure, considering changes in land use, population, and technology. A study by Costanza et al. (2014) revisited the 1997 assessment and found that due to changes in these factors, as well as improved valuation methodologies, the estimated global ecosystem services value in 2011 (in 2007 USD) ranged from $125 to $145 trillion per year. These figures, while illustrative of the magnitude of value associated with ecosystem services, should be interpreted with caution, as they represent a partial, conservative estimate. They underscore the substantial contribution of ecosystem services to human well-being and the economic cost of their degradation or loss. Traditionally and even today, these ecosystem services, which include the regulation of atmospheric composition, climate and hydrological flows, biogeochemical processes, nutrient cycling, soil formation, erosion control, waste assimilation, detoxification, water purification, and maintenance of diverse gene pools, have been implicitly assumed and ignored in societal economics. Current exigencies demand that these services be properly valued within a country's national economy and reflected in public policy. Agroecosystems built on ecological principles exhibit immense potential for the preservation of ecosystem services. Recognizing and duly rewarding farmers who apply these principles and cultivate healthy, productive agroecosystems worldwide would enhance nature conservation and the perpetuation of environmental services.

Global climate change, as indicated by atmospheric and oceanic temperature rises, accelerated snow melting, glacier retreat, sea-level rise, and increasing greenhouse gas concentrations (IPCC, 2014a, b), poses severe threats to human health. These risks encompass increased thermal stress, storms, floods, bushfires, the spread of vector-borne diseases and dangerous microbes, elevated respiratory and cardiovascular diseases due to air pollution, malnutrition resulting from loss of crops, fisheries, and livelihoods, and population displacements (McMichael, 2013; IPCC, 2014a, b). It is noteworthy that the adverse impacts of these environmental changes disproportionately affect disadvantaged communities (Woodward & Mcmillan, 2015). These rapid environmental changes necessitate environmental interventions with dual conservation and human health objectives; however, environmental services remain inadequately integrated with human health and development (Reis et al., 2013).

An approach to amalgamating human health with the natural environment suggested by researchers is through the application of ecosystem services (ESs) to human health (Keune et al., 2013). The ES approach emphasizes the benefits that humans derive from ecosystems and their correlations with human health. Further empirical studies are crucial to establish connections between environmental services and human health, given the rapid, large-scale environmental changes that have significantly undermined human health and well-being. Therefore, it is vital to incorporate environmental and human health considerations into the development of policies and practices. Identifying the ecosystem drivers of human health and incorporating them into ES frameworks could advance research and policy on environmental change and human health.

The basic assumption is that a healthy ecosystem, due to its productivity, undergirds the livelihoods of its inhabitants and its services can directly influence their health. Therefore, understanding the concept of human and ecosystem health, and the interplay between them is paramount. The World Health Organization (WHO) defines human health as a "state of complete physical, mental, and social well-being and not merely the absence of disease or infirmity" (WHO, 1998). The Centers for Disease Control and Prevention (CDC) defines human well-being as "a multidimensional concept covering physical, psychological, and social aspects of wellness, positive emotions and moods, satisfaction with life, fulfillment, resilience, and positive functioning" (CDC, 2013). Therefore, a healthy ecosystem can be considered as one that demonstrates resilience in its structure and function, delivers ecosystem services, and exhibits no overt signs of distress under external pressure. Table 5.1 below summarizes some of these ecosystem services.

The concept of ecosystem health is intertwined with the concept of ecosystem services. Healthy ecosystems generate a full range of ecosystem services. Ecosystems that are degraded and suffer from ecological stress generate low quality and quantity of ecological services, which will ultimately and negatively impact human health and well-being (Rabinowitz & Conti, 2013).

5.5 Manhattan Principles and Lessons from COVID-19

On September 29, 2004, the Wildlife Conservation Society of New York, in conjunction with The Rockefeller University, convened a symposium in Manhattan, New York. The primary objective of this gathering was to explore the extant and prospective disease transmission among human beings, domestic animals, and wildlife populations. Employing case studies of diseases such as *Ebola, avian influenza,* and chronic *wasting* disease, expert panelists underscored the necessity for an international, interdisciplinary strategy aimed at counteracting threats to terrestrial life's health. These discussions culminated in the establishment of the "Manhattan

Table 5.1 List of some ecological services

Categories of services	Definition of services	Ecosystem services
Provisioning services	Provisioning services reflect goods and services extracted from the ecosystem	Food Fodder (including grass from pastures) Fuel (including wood and dung) Timber, fibers, and other raw materials Biochemical and medicinal resources Genetic resources Ornamentals
Regulation services	Regulation services result from the capacity of ecosystems to regulate climate, hydrological, and biochemical cycles, earth surface processes, and a variety of biological processes	Carbon sequestration Climate regulation through control of albedo, temperature, and rainfall patterns Hydrological services: Regulation of the timing and volumes of river flows Protection against floods by coastal or riparian systems Control of erosion and sedimentation Nursery services: regulation of species reproduction Breakdown of excess nutrients and pollution and waste assimilation Pollination Regulation of pests and pathogens Protection against storms Biological nitrogen fixation, nutrient recycling, maintenance of organic matter and microbial ecosystems Protection against noise and dust
Cultural services	Cultural services relate to the benefits that people obtain from ecosystems through recreation, cognitive development, relaxation, and spiritual reflection	Provision of cultural, historical, and religious heritage (historical landscape or a sacred forest) Scientific and educational information Opportunities for recreation and tourism Amenity service: Provision of attractive housing and living conditions Habitat service: Provision of a habitat for wild plants and animal species

Sources: Adapted from Ehrlich and Ehrlich (1983) and Costanza et al. (1997)

Principles," a set of 12 recommendations designed to foster a more integrative approach to averting epidemic/epizootic diseases while preserving ecosystem integrity for humans, their domesticated animals, and the biodiversity upon which we all depend. Today, the importance of these principles has been heightened by the advent of the *COVID-19 pandemic*. These principles (as presented below in a box) provide a framework for integrated and interconnected multidisciplinary solutions to the challenges faced by humanity today, particularly in the context of unprecedented loss of biodiversity, wildlife habitat, and ecosystem destructions and adverse climate change.

The Manhattan Principles on "One World, One Health"
Source: Robert A. Cook, William B. Karesh, and Steven A. Osofsky. Wildlife Conservation Society, Bronx, New York, USA, 2004.

Recent outbreaks of the West Nile virus, Ebola hemorrhagic fever, severe acute respiratory syndrome (SARS), monkeypox, mad cow disease, and avian influenza remind us that human and animal health are intimately connected. A broader understanding of health and disease demands a unity of approach achievable only through a consilience of human, domestic animal, and wildlife health – "One Health." Phenomena such as species loss, habitat degradation, pollution, invasive alien species, and global climate change are fundamentally altering life on our planet from terrestrial wilderness and ocean depths to the most densely populated cities. The rise of emerging and resurging infectious diseases threatens not only humans (and their food supplies and economies) but also the fauna and flora comprising the critically needed biodiversity that supports the living infrastructure of our world. The earnestness and effectiveness of humankind's environmental stewardship and our future health have never been more clearly linked. To win the disease battles of the twenty-first century while ensuring the biological integrity of Earth for future generations requires interdisciplinary and cross-sectoral approaches to disease prevention, surveillance, monitoring, control, and mitigation as well as to environmental conservation more broadly. We urge the world's leaders, civil society, global health community, and institutions of science to:

1. Recognize the essential link between human, domestic animal, and wildlife health and the threat that diseases pose to people, their food supplies and economies, and the biodiversity essential for maintaining the healthy environments and functioning ecosystems that we all require.
2. Recognize that decisions regarding land and water use have real implications for health. Alterations in the resilience of ecosystems and shifts in patterns of disease emergence and spread manifest themselves when we fail to recognize this relationship.
3. Include wildlife health science as an essential component of global disease prevention, surveillance, monitoring, control, and mitigation.
4. Recognize that human health programs can greatly contribute to conservation efforts.
5. Devise adaptive, holistic, and forward-looking approaches to the prevention, surveillance, monitoring, control, and mitigation of emerging and resurging diseases that take the complex interconnections among species into full account.
6. Seek opportunities to fully integrate biodiversity conservation perspectives and human needs (including those related to domestic animal health) when developing solutions to infectious disease threats.
7. Reduce the demand for and better regulate the live, international wildlife and bushmeat trade not only to protect wildlife populations but to lessen the risks of disease movement, cross-species transmission, and the development of novel pathogen–host relationships. The costs of this worldwide trade in terms of

impacts on public health, agriculture, and conservation are enormous, and the global community must address this trade as the real threat it is to global socio-economic security.

8. Restrict the mass culling of free-ranging wildlife species for disease control to situations where there is a multidisciplinary, international scientific consensus that a wildlife population poses an urgent, significant threat to human health, food security, or wildlife health more broadly.
9. Increase investment in the global human and animal health infrastructure commensurate with the serious nature of emerging and resurging disease threats to people, domestic animals, and wildlife. An enhanced capacity for global human and animal health surveillance and for clear, timely information-sharing (that takes language barriers into account) can only help improve the coordination of responses among governmental and nongovernmental agencies, public and animal health institutions, vaccine/pharmaceutical manufacturers, and other stakeholders.
10. Form collaborative relationships among governments, local people, and the private and public (i.e., nonprofit) sectors to meet the challenges of global health and biodiversity conservation.
11. Provide adequate resources and support for global wildlife health surveillance networks that exchange disease information with the public health and agricultural animal health communities as part of early warning systems for the emergence and resurgence of disease threats.
12. Invest in educating and raising awareness among the world's people and in influencing the policy process to increase recognition that we must better understand the relationships between health and ecosystem integrity to succeed in improving prospects for a healthier planet.

It is clear that no one discipline, or sector of society, has enough knowledge and resources to prevent the emergence or resurgence of diseases in today's globalized world. No one nation can reverse the patterns of habitat loss and extinction that can and do undermine the health of people and animals. Only by breaking down the barriers among agencies, individuals, specialties, and sectors can we unleash the innovation and expertise needed to meet the many serious challenges to the health of people, domestic animals, and wildlife and to the integrity of ecosystems. Solving today's threats and tomorrow's problems cannot be accomplished with yesterday's approaches. We are in an era of "One World, One Health," and we must devise adaptive, forward-looking, and multidisciplinary solutions to the challenges that undoubtedly lie ahead.

http://www.oneworldonehealth.org/sept2004/owoh_sept04.html

The Manhattan Principles, put forth two decades ago, bear profound relevance to ecosystem health, its relationship with human societal well-being, and the management of this relationship. In the aftermath of the COVID-19 pandemic, these principles take on even greater importance given the vast loss of biodiversity and extensive degradation of countless species' natural habitats.

5.5.1 Lessons from COVID-19

The severe acute respiratory syndrome coronavirus 2 (SARS-CoV-2) first appeared in December 2019 in Wuhan, China, and has since spread to more than 200 countries and territories around the world. As of August 14, 2020, there have been around 20.7 million confirmed infections and more than 751,000 reported deaths. The *COVID-19 pandemic* has had a significant impact on social and economic life, from physical isolation and social distancing to global famines affecting 265 million people and the largest global economic recession since the Great Depression in the 1930s.

The virus is part of the *zoonotic diseases* passed from animals to humans, and the origin of this zoonotic transmission lies in human invasions into natural environments and habitats. This invasion is due to factors such as population and economic growth, unsustainable land use practices and degradation, deforestation, biodiversity reduction, and illegal wildlife trade. Many environmental or ecology groups, and even the Secretary General of the United Nations (UN), see in the appearance of *COVID-19* an unprecedented and clear warning shot about the consequences of the ongoing degradation and destruction of the natural environment and biodiversity (IBPES, 2020; Carrington, 2020a; GEN, 2020). Among the preconditions or factors that allowed the worldwide spreading of the *COVID-19 pandemic*, we find the almost unlimited and unrestricted mobility of people, poor health systems, ignorance or downplaying of the threats, inappropriate or delayed decisions by some political authorities, and adverse socioeconomic settings in some countries.

The outbreak's origin and transmission pathways are still unknown, but there has been a rise in zoonotic diseases, which are diseases passed from animals to humans. This increase is due to the unprecedented destruction of wild habitats by human activity. According to the UN's environment chief, Inger Andersen (Carrington, 2020a, b), and other environmental experts, "Nature is sending us a message with the coronavirus pandemic and the ongoing climate crisis." The Intergovernmental Science-Policy Platform on Biodiversity and Ecosystem Services (IPBES, 2020) report released in October 2020 predicts that future pandemics will emerge more often, spread more rapidly, do more damage to the world economy, and kill more people than *COVID-19* unless there is transformative change in the global approach to dealing with infectious diseases.

The same human activities that drive climate change and biodiversity loss also drive pandemic risk through their impacts on our environment. Changes in the way we use land, expansion and intensification of agriculture, and unsustainable trade, production, and consumption disrupt Nature and increase contact between wildlife, livestock, pathogens, and people. Post-pandemic recovery plans provide an opportunity for global, coordinated efforts to redesign our systems, aiming for transformative change within the framework of the 2030 Agenda for Sustainable Development and the Paris Agreement on Climate Change. The COVID-19 pandemic has spotlighted our society's vulnerability and emphasized the need for transformation grounded in sustainability and resilience principles.

Degradation of natural habitats, potential extinction of natural hosts, and the irreversible damage to ecosystem functionality cannot be dismissed as a potential trigger for such pathogens' encroachment into human populations. The health of an ecosystem is directly linked to the quality and quantity of the ecosystem services it generates. A robust ecosystem delivers a multitude of services, fundamental to human health and well-being. This underlines the premise that ecosystem health should be at the core of nature conservation, protection, and sustainable living. Ensuring and maintaining ecosystem health is, in essence, a means to safeguard and uphold the ecosystem's autopoietic capacity, or self-organizing ability. Understanding the concept of autopoiesis, ecosystem health, and their implications for human health and well-being is indispensable. Furthermore, there is an urgent need for restoration and reengineering efforts to restore and redevelop degraded ecosystems, reinstating their functional integrity. Such measures are critical to the survival of humanity and the continuation of living systems on planet Earth.

Chapter 6
Satisfaction of Human Needs and Environmental Sustainability

The Planetary Boundaries (PB) framework does not dictate how societies should develop. These are political decisions that must include consideration of the human dimensions, including equity, not incorporated in the PB framework. Nevertheless, by identifying a safe operating space for humanity on Earth, the PB framework can make a valuable contribution to decision-makers in charting desirable courses for societal development.

Steffen et al. (2015)

6.1 Introduction

This chapter intends to dissect the intricate task of harmonizing human needs with the imperatives of nature conservation, protection, and sustainable development. This approach recognizes that human needs largely shape and drive the patterns of human behavior. Abraham Maslow, in his seminal work of 1943, proposed a hierarchy of needs, ranking five key needs in descending order of priority: "physical," "security," "social-affectional," "self-esteem," and "self-actualization." Regardless of cultural or historical differences, fundamental human needs remain constant, with humanity's reliance on ecosystem services – encompassing material inputs, life-supporting services such as land, air, water, biodiversity, and waste receptor services – provided by natural ecosystems. The operation of human economies occurs as a subsystem within the boundaries of the planetary ecosystem. Sustainable development, as a concept, presents profound complexities and challenges due to its multidimensional nature, encapsulating ecological, economic, and social dimensions and the antagonistic interactions inherent within them. Arguably, the most significant predicament that humanity confronts in the Anthropocene is striking a balance between the ecological and social facets of sustainability. Current economic models oriented toward infinite growth, excessive consumerism, and negligence toward ecosystem health, processes, and the resilience of Earth's systems are inherently unsustainable. This chapter advances the notion of a value-based development

G. Upreti, *Ecosociocentrism*, https://doi.org/10.1007/978-3-031-41754-2_6

model. This model emphasizes the necessity of safeguarding ecosystem health, maintaining critical ecological processes and services, preserving resilience, and ensuring the sustainable utilization of natural resources while staying within the regenerative biocapacity of our planet. Such a model holds promise as a viable approach toward "sustainable development" and "sustainable living" on planet Earth.

6.2 Human Needs: The Prime Mover

We all know that human needs are the powerful drivers of human behavior and social interaction. Human needs, to a large extent, can explain dominant human behavioral patterns. About a 150 years ago, Karl Marx (1987) mercilessly wrote against the capitalist system, describing it as the cause of the inhuman demotion of poor people to a mere appendage of a factory machine. Nobody disputes the fact that the eighteenth century capitalism in Europe was extremely crude and the working conditions of the workers were miserable and far less than satisfactory. This impacted Marx's thinking then and is well-reflected in his writings at a time when he was poised to develop a radical, explanatory, and predictive socioeconomic theory. His theory of materialistic conception of history is based on man's struggle for the satisfaction of basic needs and his quest for self-actualization. Coate and Rosati's (1988) argument brings forth an insightful perspective on how human needs drive human behavior and social dynamics: "It is the existence of individual needs that makes society possible and necessary: human beings must interact in order to satisfy their needs. All politics, then, including global politics, are inextricably tied to processes and outcomes related to the satisfaction or deprivation of human needs. Thus, it is the interaction of people in the pursuit of needs in social settings that underlies and gives meaning to politics."

Marx was the most powerful philosopher whose writings were heavily influenced by the economic inequality he observed in the society, and he was the first one to point out how new higher-level human needs arise from the satisfaction of the most basic ones. In 1843, he wrote "the root is man……theory is only realized in a people so far as it fulfills the needs of the people." In 1846, Marx and Engels wrote that "life involves, before anything else, eating and drinking, a habitation, clothing., and many other things" and "that as soon as a need is satisfied …., new needs are made" (Marx 1867). According to them, the first new need was the establishment of the family, the first social relationship, and, later, many new social relations and new needs arose. Although the dictatorial communist regimes conceived and built in Marx's name have been severely criticized to have suppressed human freedom, there are people who believe that Marx's philosophy was dedicated to human freedom. As Menand (2016) argues: "Marx was an enlightened thinker who wanted a world that is rational in which human beings could be liberated from the control of external forces." Marx, as a humanist, believed that human beings transform the world around them through their labor to produce objects for the benefits of all. This is the essence of the *Homo sapiens* as a species.

An economic system that transforms this activity into labor that is bought and used to aggrandize others is an obstacle to the full realization of our humanity. The central idea of Marxism is that when a large number of people are alienated from the fruits of their labor and a small number of people use that for their self-aggrandizement, such a system cannot prolong and becomes self-destructive from its inherent contradictions. Marx predicted that the income gap between the workers and owners would increase, wages would remain at subsistence levels, the rate of profit would fall, and thus capitalism would eventually collapse. There would be revolutions in advanced capitalist countries that will pave the way to socialism. Some of these predictions like increase in the income gap, stagnant wage rate, and diminishing rate of profit are certainly true, but the prediction that capitalism would collapse by revolutions in advanced capitalist countries has not yet happened and does not seem to in the foreseeable future. Thomas Piketty (2014) argues that the extremely high level of private wealth accumulated since the 1980s and 1990s in the wealthy countries of Europe, Japan, and the United States reflects the Marxian logic. There is no doubt that the capitalist system has increased inequality. As Piketty further points out, by 1900, the richest 1% of the population in Britain and France owned more than 50% of those nations' wealth; the top 10% owned 90%. It is worth mentioning following the observation of Piketty that "in all known societies in all times, the least wealthy half of the population has owned virtually nothing," and the 10% has owned "most of what there is to own."

Menand (2016) also reinforces Piketty's observation by revealing the current level of wealth accumulation in the United Staes as the Federal Reserve informs that the top 10% of the population owns 72% of the wealth and the bottom 50% owns 2%. About 10% of the national income goes to the top 247,000 adults (one-thousandth of the adult population). Menand further explains that inequality has been in existence throughout history and that neither did industrial capitalism reverse it in the nineteenth century nor is finance capitalism reversing it in the twenty-first century. As Ryan (2015) points out, "The modern republic attempts to impose political equality on an economic inequality, it has no way of alleviating." This has been a relatively recent problem because the rise of modern capitalism coincided with the rise of modern democracies, making wealth inequality inconsistent with political equality.

If democracy is all about political equality, then democracy and capitalism (system of economic inequality) need to be compatible with each other and not contradictory. I believe that this is the biggest challenge of modern liberal democracy and political economy in the Anthropocene era. Menand (2016) suggests that the only thing in democracy that can reverse such inequality is political action aimed at changing or altering the systems. Nobody can deny the fact that human-invented social systems can be modified or changed if they do not work for the majority of the people. This is what Bernie Sanders and his coalition within the democratic party have been advocating: reforming American capitalism to make it more compatible with democracy. Marx strongly argued that man achieves self-actualization only through the process of satisfaction of his needs; as soon as he satisfies a need, new kinds of needs arise, and so on, and, thus, it is a never-ending process, always approaching a higher and more complex level of development. According to Marx, the history of mankind is the history of the self-actualization of man himself. Abraham Maslow (1943) further expanded on this and hypothesized a mechanism of how a hierarchy of needs arises. When the most prepotent need is realized, the next higher need emerges. "Thus, man is a perpetually wanting animal." Davies (1977) prioritized the basic human needs that best fit with the way people behave as follows: first, physical needs; second, social-affectional or love needs; third, self-esteem or dignity or equality needs; and fourth, self-actualization needs.

Human behavior is greatly influenced by the human value system that forms an integral part of the human cultural system. A value judgment is an appraisal of the worth of a thing, action, or entity. It has been argued that this worth must be assessed in terms of a standard of exchange, such as money; a value judgment compares the worth of one thing with that of another. Value systems are relative concepts with respect to time, geography, and circumstances. There cannot be any "absolute" value system. They may be similar in widely separated but similar environmental circumstances and quite dissimilar in societies living near but different environmental circumstances. One must search the roots of "value systems" of a society in the dynamics of its survival, whether that be conservation of matter and energy at the inorganic level, preservation of the individual or species at the level of life, or conservation of wisdom at the cultural level. As Gorney (1979) points out, it is the "human survival mechanisms that we must view values and value system." Irrespective of what they may be, value systems have evolved as man's most unique adaptation to a specific set of realities, and these realities are nothing other than man's basic needs, which are the same for all humans everywhere. The flexibility of human beings and the specific circumstances that they encounter with respect to the satisfaction of their basic needs permit a great deal of variation in the value system that is comparable to survival. It is in this respect that human value systems function as the matrix of human survival mechanisms or otherwise people subscribing to it will not survive. The concept of development loses its social meaning if it is not directed toward the satisfaction of fundamental human needs of the larger population. It is also equally important to make a distinction between fundamental human needs and needs in general. As Max-Neef et al. (1989) point out, fundamental human needs are the same in all human societies and they function as a system and are interactive and interrelated. Self-actualization of human individuals is possible only through the fulfillment and gratification of human needs. When we talk about development, it should be clear that we talk about the development of people and not objects (means) as some development professionals seem to equate development with the growth of physical objects like buildings, roads, cars, ships, etc. Physical infrastructures and artifacts are the means, and, if used appropriately, they can help realize human potentials. From this perspective, development must be construed as a process of imparting meaningful improvement to people's quality of life. Mainstream development professionals equate development with mere economic growth, but this is true only when economic growth brings improvement to people's quality of life, social justice, equity, and realization of human potentialities. Therefore, development must be defined as a process that brings improvement to the quality of life of masses of people through improvement in human conditions and fulfillment of basic human needs. The goal of development should be the elimination of conditions that cause human misery, which results from either stagnated or overextended development. The main cause of stagnated development is poor and inequitable use of resources or scarcity of resources to satisfy the basic human needs. The larger part of human misery currently manifested in different parts of the world has resulted from the over extension of the carrying capacity of the resource base. As Dombois (1983) points out, growth is normally associated with the linear order of three steps: "access roads," "human settlement," and "conversion of natural habitats" for material needs. In the beginning, the use of resources from natural habitats or ecosystem conversion enhances human happiness; however, thoughtlessly accelerated conversion and expansion ultimately leads to degradation and destruction of the natural environments and that of human happiness as well. "The success or failure of current development paradigm" as G. Upreti (1994) correctly states "primarily rests on solving basic human needs and reversing ecologically hostile consumerism while maintaining the productive capacity of planetary ecosystem. Environmental conservation and the satisfaction of human needs should not be viewed as the antithesis of each other whereas ecologically hostile consumerism (consumption pattern in Western world) is the very antithesis of environmental conservation." It is a herculean task for development thinkers, political elites, and economists to design a development framework that entails ecologically enabling policy instruments to the effect that people can satisfy their basic needs and reverse ecologically hostile consumerism with ecological consciousness to maintain the environmental resource base in a productive state. When people obtain their sustenance from an environmental resource base with the knowledge and ecological consciousness that degradation of their resource base poses an existential threat, then this constitutes the foundation for sustainable development. Without ecological wisdom consciousness, sustainable development is like a water "mirage" in the sand dunes of a desert.

6.3 Ecosystem Protection and Basic Human Needs

Human existence is dependent upon the environmental goods and services (material goods, life support services such as air and water, and waste receptor services) of the planetary ecosystem. Planet Earth's systems serve as both the resources and the sinks for all waste throughputs produced by human socioeconomic technometabolism. The human socioeconomic system (sociosphere) must operate within the boundary of the planetary ecosystem, which contains finite material goods and generates limited ecological services. The material goods and services from Earth's systems enter into the production process of the human socioeconomic system from which man-made goods and services enter into the consumption process and are eventually released into the environment as a waste throughput (Daly, 1990). First, we need to realize that the finite material inputs and environmental services of the planetary ecosystem can sustain human existence and needs only if the "modus operandi" of the extraction and uses of these goods and services can maintain the health, integrity, and stability of the planetary ecosystem. Excessive consumption of finite environmental resources and services that destroys environmental health, resilience, and ecosystem stability would be a highly destructive pathway for both mankind and other living systems on planet Earth. Therefore, ecosystem protection, its sustainable uses for the satisfaction of basic human needs, and its conservation should become the central axis of the modern development approach that can be considered sustainable.

6.3.1 People and Ecosystem Protection

Clark and Zaunbrecher (1987) have defined ecosystem management, and it is critically important to understand this concept in the context of sustainable development. For them, ecosystem management refers to the following set of principles and activities:

> Maintenance of all existing plant and animal populations and restoration of species that have been eliminated by humans. Monitoring of major ecological processes such as air and water quality and wildlife population trends
>
> Integration of local people's needs and long-term human economies within the management and conservation framework
>
> The ecosystem management concept acknowledges the intrinsic value of all the components of the ecosystem whether or not they are directly useful to humans, whereas traditional management values resources primarily on the basis of their direct usefulness for human purposes.
>
> Ecosystem management emphasizes maintaining ecosystem health and the long-term viability of ecological processes; traditional management focuses on managing target species allowing the extraction of other valuable species.
>
> Ecosystem management encompasses all species, whereas traditional management includes only those species with high aesthetic or known commercial values.
>
> Ecosystem management intends to benefit all biotic communities and species, including humans, whereas traditional management is intended to exclusively benefit humans.

There has been increasing realization among development planners, natural resource managers, social scientists, and ecologists that there is a fundamental need to develop appropriate policy instruments and strategies that can effectively involve local people's participation in the planning and management of natural ecosystems (natural resources, parks, and protected areas). This realization has stemmed from two important considerations:

- First, local people who constantly interact with natural ecosystems possess profound wisdom and knowledge of their ecosystems. The cultural knowledge and social system of local people has come into being as a result of hundreds of years of their interactions with natural systems. Such knowledge systems serve as the repositories of the adaptive mechanism and survival values of the community.
- Second, the knowledge system of local people combined with the science based knowledge and understanding of ecosystem functioning can be a valuable

tool that can be effectively applied in the management and sustainable use of a natural ecosystem.

Modern ecosystem conservation and management must effectively integrate the basic needs of the local people, particularly in developing countries. Such integration becomes effective only when people are encouraged to participate in the planning and management of the natural ecosystem of which they themselves constitute an important component. Understanding local people's needs, knowledge, and the sociocultural system is as important as the ecological understanding of the ecosystem itself from the perspective of developing appropriate policy instruments and strategies for nature conservation and sustainable living.

There is no substitute for the effective participation of local people in the planning and management of ecosystems for this is the only way to develop a mutually reciprocal and positive relationship between people and Nature, thus ensuring a greater opportunity for all, including the biotic community, to flourish. The indigenous human knowledge systems combined with the knowledge and understanding of ecosystem functioning in terms of generating life support services, perhaps, holds the most potential to be used in the planning and management of ecosystems and protected areas. "The most important task for development thinkers and planners, economists, ecologists, resources managers, and conservation biologists is to find out how natural ecosystems function, what indigenous knowledge systems are, how they influence the behavior of people, and, ultimately, how they determine how such knowledge systems can be incorporated into the long-term sustainable use and management of the ecosystems." This is a kind of management system that ensures ecological sustainability of the biophysical resources while meeting the basic needs of the local inhabitants. The traditional knowledge system of resource management existing at the local level deserves special attention and scrutiny because they are not only well-accepted by the people but are also well-defended against outside interference. Equally important is the recognition of the fact that they represent and reflect century-old local systems of social organization and customs, which, if incorporated into modern management principles and practices, ensures a greater chance for successful management.

The loss of genetic diversity and the extinction of species pose a direct threat to life on Earth. The loss of human cultural diversity and extinction of social systems and indigenous knowledge is also equally appalling. In natural and human cultural systems, "diversity" and "autopoiesis" a value for themselves and must be respected and protected from being obliterated. Both natural and cultural diversity possess a creative self-organizing property called "autopoiesis," which is the essence of an organizational system. Life becomes impossible when this self-organizing property of biological and social systems is damaged and destroyed beyond a certain threshold. The higher the amount of biological and cultural diversity in Nature and society, the greater is the ability of natural and social systems to cope with disasters and crises. By maintaining and preserving an ample amount of biological and a broad range of cultural diversity, natural and social systems retain far greater organizational options and adaptive solutions to crises. In other words, the greater the amount of ecological and social autopoiesis, the greater is the capacity of the ecological and social systems to cope with adverse changes in the environment and adapt to adverse environments.

Local people's participation in the planning management of natural ecosystems, parks, and protected areas constitutes a large and complex set of activities. It involves, among other things, the need and ability to establish effective communication between people and the park resource managers and conservationists. The ability to integrate people's needs, knowledge systems, and their participation in the planning and implementation of ecosystem management is a fundamental requirement for the long-term success of any conservation and sustainable development strategy. A sincere attempt must made to involve local people and their indigenous knowledge system in the planning process and implementation of ecosystem management and conservation programs. Messerschmidt (1985) has outlined three distinct time phases for the local people's participation: "planning," "management," and "conflict resolution."

6.3.2 Sustainable Uses of Ecosystem Resources and Services

People's participation in ecosystem planning and the sustainable uses of ecosystem services is based on the premise that peoples are the repositories of century-old conservation knowledge and practices and their participation in planning is essential for the success of any nature conservation program. The first step in the process of people's participation involves conservation education in the community. This must be a two-way process. Both conservation experts and people should have the opportunity to learn from each other in this process. The objective and goal of such conservation educational programs is to enhance people and expert's knowledge of the management issues and environmental concerns. This will not only help them broaden their cultural perspectives on conservation but also provide an adequate base to interpret change and induce appropriate behavioral responses. Glick (1980) eloquently points out why such participation is essential: "For millennia, native peoples .... have successfully coexisted with their fragile natural environment – an environment which today is being ravaged by many instances of inappropriate development. Their sustainable land use practices are the result of years of trial and error, culminating in a vast storehouse of ecological wisdom."

Development of locally appropriate technologies and management systems is fundamental for the conservation and preservation of genetic resources and ecosystems. Such technologies and management systems must evolve from the interaction of the rich historical and cultural heritage of local people and modern conservation science. It is for this reason that people's participation becomes indispensable to the planning phase of natural resource management. The methods and mechanisms for participatory involvement must be sensitive to social, cultural, economic, and political circumstances on the microlevel and should be flexible enough to accommodate factors pertaining to temporal, spatial, and cultural dimensions. It is generally believed that the involvement of people right from early initial data collection point onward throughout the planning phase greatly enhances their interest, trust, and commitment to long-term management and sustainable uses of the natural resources.

Having gone through the planning process, local people can be encouraged or trained to have positive inputs in the day-to-day management practices of national parks. Local people can fill many management and administration roles depending on their own initiatives, interests, and educational training. They can participate in various capacities such as advisory and management committees or councils. Some may form cooperatives for investment in the necessary support services, e.g., running inns and guest houses, raising and delivering the supplies necessary for the day-to-day operation of the park institutions, etc. It is also conceivable that local men and women can be trained to take up jobs ranging from park guards to tourist guides. The most important involvement of local people in the management of park resources constitutes following "practicing the sustainable use of park resources, ensuring the equitable distribution of benefits from the park resources, and monitoring biological or ecological changes taking place in the park ecosystem."

Once people realize that the preservation and sustainable uses of ecosystems or natural resources enhance and sustain their livelihood, they become the custodian and protector of Nature. Conservation education, workshops, and training by ecologists, conservationists, and environmental educators become effective means to empower peoples' capability and understanding about sustainable uses and preservation of ecosystem resources and services for sustainable living.

6.4 Neoclassical Economics and Environmental Sustainability

One of the most difficult problems in the development discourse today is the lack of appropriate analytical methodologies and techniques to appraise sustainable development projects. It is true that measurement of the biophysical conditions of an environment is the essential first step in determining whether a given use or exploitative practice is sustainable or not. Uncertainties in these measurements and lack of understanding of ecosystem processes hamper the guidance of economic planners and managers. It is especially difficult to detect and predict the shift to unacceptable degradation in a highly productive managed ecosystem, such as intensive agriculture or forestry. Opportunities for improved measurement are to be found in new approaches such as restoration ecology, long-term ecological research, and the concept of "keystone species." Tremendous difficulties lie ahead in the incorporation of environmental objectives in four major techniques of economic planning: social cost–benefit analysis (CBA), resource accounting, macroeconomic policy, and applied project or sectorial research. These difficulties arise from two main factors.

First, environmental impacts are usually difficult to measure, value, and predict as they are not marketed but have intangible attributes and are enjoyed or consumed collectively. Second, the basic tenets of conventional economics, as they are tailored to maximize both profit and growth, are fundamentally contradictory to the requirements of sustainable development. Hence, fundamental reorientation in the objectives and technique of economic analysis is therefore necessary. How to internalize environmental costs into the social cost–benefit analysis has become the most pivotal issue in environmental/ecological economics and demands a great deal of concerted efforts and consensus among experts in this particular area of inquiry. The purpose of any such methodology would be first to identify and measure the environmental effects and then translate them into monetary value for inclusion in the formal project analysis (Barbier, 1987). Rees (1990) points out the following inherent shortcomings of neoclassical economics in an attempt to show why this economic model, in association with any development project, cannot be used as a guide for sustainable development in general and measurement of environmental effects in particular. A brief discussion of some of these shortcomings will be useful for understanding the complexity of environmental sustainability and sustainable development approaches.

6.4.1 Gross National Product (GNP) and Human Well-Being

Neoclassical economics evaluates development projects on the basis of their projected contribution to a country's gross national product (GNP), and it does not consider the fact that GNP is only a partial measure of those conditions that contribute to human happiness and well-being. It does not say anything about the distribution of wealth. It does not measure nonmarket transactions and, therefore, undervalues both environmental services and nonmarket sources of natural capital resources. GNP does not have any provision to include in the analysis of the economic benefits of properly functioning ecosystems or their degradation because such processes do not interact with the market and are not integrated in the analysis.

6.4.2 Validation of Neoclassical Economic Assumptions

The fundamental assumptions of neoclassical economics remain virtually untested. Empirical tests to validate economic models have not been undertaken in either developed or less developed countries in the world. Objective economic analysis such as the cost–benefit analysis of a project, total cost, and the least cost of a project based on arbitrary and convenient assumptions may produce logically and mathematically coherent, but not necessarily correct, models. Classical and neoclassical theories, developed on the concepts of markets that existed in agrarian societies, have been transferred more or less unchanged to the application in the modern industrial capitalist world without considering the industrialization, development of large corporation and institutions or impacts of advertising, and the consequences of the power of money itself, each of which characterizes the contemporary society and the markets where we buy and sell.

6.4.3 Neoclassical Economics and Destruction of Natural Capitals

Critics argue that the fundamental premises of neoclassical economics lead to the destruction of natural ecosystems and natural capitals since market prices do not reflect ecosystem services (Hall, 1990; Rees, 1990). In reality, neoclassical economics destroys the real natural wealth on which depend man-made capital wealth. The economic policies driven by neoclassical economics have devastating effects on the natural resource base and the economics of developing countries because such policies encourage excessive burrowing from developed countries, hence subjecting them to growing debt servicing. This, in effect, creates pressures on these nations to mine their natural resources to get a quick return on the investment so that lending agencies can get their cash return. This is the main cause for the enormous amount of capital flights that takes place every year from developing countries to the developed ones in the form of debt servicing.

Another factor that directly contributes to the destruction of natural capital resources is the use of discount rate in the economic analysis. A discount rate reflects the cost of borrowing money from commercial banks and changes with monetary policies and other factors. A high discount rate heavily discounts the future. As Rees (1990) states: "Since many of the direct benefits of natural ecosystems are gained at low rates (as measured in dollars) but over very long, even indefinite time scales, their value tends to be heavily discounted." Present decisions are made by discounting the future, but this procedure has an extremely serious repercussion on certain aspects of natural ecosystems, especially the vital life support ecosystem services. A negative discount rate for natural capital resources and ecosystem services would be well-justified rather than a normal positive discount rate from the perspective of sustainable development. This could lead to the conservation and protection of natural resources and ecosystems. Hall (1990) articulates the consequences of discounting the natural capitals in the following passage very well: "Due to depletion, a barrel of oil is likely to be more, not less valuable in the future. The same is true for a ton of soil or a hectare of forest. The use of positive discount rate makes any of the resources essentially worthless in a decade or two. Society cannot afford to discount the future. If forests are destroyed, the rainfall, and hence agricultural production, of a region may be diminished. If discounting is used in economic analysis, the value of agricultural loss would appear negligible. Much of the Levants was forested and farmed in Biblical times, but is now desert, probably due to largely human activities. The money gained from that original deforestation was almost certainly trivial, even if invested, compared to a loss of thousand or more years' agricultural production."

It is unfortunate that the underlying facts expressed in the above statement of Hall have not caught the attention of mainstream economists in the light of the rapid degradation of ecosystem services and adverse climate change. The assumptions and "modus operandi" of neoclassical economics with respect to the uses of natural capital resources of Earth must be revisited and modified to make them compatible with the sustainable uses of these resources if we are truly committed to sustainable living on planet Earth.

6.4.4 The Market Yardstick and Large-Scale Economic Analysis

Hall (1990) and Rees (1990) consider that market is a wrong yardstick for large-scale economic analysis. They point out that the economic decisions of an entire nation are made on the basis of individual consumers' behaviors on the assumption that the consumers rationally allocate their monetary resources in a way that is best for them. Neoclassical economics assumes that people' wants and needs are best expressed by their purchasing behavior in the marketplace, and this behavior is a direct function of how much money one possesses rather than the function of one's wants and needs. It follows that people, who do not have enough money, do not display purchasing behavior in the marketplace and, consequently, do not have wants and needs, which is not the case. It is interesting to note that neoclassical economics does not say anything about the needs and wants of those people who do not have money to purchase from the market. Hence, neoclassical economics cunningly avoids the necessary discussion about economic means and ends and replaces it with simplistic objectives based on short-sighted and manipulated human greed (Hall 2004).

6.4.5 Price and Scarcity

Economists regard price as the best measure of scarcity, but it reflects many important aspects of scarcity very poorly. A large contingent of scientists from environmental science to economics have argued against the perspective of Barnett and Morse (1963) that inflation-corrected price changes are the only relevant measure of scarcity. Daly (1990) demonstrated that as all resources become scarcer, then the prices of all goods, including resources, will inflate as a general trend, and inflation-corrected values for all materials will not increase. Barnet and Morse's analysis (they found no indication of increasing scarcity of raw materials as reflected by their prices) was incomplete because the decreasing price of energy and its increasing use masked the consequences of resource depletion. Only development that entails social, economic, and ecological sustainability can be called "sustainable development"in the true sense. Ecological sustainability is the basis for social and economic sustainability, which is totally ignored by such analysis.

The current growth development paradigm advanced by neoclassical economists under the rubric of "sustainable development" is simply a self-defeating proposition. An economic system with "infinite growth" in a finite resource base of planet Earth is simply inconceivable to be sustainable. In order for an economic system to be sustainable, it must remain within the regenerative capacity of the planetary ecosystem, and, by default, such a system is bound to be a "steady-state economy" as Daly and Townsend (1993) have been advocating over the last three decades. It remains to be seen how long it is going to take for mainstream neoclassical economists to come to terms with "steady-state economy," which seems inevitable for the existential preservation of humanity and other living systems on planet Earth. The longer they take to recognize this "inconvenient truth" and revisit their basic assumptions, the harder they inflict misery and suffering on humanity and living systems on planet Earth. They may think that they are saving the world, but, in reality, they are pushing the world to the brink of planetary catastrophes.

6.5 Strategies for Environmental and Social Sustainability

Goodland and Ledec (1987) argue that sustainable development is a pattern of social and structural economic transformation, which optimizes the economics of resource use and societal benefits available at present without jeopardizing the likely potential for similar benefits in the future. There are two important dimensions of sustainable development: the first refers to our desire for economic improvement and the maintenance of the ecology of the natural resource base. The second refers to the notion that equality, social justice, and cultural diversity promote human well-being. The sustainable use of the natural resource base and its maintenance and the equity and opportunity for everybody in society to satisfy their basic needs necessary for the actualization of human potential are the basic components of sustainable development. It is impossible to keep ecosystems intact and in pristine state without fulfilling the basic needs of people, and it is impossible to solve basic needs problems without looking into the nature and mode of relationship between human beings and resource use, their means of production, and their distribution. Without promoting equity, social justice, and maintaining ecosystem health, sustainable development is inconceivable. Hence, it is important to know the basic parameters of the equation of environmental and social sustainability if one were to devise and work out strategies to achieve development with environmental and social sustainability.

The following parameters constitute the basic equation of the development with environmental and social sustainability and must be considered in all development projects, programs, and practices if they are to be truly sustainable in both spheres. Development with environmental and social sustainability can be considered as a function of "ecological stability," "social equity," and "sustained economic growth" (Panayotou, 1990). The equation can be written as:

$$\text{Sustainable development} = f\left(\begin{array}{l}\text{ecological stability} + \text{social equity}\\ +\text{sustained economic growth}\end{array}\right)$$

Panayotou (1990) suggests five conditions for all development projects, programs, and policies to be sustainable and achieve environmental and social sustainability as defined by the equation above:

Limit population growth rates to below the sum of the rates of capital accumulation and technical change.
Alleviate poverty and reduce income disparities.
Maintain ecological balances, a renewable natural capital base, cultural stocks, human-made capital stocks, and the assimilative capacity of the environment.
Avoid irreversible changes to the environment and any foreclosure of options unless near-perfect substitution exists in order to provide for the next generations.
Attenuate development activity (i.e., reduce the scale, scope, and speed of development) to allow the gathering of needed information in cases in which there is considerable uncertainty or in which irreversible changes or major adverse impacts are suspected.

Development professionals and policy analysts (Milbrath, 1989; Repetto et al., 1989) argue that the objective of any development is to achieve a reasonable and equitably distributed level of economic well-being that can be continually sustained for many human generations, and this can be made possible only through transition from prevailing economic growth based on the depletion of nonrenewable resource stocks to an economic system based on renewable resources over time. From an ecological and basic needs perspective, Repetto et al. (1989) argue that development is living on the "planet's income instead of depleting nature's capital for meeting the needs of today's population without compromising the ability of future generations to meet theirs, and managing of natural, human, and financial assets so as to increase long-term wealth, and well-being." This definition captures environmental and social sustainability more closely than the Brundtland Commission's definition but is still inadequate in terms of clarity and emphasis.

6.5.1 Dimensions of Sustainability

Sustainability is a multidimensional construct when it comes to the study of the interaction of the human social system (political economy) with the environmental system. Primarily, environmental and social dimensions of sustainability can be conceptualized from the local to global level. The local and regional dimensions of environmental and social sustainability are embedded in the global dimension and are much affected and impacted by what happens at the global level. A brief discussion of these dimensions will bring about some perspectives that will help us understand the complexity of the issues that are intertwined with environmental and social sustainability.

6.5.1.1 Environmental and Social Sustainability

It may be difficult for us to predict as to what constitutes the future generations' interests; it would be prudent on our part to assume that their needs for natural capital resources would be at least not less than ours. This directly implies that the renewable natural capital resources must be used in a manner that does not deplete or degrade them permanently so that the future generation could also use them to satisfy their needs. There are two important areas in which the concept of sustainability needs to be thoroughly investigated and applied. These are environmental sustainability and social sustainability.

"Environmental sustainability" refers to the functionally productive state of natural capital resources (land, water, forest, biodiversity, minerals, and all the material resources of nature) that can sustain the consumptive uses of humanity and living systems on planet Earth. This means harvesting renewable capital resources on a sustained yield basis rather than mining them to extinction. Forests, biodiversity, freshwater, lands, whales, and coral reefs are some of the renewable capital resources that can be harvested or used on a sustainable basis. In other words, harvesting or uses of natural capital resources must be within the regenerative biocapacity of nature. Environmental sustainability also implies using the nonrenewable resources at a slow rate so as to ensure a smooth transition to renewable energy sources (solar, wind, wood, biomass, hydroelectric, and other water-based sources) and a long-term planning to guide the transition to a renewable energy source. With respect to agriculture and other biological production systems, environmental sustainability implies the permanent maintenance of biological productivity of the landscape ecosystem. The environment has certain limits in terms of physically and biologically supporting the human political economy with respect to food and industrial needs and also processing the human wastes that enter into the environment. This biocapacity of the environment to provide "life support system" and "services" directly depends on the demand made by the human political economy on the natural systems' biological and physical processes. Without some understanding of the basic biological and ecological processes of natural ecosystems (physical environment), the strategies for the use of natural capital resources and food production on a sustainable basis cannot be worked out.

"Social sustainability" refers to the economics, social–political organizations, and control of resources and income of a society collectively called "political economy." Moreover, it specifically refers to the relationship of human populations with the means of production on which their sustenance is directly dependent. In most of the societies, the means of production and natural capital resources and political institutions are controlled and owned by a small minority of the population, as a result of which the majority of the population are forced to live under utmost misery and dehumanized conditions. This situation is highly pervasive in the developing world. In developed countries, the majority of people do not have to suffer from the problem of meeting their basic human needs. Unless there is a strong political will and determination in the ruling elites of the developing world and the developed countries to deal with the problem of income distribution, equity, and social justice along with the appropriate population control measures, the societies (social institutions) in these countries may not be able to sustain their functions as stable institutions for human development in the "sociosphere." The gross disparity between the haves and have-nots may lead to social instability, and the effects and consequences of such instability will ultimately be reflected in greater environmental impoverishment and degradation.

There is a direct causal relationship between "environmental sustainability" and "social sustainability," and, sometimes it is difficult to identify which is the cause and which is the effect. However, the fact is that both can be the cause and the consequence of each other, and, for this reason, these two must not be isolated while dealing with either social or environmental problems. Environmental and social sustainability are the two fundamental dimensions of sustainable development required for present and future generations. The nature of the dynamical and dialectical relationship between these two dimensions determines the future trajectory of human civilization and the destiny of the *Homo sapiens* as a species.

6.5.1.2 Challenges of Global Sustainability

It is interesting to note that Western scholars emphatically talk about resource conservation and global sustainability but do not intend to inform the outside world that the United States and Europe together constitute just 12% of the world population but consume nearly 60% of the world's resources, whereas the rest of the world with 88% population has to depend on the rest of the world's resources. The same question can be asked to other members of the group of seven industrial nations. Why is there always a closed and open-door meeting first among them and then between the northern and southern countries (north-south dialogue)? Is it not about the control of the resources, whether it be in the form of raw materials or finished products or import or export of material goods? The only thing that one can be sure about their meeting is that they do not meet to work out an economic arrangement to eliminate poverty from the world or to develop a strategy to fulfill the basic human needs and self-actualization of humanity on planet Earth, despite the fact that, with a little bit of conscientious assistance from the big brothers of the world, such essential needs could have been met in the developing world a long time ago. Compared to the investments of powerful or industrial nations in weapons of mass destruction, the investment in the elimination of poverty from the world and a dignified life is trivial. Until now, there was a powerful argument that the world was divided between the capitalist and socialist systems and every nation on Earth had to go through the ordeal of identifying itself with the existentially powerful system on land and did not have time to think about the apolitical problems of the world.

Now, the global situation is different. Tensions between the socialist and capitalist systems are effectively dead after the collapse of the Soviet Union. The United States has already initiated a New World Order policy, which was first experimented in the Persian Gulf, and it remains to be seen how it proceeds and progresses after its experimentation in Afghanistan to other parts of the world. The most likely scenario is that the world is going to increasingly experience a greater American involvement, intervention, and control of the world, specifically the world market and resources. Tensions may arise among the competing countries, especially between America and China, Japan, and India for control of the world market and resources. Tensions may also arise among the European Economic Community (EEC), China, Japan, and the United States from their competition for control of the world market and resources. It will be extremely difficult for China and Russia or any other country to directly challenge the American political and military offensive.

It is difficult to say how the relationship between the south and north, including the United States, will emerge in the foreseeable future, but northern countries are likely concentrating their effort on maintaining and strengthening their status quo through the use of the power of international financial and trade agencies, mainly the World Bank, International Monetary Fund (IMF), and World Trade Organization (WTO). The policies, terms, and conditions of these powerful agencies almost invariably have resulted and will result in strengthening and expanding the interests of the northern countries. Very few southern countries may resist the undue pressure from the north because they may not want to be singled out for punitive action and be alienated from the international economic system. Western countries (the United States and the European Union (EU)) and Asian countries, including China, India, and Japan, will compete for access to and control of resources, especially petroleum, forests, coals, mines, and rare elements in the southern countries. After all, in practice, politics is nothing more than a strategy of the manipulative exploitation and control of resources. The history of colonialism is the history of the political control and exploitation of the resources of the weaker nations by the more powerful ones. Countries that have limited resources and several human problems, namely, poverty, malnutrition, starvation, no access to drinking water, and poor health, will probably be forgotten with some insignificant token supports. This is simply because of the uncontrolled liberal market economy that does not recognize values other than profit and accumulation of wealth in the hands of few powerful political and corporate elites. Hence, poor nations will not have a visible existence in the world economic integration.

However, it does not need to be that way. The world has enough resources to fulfill basic human needs and aspirations, and if these resources are used on a sustainable and equitable basis, the needs and aspirations of the future generations will also be fulfilled without putting much pressure on the planetary ecosystem (biosphere). The question is: will developed countries, which constitute less than 12% of the world's population but control and own more than 80% of the world's resources, have a change of heart and be able to allocate a small portion of their wealth or income to help solve the basic human problems of the developing world? Global sustainability will make sense and have a chance only if people in the developing world get a fair chance to sustain their life with human dignity, a life free from the plight of poverty, malnutrition, diseased conditions, and insecurity. No country or organization in the world has a moral right to tell the poor farmers in developing countries to stop cultivating marginal lands for their survival, unless there are some alternative programs tailored to address their needs. It is ironical that very few Nature conservationists, ecologists, economists, anthropologists, or political economists raise the issues of resource consumption or income distribution or see it necessary to raise such issues. It is not surprising that they have a political reason for not doing so, but the fact is that without addressing the excessive consumerism of Western society, equity, and distribution issues in developing countries, it is not even worthwhile to talk about conservation of biodiversity and ecosystem protection at the local, reginal, or global level.

There are two powerful motives behind environmental destruction, and these pose challenges to global sustainability; these are global corporate economic motives with greed and people's survival motives to satisfy their basic needs. It is important to analyze these motives in proper contexts to see the interconnection between social and environmental sustainability.

6.5.1.2.1 Greedy Economic Motive (Greednomics)

The biggest challenge to global sustainability is the driving force of a "greedy economic motive." In more acceptable terms, it can be described as the principle of "profit maximization," the higher the profit, the better the economic enterprise. I would like to call this "greednomics," the economics driven by greed rather than by need. Global Corporate behavior primarily motivated by profit maximization is the prime cause of environmental problems. Unless the reckless desire of global corporate capitalism for profit or capital accumulation driven by greed is controlled or regulated, the breakdown of the planetary ecosystem cannot be stopped. The principle of profit maximization inevitably leads to the maximization of control over natural capital resources. Natural capital resources are the inputs that enter into the production process through which they are converted into the output (finished product) that enters into the market as consumer goods and generate capital. Hence, it follows that whoever controls the greater portion of resources, also controls production processes, and the market, and, consequently, maximizes capital accumulation. This ideology of capital maximization and control over resources was the driving force behind "colonialism" in the past, and it became the heart and soul of modern global "corporate capitalism." If we look into the history of the past world wars in general and any war in particular and try to analyze what was the motivating cause behind these wars, we find only one motive. That is the control and ownership of resources because such control enables a nation to occupy a more dominant position relative to other nations. It was basically the reason behind Iraq's invasion of Kuwait and United States' declaration of war on Iraq. Control over oil resources was the primary motive behind both Iraq and the United States' involvement in the Persian Gulf war. The rest is just the story of rationalization of one's involvement.

This approach of control and exploitation of the resources is highly dangerous for humanity because this does not recognize other human values such as common people's right to resources for the satisfaction of basic human needs, equity, and social justice. It does not recognize any value other than the value of profit maximization, and, certainly, such an approach cannot do justice to Earth's systems. The biggest threat to the breakdown of the planetary ecosystem (biosphere) or that of environmental sustainability comes from the current dominant paradigm of reckless profiteering and wealth accumulation propelled by the engine of neoliberal global corporate capitalism, which is "greednomics" in essence.

6.5.1.2.2 Survival Motive

Anthropogenic activity driven by survival motive has also caused environmental degradation but to a much lesser extent compared to corporate greed. This is the fundamental motive that was pervasive right from the beginning of biological and cultural evolution of the *Homo sapiens* throughout history. In today's world, resource-poor farmers are forced to cultivate marginal lands to produce food for their sustenance and survival through encroachment upon forest and marginal lands. The urge to survive is naturally stronger than any other urge. Population pressures have expedited this process in most of the developing countries. The societies that recognize every individual's right to natural capital resources to satisfy the survival needs have developed social systems based on sharing and distribution of the resources to meet those needs. Once resource sharing is achieved, scientific approach and strategies based on ecological principles can be used in the management of these natural capital resources so that the environment can sustain the optimum production needed to support human needs. Sustainable development must entail following strategies that promote both environmental and social sustainability.

6.5.2 Major Strategies

Mainstream economists have defined economic development as the process whereby the real per capita income of a country increases over a long period of time provided that the number below the "absolute poverty line" does not increase and that the distribution of income does not become more unequal (Meier, 1976). It has been argued that the development of a society involves not only changes in the economic activities but also sociopolitical and cultural transformations, and the per capita income used as the overall index of economic development may not necessarily represent the development of a society because there are important qualitative dimensions to development that distinguish it from economic growth. Development with environmental and social sustainability results from the interaction between environmental and social system goals, namely, biological–ecological system goals and social system goals. Biological–ecological system goals entail "biodiversity," "resilience," and "biocapacity." Social system goals entail basic "human needs" (useful goods and services), "equity," "cultural," and "institutional sustainability."

Environmental and social sustainability must entail a process of trade-offs among these system goals. It is impossible to maximize all these system goals all the time. We know that political economy is dependent on natural capital resource use, and even increasing useful goods and services may directly conflict with the productivity and stability of the ecological (biological and physical resource) system. Based on the relative weight of environmental and social system goals and their adaptive trade-offs over time and space, one has to make the choice as to which system goals should receive priority and greater weight in the development strategy. Taking system goals into account, sustainable development can be appropriately defined as a development approach that raises social welfare with the maximum amount of resource conservation and the minimum amount of environmental degradation allowable within given economic, social, and technical constraints. Environmental sustainability is the basis and foundation of social and economic sustainability, and, under certain circumstances, environmental sustainability overrides social sustainability. Only development that entails strategies to achieve environmental and social sustainability system goals can be sustainable development in the true sense.

Here, I want to describe some strategies in an attempt to highlight some inherent features of environmental and social sustainability and develop a perspective that will help us better understand how we can approach the difficult tasks of achieving environmental and social sustainability goals. These strategies will certainly help us visualize the tremendous constraints and difficulties in designing a realistic framework and pathways for sustainable living.

6.5.2.1 Limit to Growth

The neoclassical economic theory assumes that there is no limit to growth in the physical scale of production and consumption, but this has been proven utterly wrong by a great deal of evidence that has surfaced over the past many decades. The recent trend in forests, fisheries, croplands, grasslands, and biodiversity has indicated that the productivity of these renewable resources has been rapidly declining worldwide. Goodland and Ledec (1987) point out that the marginal cost of discovering and exploiting new mineral and fossil fuel deposits has been increasing exponentially. In the neoclassical microeconomic theory, growth in production or consumption is possible only to the point where marginal benefit (revenue) equals marginal cost. The macroeconomic theory does not recognize the optimum size of an economy. It rather promotes the idea of bigger is always the better. It negates the environmental and social costs associated with high and growing rates of per capita natural resource consumption (throughput). The limit to growth becomes clear only when these costs are considered and included in the economic calculus. Even with technological advances and innovation, the growth in natural resource consumption and production is constrained by the physical laws of thermodynamics and the finite size of planet Earth. The current developmental strategies inevitably lead to environmental degradation and ecosystem instability. Policymakers must recognize the facts of limit to growth and adopt policy measures to reduce resource throughput to sustainable levels within the regenerative capacity of Earth's systems if they wish their developmental strategies to be sustainable.

6.5.2.2 Safe Minimum Standard (SMS)

Goodland and Ledec (1987) argue that "safe minimum standard" (SMS) analysis can be used as a strategy and methodology to address ecological concerns, which hitherto have not been given any attention in economic cost–benefit analysis (CBA). SMS, as they explained, is a time-tested standard operating procedure (SOP) widespread throughout engineering design, health planning, and industrial worker safety. For example, a bridge is commonly designed with a safety factor of three or more to accommodate unexpected and unknown future events. Since there is no well-established agreed-upon methodology associated with measurements and valuation, discounting, and irreversibility of ecosystem services, the concept of SMS becomes the most appropriate instrument to deal with ecological concerns, which would otherwise go unnoticed or undervalued in conventional CBA. Mainstream economists resist the use of the SMS approach because they argue that the SMS criteria are established subjectively and arbitrarily without any reference to what is economically efficient or optimal. In response to such criticism, Goodland and Ledec (1987) argue that reliance on marketplace as the yardstick to measure the social well-being (as is done in CBA) is no less arbitrary and no more objective or optimal than SMS analysis and, hence, there is no reason why it cannot be used. Even the World Bank, which operates under the theoretical framework of neoclassical economics, has pushed the use of SMS strategy considerably in the appraisal of the proposed developmental projects.

6.5.2.3 Sustained Yield

There are two principles that can be applied to the management of renewable natural capital resources. The first principle states that the harvest rates of such resources must equal the regeneration rates. The second principle states that the total waste emission rates (throughputs) must equal the natural assimilative biocapacity of the ecosystem into which wastes are emitted. David Pearce (1988) argues that the regenerative and assimilative capacities of ecosystems must be treated as the natural capital, and failure to maintain these capacities must be treated as capital consumption and therefore not sustainable. Herman Daly (1990) argues that both man-made and natural capitals can be maintained at various levels, but the idea is to maintain them intact at the optimum level. It has long been recognized that, for a renewable natural capital resource, there exists a stock size that gives maximum yield per time period and that this biological maximum coincides with the economic optimum only in the case of constant cost of harvest. Hence, the optimum ecological and economic criteria can be applied to choose the level at which the natural capital can be maintained intact and the sustained yield can be harvested.

6.5.2.4 Complementarity

One important question that is often asked is: are man-made and natural capitals substitutes or complements in the process of production? Neoclassical economics assumes that man-made capital is nearly a perfect substitute for natural resources and, consequently, that man-made capitals substitute the flow of natural resources generated by natural capitals. Herman Daly (1993) argues that man-made and natural capitals are complementary and not substitutionary. In the production process, the flow of matter and energy from nature is transformed into a flow of finished products by labor and capital. Capital and labor are substitutable for each other to a considerable degree because their functions in production are qualitatively the same; they are both agents of transformation of raw materials into the finished product. However, the qualitative roles of natural resources and man-made capitals are totally different and cannot substitute each other. Similarly, there is considerable substitutability among different resources with similar qualitative roles, for example, stone for wood or aluminum for copper, because their role in the production is qualitatively similar. However, the substitutability of man-made capital for natural resources is entirely and qualitatively a different matter and is extremely limited. As Daly (1993) points out, increased sawmills cannot substitute for diminishing forests nor can refineries for depleted oils nor larger nets for declining fish populations, but, on the contrary, they will all cause a corresponding further decline of those natural resources. Form this, we can clearly see that natural capital, as a provider of raw material and energy, is complementary to man-made capital. When one recognizes the complementarity of natural and man-made capital, then it becomes clear that development is limited by the one in shortest supply. In the past era of "empty-world economics" (to use Daly's phrase), man-made capital was limitative but now we are entering an era of "full-world economics" in which natural capital will be increasingly limitative. This is a clear demonstration that development with environmental and social sustainability cannot be pursued without keeping the natural capital intact and within their regenerative capacity to ensure the constant flow of material resources and energy.

6.5.2.5 Sustainable Replacement

Nonrenewable resources cannot be maintained intact, but it is possible to use them in a "quasi-sustainable" manner by limiting their rate of depletion to the rate of creation of renewable substitutes. This requires that any investment in the exploitation of nonrenewable resources must be paired with a compensating investment in a renewable substitute. For example, oil extraction must be paired with tree plantation for wood alcohol. When granting the renewable substitute for a nonrenewable resource, one should not forget to assure the continued existence of the complementary natural capital, such as the ecosystem's capacity to absorb wastes. For example, in the case of coal extraction, the paired renewable investment should therefore be in expanding the sink capacity of the ecosystem. Hence, planting trees serves both as a sink for carbon dioxide (CO_2) and as an alternative source of energy, but the sink function dominates.

6.5.2.6 Efficiency Innovation

Some economists and scientists believe that it is technology that will ultimately bail out humanity from a crisis and that technology can transform a thing that has no use value today into a thing of enormous use value in future. Theoretically, such possibility cannot be ruled out, but, in practical terms, it is unlikely that technology can solve all environmental and human social problems. Technology cannot create new matter and energy. It cannot create new genes and gene pools. All it can do is assemble, reassemble, arrange, and transform energy and matter from one state to another in more efficient ways. However, there lies a tremendous opportunity in developing a technology that is more energy-efficient and sustainable. New innovative technology can be designed that can increase resource productivity, the amount of value extracted per unit of resource, rather than increase the resource throughput. For example, production of more efficient light bulbs rather than more power plants must be emphasized. Innovative technology can be developed, which can generate products and processes that can facilitate material recycling and reuse of both within the economy and natural ecosystems. Technological innovation can make significant impacts on the functioning of natural ecosystems, but technology cannot replace the vital environmental goods and services generated by natural ecosystems that were created through billion years of evolutionary processes.

6.5.2.7 Sustainable Economic Scale

Usually, the scale of an economy refers to the total per capita resource use of a population (population times per capita resource use). This scale must be within the limit of the carrying capacity of the natural ecosystem so that it can be maintained without resorting to the overconsumption of natural capitals. This implies that ultimately there is a limit on the total scale of resource throughput, which, in turn, implies limits on and trade-offs between population size and per capita resource use in the bioregion. Poor countries that have extremely low per capita resource use will have to concentrate their efforts on population control, and the countries that have high rates of per capita resource consumption must concentrate their efforts on the reduction of excessive consumption. Development with environmental and social sustainability requires that excessive consumption of goods and services and growth in human population must ultimately end because an optimum quantitative growth will give rise to a qualitative improvement in the life of both people and natural ecosystems.

Developed countries have found that development is the best means of population control. Rising income levels, improved education, urbanization, and the empowerment of women have about brought negative rates of population growth in West Germany and Sweden and most developed countries. Many developing countries, with some exceptions, have been caught up with a high population growth rate. The increased population pressure will exacerbate environmental and social problems (poverty, deforestation, land degradation, water pollution, etc.) and undermine the goals of sustainable development. Scientists and development experts tend to link unsustainability to the high population growth rate prevalent in the developing world, whereas it is not the population growth per se but the scale of economy, the per capita consumption of the resources that is the ultimate cause of environmental and social unsustainability. What is needed is a development strategy that can maintain human population at an optimum level in developing countries and reduce the excessive per capita consumption of goods and services in developed countries. The Western consumeristic culture that thrives on market capitalism has encouraged deforestation, species loss, and desertification in developing countries. American and European transnational corporations operating in developing countries are causing a massive destruction of tropical forests and other natural resources. There is a direct connection between deforestation in Brazil and the dairy and beef industries in the United States. The products of these industries are being sold in US and European markets. Most of these industries are owned by American and European companies. The crucial question to ask is can the people and the government of industrialized countries help design policies that can control and regulate the behavior of their multinational companies to protect the environment? Until now, the world has witnessed just the opposite.

In poor developing countries, poverty can be eliminated only with economic development, population control, and income redistribution. The Brundtland Report (1987) envisaged a 5-to-10-fold increase in the size of developing world economy in order to be able to eradicate poverty and solve basic human needs problems. It is highly unlikely that developed countries will cut down on their current rate of consumption of goods and services. Developing countries are also unlikely, with few exceptions, to increase their economic growth to a level necessary to alleviate poverty and create infrastructures for a more humane and dignified life without the assistance of developed countries. The biggest challenge of the twenty-first century is to develop environmental and social (economic) governance systems that can end poverty and achieve levels of consumption and population while securing the environmental life support services underpinning current and future human well-being, which is impossible without incorporating natural capital and ecosystem services into decision-making (Guerry et al., 2015; Upreti, 1994, 2020). Unless the world's nation states come to mutually agreeable terms as to what, how much, and in what manner to produce, distribute, and consume goods and services without destroying or degrading planetary Earth systems, the world will experience an ever-increasing economic growth, which will have surpassed the optimum scale beyond which "economic growth" will be impossible due to the breakdown of Earth's systems. As soon as the economic scale surpasses the biocapacity of the planetary ecosystem, ecological laws and the law of "thermodynamics" (the "entropy law") will begin to govern and regulate life on this planet. It is impossible to reverse "entropy law," but it is very much in the hands of humanity to slow it down and prolong human civilization. A steady-state economy with stable human population and reduced per capita consumption is necessary to keep the planetary ecosystem functioning. It is possible only if the nation states embrace and adopt these strategies discussed above in developing, restructuring, and redirecting economic, social, and governance systems with innovative technologies.

Chapter 7
Climate Change and Its Threat to Humanity in the Anthropocene

There are very strong indications that the current rate of species extinctions far exceeds anything in the fossil record... Never before has a single species driven such profound changes to the habitats, composition, and climate of the planet.

Royal Society of London (2010)

7.1 Introduction

This chapter intends to explore how *anthropogenic* activities are causally interconnected with the degradation of the planetary ecosystem, thus accelerating climate change with dire consequences on living systems. The intrinsic interconnectedness of planetary ecosystems and climate change constitutes an unparalleled area of scientific exploration, delineating the profound and intricate relationship that our planet's biological and physical systems share. This symbiotic relationship has sparked considerable discourse across various academic fields, from environmental science and ecology to geophysics, climatology, and astrobiology. Planetary ecosystems, intricate matrices of biotic and abiotic components, provide essential goods and services that fortify Earth's habitability. These diverse systems, however, face significant destabilization threats due to *anthropogenic* climate change. This perturbation, characterized by accelerated global warming, erratic precipitation patterns, intensified storm events, and sea-level rise, propagates wide-ranging effects across global ecosystems, thus jeopardizing the biosphere's stability.

Degradation in the functioning of the planetary ecosystem has accelerated the climate change with adverse impacts on the living systems on planet Earth. It is essential to consider the reciprocal nature of this relationship. Although ecosystems suffer from climate change's adverse effects, they also contribute to its progression, either as carbon sinks or sources. For example, carbon sequestration in forests and oceans mitigates climate change, yet deforestation and ocean acidification can amplify greenhouse gas concentrations, accelerating global warming. The

G. Upreti, *Ecosociocentrism*, https://doi.org/10.1007/978-3-031-41754-2_7

catastrophic effects of climate change also transcend the biological sphere, impacting socioeconomic systems and human health. Consequently, to grasp climate change's full magnitude and implement effective mitigation strategies, it is essential to adopt a holistic, interdisciplinary approach, converging insights from Earth science, ecology, economics, public health, and policymaking. Scientific scholarship should be directed toward high-resolution modeling, remote sensing technology, and rigorous field studies to elucidate the complex dynamics between planetary ecosystems and climate change. This critical understanding will subsequently inform the development of sustainable, resilient systems that preserve biodiversity, mitigate climate change, and secure human well-being for generations to come.

7.2 Planetary Ecosystems and Climate Change

Earth can be conceived as an intricate system of reciprocally interactive components, namely, the *geosphere, hydrosphere, atmosphere*, and *biosphere/ecosphere*. These four interconnected components persistently interact to ensure the planet's optimal functionality. These four facets, often referred to as "spheres," are not isolated entities but function in a concerted manner. All spheres within Earth (geosphere, hydrosphere, atmosphere, and biosphere) are engaged in continual interactions and are interlinked in an intricate web of relations. The *biosphere/ecosphere* the sphere encompassing all life-forms, was borne out of the interplay between the *geosphere, hydrosphere*, and *atmosphere*, propelled by solar energy. This interaction, underpinned by the primary catalyst of solar energy, is what enables the interconnectedness of these spheres. Each sphere bears its unique function and perpetually transforms through a sequence of processes known as cycles. The biosphere acquires gases, thermal energy, and solar energy from the atmosphere, whereas the hydrosphere contributes water, and the geosphere offers a suitable medium for life, thus making it possible for the biosphere */ecosphere* to harbor all life-forms that have evolved on Earth. We are cognizant of the fact that the current state of the biosphere/*ecosphere* is the outcome of billions of years of evolution and intricate interactions.

We inhabit an era defined as the *Anthropocene epoch*, hallmarked by humans emerging as the dominant driving force behind planetary change. This shift carries significant implications for humanity and its impacts, the depth of which we are just beginning to fathom. It is essential that we acknowledge and understand that the *sociosphere* (our socioeconomic–cultural system) did not evolve independently of, but rather within and as an integral part of, the *ecosphere* (biosphere). The existence of humanity, along with its socioeconomic, cultural, and technological superstructures, cannot be conceived as separate and independent. Our collective actions, whether socioeconomic, cultural, or technological, have far-reaching impacts on the ecosphere/biosphere and have contributed to the degradation of the planetary system, creating conditions detrimental to human well-being and the overall well-being of life-forms in the *Anthropocene biosphere*.

An ecosystem encompasses living organisms (plants, animals, and other organisms) in a particular area, interacting not only amongst themselves but also with their abiotic environment (including weather, Earth, Sun, soil, climate, atmosphere). The life-sustaining services offered by these ecosystems, often termed as "ecosystem goods and services," are fundamental for the survival of human beings and are the bedrock for future economic and social development. Ecosystems provide a multitude of resources, including food, water, timber, biodiversity, fossil fuels, air purification, oxygen, soil formation, and pollination. Without these services, human survival would be compromised. Human social, economic and technological systems are essentially subsystems of the planetary ecosystem (biosphere). Ecosystems form the core of the biosphere and determine the health and functional status of the planetary ecosystem as a whole. Unfortunately, human activities have wreaked havoc on these ecosystems, causing substantial damage to a myriad of ecosystems and biomes. Scientists have raised the alarm regarding significant alterations in the world's climate, impacting both living and nonliving entities. Places previously warm are now experiencing cold weather, and traditionally colder regions are seeing extremes in temperature fluctuations (Specktor, 2019; Surpran & Achakulwisut, 2019; Fleming et al., 2018; Hayhoe et al., 2018).

Between 1901 and 2012, it is postulated that Earth's temperature increased by 0.89 °C, and rainfall amounts have risen in the northern hemisphere's midlatitudes since the start of the twentieth century. Concurrently, sea levels have risen by approximately 19 cm worldwide, accompanied by significant melting of glaciers in the Arctic, Antarctica, and high-altitude regions such as the Himalayas (Fleming et al., 2018; Lal et al., 2018). Dirzo et al. (2014) posit that we are experiencing an epoch of human-induced biodiversity loss: characterized by species extinction, population reduction, and declines in local species abundance. They report that 322 terrestrial vertebrate species have been obliterated since 1500 and that the remaining species are experiencing a 25% average decline in abundance. Similar trends are observed in invertebrate species with a 45% mean abundance decline in 67% of the monitored populations. It is irrefutable that these animal declines will trigger cascading effects on ecosystem functioning and human well-being. This *Anthropocene* defaunation is evidently a widespread component of the planet's sixth mass extinction and is also a primary driver of global ecological change.

Climate scientists have long been sounding the alarm that the current patterns of resource extraction, production, and consumption are inflicting escalating levels of harm upon our planet. They caution that if we do not promptly revise our behaviors, the inflicted damage may soon reach a point of no return. Yet, even in the face of such dire predictions, the prevailing power dynamics and their impact on our collective consciousness often supersede a pragmatic outlook, a path that leads to universal suffering. The cascading effects of an unsustainable and environmentally detrimental consumer economy are no longer obscured in the abstract. The damage imprinted on the atmosphere (air), lithosphere (land), hydrosphere (water), and biosphere (living system) are now evidently perceptible to a wider community of people than ever before.

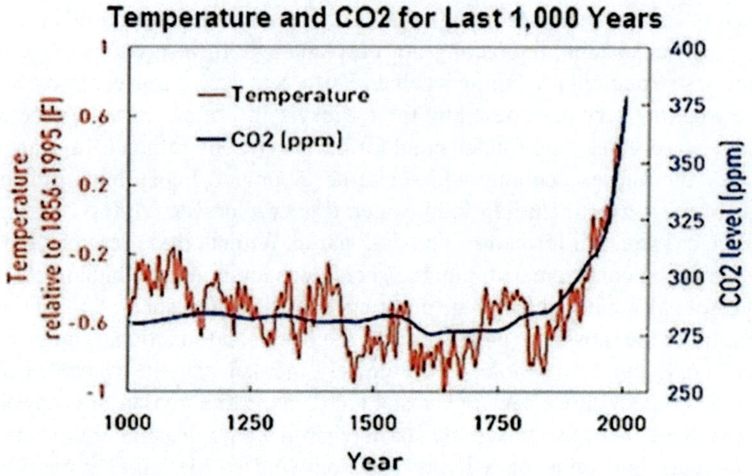

Graph 7.1 Association of carbon dioxide (CO_2) in the atmosphere with the heating of the planet. (Source: Adapted from Mellilo, J. M., Richmond, T., & Yohe, G. (2014). Climate Change Impacts in the United States: The Third National Climate Assessment Report. *US Global Change Research Program, Washington DC*, 2014)

7.2.1 The Danger of Tipping Points

Researchers are sounding the alarm, predicting that the world is on a trajectory toward a tipping point marked by extinctions and unpredictable changes on a scale unprecedented since the retreat of the glaciers 12,000 years ago (Pappas, 2012). Professor Anthony Barnosky et al. (2012) from the University of California, Berkeley, along with 17 other scientists, caution that planet Earth may transform into a less hospitable abode for life. As per Barnosky et al. (2012), "*There is a very high possibility that by the end of the century, the Earth is going to be a very different place. You can envision these changes as a fast period of adjustment where we get pushed through the eye of the needle. As we are going through the eye of the needle, that is when we see political strife, economic strife, war, and famine.*"

The smooth curve in blue shows northern hemisphere temperatures over a period of 1000 years. Its uncertainty range is in light blue, which is overlaid with green dots showing a 30-year global average. The red curve shows the measured global mean temperature since the Industrial Revolution.

For decades, climate change deniers have vigorously attempted to discredit the hockey stick graph (Graph 7.2) developed by Michael Mann and his colleagues. However, in a remarkable turn of events, more than 25 years later, numerous independent reconstructions of past temperature changes have emerged, all validating Mann and his colleagues' original findings. Recently, during his acceptance inaugural speech for the Humanist of the Year award by the American Humanist Association (AHA), Michael Mann had eloquently articulated the significance of these validations. He had highlighted that the latest studies suggest that the warming of our planet in recent decades is unparalleled in tens of thousands and even hundreds of thousands of years (Mann, 2023).

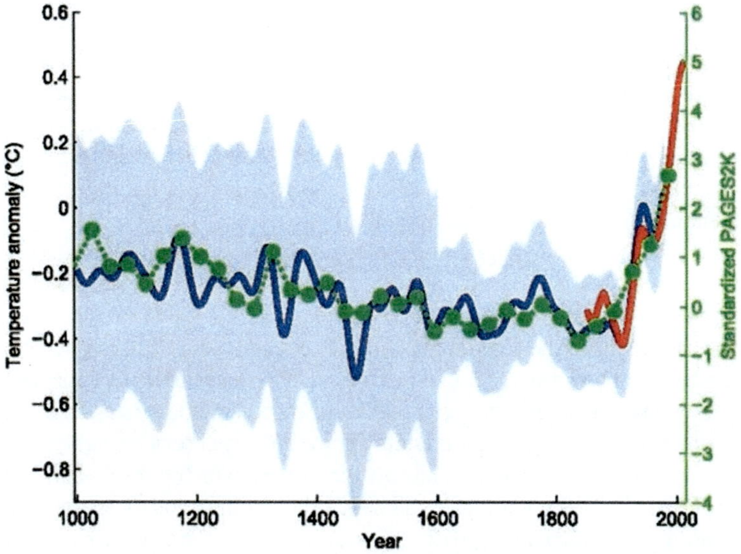

Graph 7.2 The famous hockey stick graph developed by Michael Mann, Ramon Bradley, and Malcom Hughes first published in *Nature* in 1999. (Source: Adapted from Wikipedia: https://en.wikipedia.org/wiki/Hockey_stick_graph_(global_temperature))

This emphasizes the unprecedented nature of the situation we find ourselves in today, as we are involved in an uncontrolled and unparalleled experiment with the only planet we currently call home—a planet that sustains us and countless other life-forms. Mann and his colleagues have significantly contributed to our understanding of the urgent need to address climate change and its consequences. As we continue to navigate the complexities of this global issue, Mann's words serve as a reminder of the critical importance of safeguarding our planet for the benefit of current and future generations and living systems on planet Earth.

Alan Buis (2011) predicts that by 2100, anthropogenic climate change will induce shifts in ecosystems and plant communities across almost half of Earth's land surface. This will result in nearly 40% of land-based ecosystems transitioning from one major ecological community type to another, such as forests, grasslands, or tundra. Scientists have been raising concerns about the ecological repercussions of global warming of even a few degrees. Should major greenhouse gas-emitting countries fail to drastically reduce their emissions and limit the global temperature increase to below 1.5 °C by 2050, all terrestrial ecosystems on planet Earth will be subjected to major transformations that will drastically alter the world's biomes. This will have far-reaching consequences for everything from water and food security to public health and biodiversity loss. Unchecked climate change could fundamentally transform our terrestrial ecosystems, posing immense risks to the diversity of our planet. It is likely to result in dramatic global landscape changes, equivalent to an ecological transformation occurring over one or two centuries, akin to the transformation that took place over 10–20 thousand years at the end of the last deglaciation period 21,000 years ago (E360 Digest, 2018). The changes in the atmosphere and oceans can profoundly impact the biosphere, the delicate layer of life on Earth that is intrinsically interwoven with the atmosphere and hydrosphere and provides the life-supporting matrix within which human societies exist. Yadvinder Malhi et al. (2020) posit that the degradation or restoration of parts of the biosphere will likely have regional or planetary consequences, including anthropogenic greenhouse gas emissions, which drive both climate change and ocean acidification, increasingly threatening the viability and resilience of natural ecosystems and the human societies that depend upon them.

Cultivation of land through agriculture, along with millennia of deforestation, has potentially injected hundreds of billions of tons of carbon into Earth's atmosphere. As we transitioned into the industrial era, our reliance on carbon-laden sources of energy further exacerbated this issue. The combustion of coal and natural gas in power plants that illuminate our homes and the use of petroleum in various modes of transportation have collectively bolstered the net accumulation of carbon dioxide (CO_2) in the atmosphere. At present, an average person emits approximately 5 tons of carbon dioxide annually, a quarter of which will endure in the atmosphere for over a millennium (Hsiang et al., 2017). Greenhouse gases like CO_2 disrupt the planet's energy equilibrium. According to Hsiang and Kopp (2018), the escalation of greenhouse gases in the atmosphere obstructs some of this reradiation, rechanneling energy back toward Earth. An increase of 1% in atmospheric CO2 concentrations translates to roughly 27 trillion watts, or 0.05 watts per square meter. This is equivalent to the energy output of a Hiroshima-scale atomic bomb disseminating across Earth's surface every 2.3 s. The concept that anthropogenic activities could modify the climate goes back to nearly two centuries. Empirical studies conducted in the latter half of the twentieth century have cemented the understanding that human actions have been pivotal in reshaping the climate (Stocker et al., 2013; UGCRP, 2017). Given the weight of scientific evidence, the hypothesis positing no human influence on global climate has been unequivocally refuted (Hegerl et al., 2007).

Researchers have reviewed extensive studies on climate change, ecology, and Earth's tipping points, revealing thresholds that, when surpassed, can incite an environmental domino effect. Excessive pressure on the environment can catalyze these inflection points, triggering major global transitions (Barnosky et al., 2012). The most recent of such a transition has been the conclusion of the last glacial period when, within a span of 3000 years, Earth transitioned from being 30% ice-covered to its current near ice-free state. A substantial portion of the extinctions and ecological shifts occurred within a mere 1600 years. Between 1970 and 2010, our planet has experienced a staggering loss of 52% of its biodiversity, thereby heightening concerns about the revival of Earth's lost biodiversity (Lipton et al., 2018; Specktor, 2019).

In the present day, human-induced alterations to the environment occur at a pace exceeding natural ones. The Industrial Revolution, instigated by a 35% increase in atmospheric carbon dioxide, has induced global temperature increases that outpace those observed in preindustrial times. Concurrently, humans have extensively modified 43% of Earth's land surface for urban and agricultural purposes, compared to the 30% land surface transition that occurred at the close of the last glacial period. With the human population burgeoning, exerting increasing pressure on existing resources (with a current population of eight billion), the consequences are challenging to predict, as these tipping points are propelling the planet into uncharted territories.

A departure from the current trajectory necessitates robust political determination and global collaboration. By 2025, human activities are projected to utilize 50% of Earth's surface. With an inevitable population boom to nine billion by 2050, efficiency in resource utilization is crucial for sustainability. This necessitates more efficient energy consumption and production, a heightened focus on renewable resources, and an immediate imperative to conserve species and habitats for future generations. Scientists (Specktor, 2019; Pappas, 2012; Carrington, 2019) urge us to acknowledge that we are at a critical juncture. Should we opt for inertia, we risk encountering these tipping points and an unimaginable future dystopia for our descendants. Climate change is perceived as the foremost threat to our planet. It could intensify extreme weather events, precipitate droughts in some regions, destabilize rainfall patterns, accelerate glacial melting in the Himalayas and Antarctica, alter the global distribution of animals and diseases, and inundate low-lying areas due to rising sea levels (Lipton et al., 2018; Hausfather, 2017; Kolasi, 2017).

An alarming policy paper from an Australian think tank contends that the risks posed by climate change are more severe than generally anticipated (Specktor, 2019). This paper underscores that climate change presents a near-to-midterm existential threat to human civilization and that societal collapse by 2050 is a tangible risk if substantial mitigation efforts are not implemented in the ensuing decade. It warns that climate scientists have been overly conservative in their predictions about climate change's impact on the planet in the near future. The ongoing climate crisis surpasses any previous human encounters in its magnitude and complexity. According to Hayhoe et al. (2018), global average temperature has escalated by about 1.8 °F from 1901 to 2016. As a result of the precarious shifts in climate and weather owing to the relatively small changes of one or two degrees in the planet's average temperature, numerous global regions have experienced changes in rainfall patterns, culminating in more floods, droughts, and intense rain as well as more frequent and severe heat waves. The planet's oceans, as well as the glaciers and ice caps in high mountainous regions and Antarctica, have witnessed significant alterations; oceans are warming and acidifying, glaciers and ice caps are melting rapidly, and sea levels are rising (Fleming et al., 2018). Climate scientists and researchers warn that when these and other adverse changes become more pronounced in the coming decades, they will pose formidable challenges and threats to humanity and the environment that we once perceived to be safe and secure. These adverse climate changes will impact all aspects of human, natural life, and environmental services. Researchers engaged in impact studies of climate change have summarized these impacts as follows:

- Climate change-induced warmer temperatures have escalated the frequency, intensity, and duration of heat waves, posing health risks, especially for young children and the elderly (Vose et al., 2017).
- Climate change has deleterious implications for human health by worsening air and water quality, facilitating the spread of certain diseases, and altering the frequency or intensity of extreme weather events (Fann et al., 2016).
- Rising sea levels induced by climate change pose threats to human communities and ecosystems in coastal regions worldwide (Fleming et al., 2018).

- Changes in rainfall patterns and streamflow timing and amount, resulting from climate change, impact water supply and quality and hydroelectricity production in many parts of the world (Lall et al., 2018).
- Changes in ecosystems induced by climate change have influenced the geographic ranges of numerous plant and animal species and the timing of their life cycle events, such as migration and reproduction (Lipton et al., 2018).
- The increased frequency and intensity of extreme weather events due to climate change, such as heat waves, droughts, and floods, lead to property losses, societal disruptions, and decreased insurance affordability (Ebi et al., 2018).

7.2.2 *Emissions of Carbon Dioxide*

Cumulative carbon dioxide (CO_2) emissions derived from the combustion of fossil fuels and the production of cement between 1751 and 2014, as presented in Table 7.1, elucidate that the United States is the principal emitter, contributing 26% of emissions. China trails with 12%, and Russia, inclusive of the former Soviet Union, contributes 11%. Together with Germany (6%) and the United Kingdom (5%), these nations are responsible for 60% of historical emissions. However, a shift in paradigm is witnessed when scrutinizing contemporary emission trends as opposed to historical stocks; China predominated in emissions in 2014, accounting for 30%, with the United States (15%), India (7%), Russia (5%), and Japan (4%)

Table 7.1 Historical and current emissions of carbon dioxide from fossil fuel combustion and cement production—top 15 current emitters

Country	Cumulative 1751–2014 (gigatons CO_2)	Percentage of global	Emissions 2014 (gigatons CO_2)	Percentage of global	Emissions per capita (tons CO_2), 2014
China	174.7	12%	10.3	30%	7.5
United States	375.9	26%	5.3	15%	16.2
India	41.7	3%	2.2	7%	1.7
Russia/USSR	151.3	11%	1.7	5%	11.9
Japan	53.5	4%	1.2	4%	9.6
Germany	86.5	6%	0.7	2%	8.9
Iran	14.8	1%	0.6	2%	8.3
Saudi Arabia	12.0	1%	0.6	2%	19.5
South Korea	14.0	1%	0.6	2%	11.7
Canada	29.5	2%	0.5	2%	15.1
Brazil	12.9	1%	0.5	2%	2.6
South Africa	18.4	1%	0.5	1%	9.1
Mexico	17.5	1%	0.5	1%	3.8
Indonesia	11.0	1%	0.5	1%	1.8
United Kingdom	75.2	5%	0.4	1%	6.5
World	*1434.0*	*100%*	*34.1*	*100%*	*4.7*

Source: Adapted from Boden et al. (2017). Global, Regional, and National Fossil-Fuel CO_2 Emissions, Carbon Dioxide Information Analysis Center (CDIAC), Oak Ridge National Laboratory (ORNL)

following suit. Germany is the most significant emitter within the European Union (EU), responsible for 2.1% of emissions, whereas the EU-28 collectively contributes 10% to global CO_2 emissions (Janssens-Maenhout et al., 2017). Globally, the per capita emission rate is 4.76 tons/year, with the United States leading with 16.2 tons/year, over three times the global average. China trails with 6 tons/year.

Seven countries, the United States, China, Russia, Germany, the United Kingdom, Japan, and India, account for more than two-thirds (67%) of total CO_2 emissions and are chiefly responsible for contemporary global warming and consequent climate change. Unless these nations assume moral accountability for the degradation of the planetary ecosystem and replace polluting activities with environmentally enabling policies and actions, addressing climate change seems unattainable. Unless these countries take the moral responsibility for the degradation of the planetary ecosystem (biosphere) and are willing to correct and replace their polluting activities with environmentally acceptable and enabling policies, programs, action plans, and activities, it seems impossible to bring desirable changes to the drivers of climate change. They need to seriously consider for coming to an agreement to alter their national and international policy instruments to tackle the issues of climate change and environmental problems before they run out of time.

The adaptation to some milder effects of climate change will undoubtedly ameliorate impacts on ecosystems and human well-being. However, there will be a high price to pay, including immense pain and suffering from the drastic adverse climate change, should the current ecologically destructive development model persist. Increased temperatures exacerbate drought conditions and are contributing to droughts in Africa, Asia, and South America, impacting farming communities directly.

Higher air temperatures not only accelerate drought conditions but also intensify them. Climate change with increased average temperatures driven by human-generated emissions of heat-trapping greenhouse gases is contributing to droughts in Africa, Asia, and South America, thus directly impacting the farming communities (Fig. 7.1).

7.2.3 *Rapid Deglaciation in Nepal Hindu Kush Himalaya (HKH)*

Rapid deglaciation in the Hindu Kush Himalayan region is of urgent concern. Glaciers are receding rapidly due to global warming and climate change impacts. The negative repercussions of climate change and global warming in this region are predicted to be more severe than elsewhere due to even minor temperature increases, leading to a hastened melting of snow and subsequent recession of glaciers. This will lead to rivers drying quickly, severely impacting agriculture and human settlements downstream (ICIMOD, 2020, 2021; Wester et al., 2019; Agrawal et al., 2019). Global warming and climate change's repercussions, primarily due to greenhouse gas emissions, include the rapid melting of the Himalayan glacier environment. This has far-reaching implications for biological and ecological systems, agriculture, and human habitats in the region, such as increased flooding, freshwater scarcity, threats to flora and fauna, and irrigation water shortages for agriculture. These glaciers have traditionally been perennial sources for Asia's major river systems, supplying water to more than one

Fig. 7.1 With increased average temperatures, drought has become a common phenomenon. (Source: Getty images. https://www.gettyimages.com/detail/photo/textured-cracked-mud-landscape-iceland-royalty-free-image/1176594622?adppopup=true)

billion people (Wester et al., 2019; ICIMOD, 2020). Nepal, occupying the eastern central region of the Hindu Kush Himalayan range, has experienced an accelerated rate of deglaciation, leading to moraine formation, which is unstable and can cause severe flooding downstream due to glacier lake outburst floods (Fig. 7.2).

Nepal has been on the front lines of climate change for several decades, enduring some of the most severe impacts. These effects have escalated dramatically in recent years. The rate of glacial melt in the Nepal Himalayas is particularly alarming, having accelerated by 65 percent in the last decade compared to the previous one. Equally troubling are the unpredictable shifts in weather patterns, reduced snowfall in the mountains, and increased frequency of glacial runoff and retreat, phenomena once considered rare. The annual monsoon season now routinely unleashes catastrophic floods, wreaking havoc on people's lives and their means of sustenance. Notably, Nepal's mountainous regions have experienced a temperature increase of 1.8 degrees Celsius, significantly higher than the global average warming of 1 degree Celsius. This stark disparity highlights the disproportionate impact of climate change on countries like Nepal, which contribute minimally (virtually zero) to global carbon emissions yet bear a substantial burden of the consequences.

The global community's efforts to mitigate this climate crisis in nations such as Nepal are grossly inadequate. This inadequacy was underscored by recent observations made by UN Secretary-General António Guterres. During his visit to the Mt. Everest region, Guterres was confronted with the stark realities of the melting glaciers and their implications for the livelihoods of the Nepali people. Utilizing the platform X for global outreach, he poignantly stated, "...Glaciers are retreating; we cannot retreat. We must move forward with climate action" (TKP 2023).

The Khumbu glacier in Nepal's Khumbu region (Everest region) is experiencing rapid melting and deglaciation. Projections indicate that even if carbon emissions are curtailed substantially, limiting global warming to 1.5 °C, 63% of the glaciers in the

Fig. 7.2 Glaciers receding in Nepal's Khumbu region (Everest region) due to global warming and climate change impacts. (Source: Adapted from ICIMOD (2021). HKH2Glasgow: An urgent call for climate action for the Hindu Kush Himalaya)

Hindu Kush and Himalayan range will have disappeared by 2100. Should emissions continue unabated, deglaciation could rise above 67% (ICIMOD, 2021). These glaciers in the HKH region are critical water storages for the 250 million people living in the region, and 1.65 billion people rely on the rivers flowing from this region. Even in the best-case scenario, one-third of these glaciers will be lost, an alarming revelation for the region's inhabitants. Despite its high population, the HKH region has received less attention compared to other vulnerable areas such as low-lying island states and the Arctic. The imperative to limit global temperature rise to 1.5 °C above preindustrial levels necessitates reducing emissions to zero by 2050. Although this is a highly optimistic target, it is unlikely to be achieved. Even if global temperatures rise by just 2 °C, two-third of the glaciers are projected to melt by 2100 (ICIMOD, 2020).

The HKH region, despite being far more populous, has received far less attention than other places, such as low-lying island states and the Arctic. As Bristol University professor Jemma Wadham (Wester et al., 2019) rightly pointed out, "*This is a landmark piece of work focused on a region that is a hotspot for climate change impacts*." To limit global temperature rise to 1.5 °C above preindustrial levels requires cutting emissions to zero by 2050. This is extremely optimistic but unlikely to happen. The HKH range (the fresh water tower of Asia) is 3500 km long and is the second largest reservoir of glacier snows besides the Arctic and Antarctic, but, in terms of the magnitude of impacts of glacier recession on human population, it is going to be the largest by any measure.

7.2.4 The Climate Change Policy and Trump Presidency

The Trump presidency marked a significant regression in US environmental policy, particularly with respect to global warming and climate change. Despite irrefutable scientific evidence and consensus regarding the detrimental impacts of anthropogenic activities, especially the consequences of fossil fuel consumption on global warming, President Trump's rejection of climate change as a hoax was deeply concerning. President Trump declared that the Paris Climate Agreement would "undermine" the U.S. economy, and put the U.S. "at a permanent disadvantage". The United States' withdrawal from the Paris Climate Agreement was an unfortunate departure from its commitment to global environmental responsibility. The Environmental Protection Agency (EPA), a credible and professional agency tasked with enforcing the federal government's environmental programs, has historically been lauded for its role in formulating and implementing environmental policies. However, the Trump administration enacted numerous rollbacks of essential environmental regulations, thereby seemingly prioritizing and acting in the interests of the fossil fuel industries.

Under President Trump's administration, US politics experienced an unprecedented shift in behavior, characterized by anger, arrogance, deceit, discrimination, and a prolificacy of false information. This shift was particularly evident in the manipulation of public opinion, particularly among lower middle-class white Americans who, ironically, are among the most adversely impacted by Trump's policies. Mulholland (2019) aptly highlights that the influence of large fossil fuel corporations and lobbyists has distorted the climate policy debates in the United States. While delivering his acceptance speech for the Humanist of the Year award at the 82nd annual conference of AHA, Denver, Colorado, in May 2023, Prof. Michael Mann's following remarks are worth pondering: "*We still face a multipronged strategy by polluters, and their enablers in the media and pundit class to distract, deflect, attack, divide, and delay. Among their preferred tactics is the promotion of risky, unproven strategies, such as geoengineering or massive carbon capture and sequestration, and the promise of future action as an excuse for business-as-usual fossil fuel burning today.*"

Regrettably, the Trump administration represented a period of significant environmental regression for the United States, marked by the denial of anthropogenic causes on climate change and a disconcerting indifference toward climate science. This resulted in significant rollbacks in environmental regulations, promotion of fossil fuels, and minimization of climate change impacts on government websites, all of which threaten global warming mitigation efforts. The administration dismissed and even contradicted expert research on climate change's implications for human security, failing to articulate a clear perspective on this interrelation. The Trump administration's omission of climate change as a significant threat in its inaugural National Security Strategy signaled a concerning disregard for the growing evidence and increasingly frequent climate-related incidents worldwide.

7.2.5 Climate Change, Conflicts, and Security

Researchers studying the impacts of climate change assert that it will heighten the risk of both domestic and international conflicts. Evidence points to a strong correlation between societal stability and climate, with research indicating that the risk of civil conflicts in African nations has increased by 11% since 1980 due to climate warming. Economically underdeveloped regions, already vulnerable, are likely to be further destabilized by climate change, leading to increased societal and psychological insecurities. The combined impact of climate change on sectors such as agriculture, energy, labor, health, crime, and coastal communities could cost the United States nearly 2% of its gross domestic product (GDP) for each 1 °C increase in global temperature. Furthermore, increased mortality rates, suicides, sexual assaults, murders, and birth-related harm are all projected to rise significantly. Given that carbon dioxide is the main greenhouse gas driving climate change, primarily from burning fossil fuels, the economic cost of carbon is set to soar.

The effects of climate change on international conflicts are not entirely understood, but there is a high likelihood of escalated violence over scarce resources such as water and land. The increasing scarcity of water for drinking and irrigation purposes will likely spur intra and interstate conflicts. The Hindu Kush Himalayan region could become a potential flashpoint for water-related conflicts in the future. Additionally, climate change-induced migration could create tensions between affected countries, potentially increasing both direct migration due to extreme weather events and indirect migration due to internal conflicts.

7.2.6 Insights from Climate Scientists

To ascertain the progression of global warming and to identify human contributions to this phenomenon, it is necessary to navigate the complexities and apparent ambiguities of Earth's climatic conditions. The historical patterns of climate change confirm that our climate's response to variations in energy balance is significant, implying that an excess of heat accumulation over heat loss can escalate global temperatures. The cause of this energy imbalance, however, may differ. The present circumstances indicate that an enhanced greenhouse effect, principally driven by increasing CO_2 concentrations, is dictating this energy imbalance. This is further corroborated by historical climate change data that provide evidence of our climate's sensitivity to CO_2 concentrations (Hausfather, 2017; Pappas, 2012).

It is critical to note that single data points of local temperatures do not provide an accurate representation of global warming's long-term trend. Scientists emphasize that climatic patterns can only be discerned by observing alterations in weather conditions over an extended period. A careful examination of high and low temperature records from recent decades has demonstrated that new record highs are almost twice as frequent as new record lows. For instance, a 2014 study published in *Geophysical Research Letters* revealed that over the past decade, daily records of high temperatures in the continental United States were twice as prevalent as record lows (Abatzoglou & Renaud, 2014). According to Skeptical Science (Cook, 2010), the hottest recorded decade was 2000–2009. It is a common misconception that instances of heavy snowfall or unusually cold weather dispute the existence of global warming. However, these occasional cold weather phenomena do not contradict the broader trend of global warming, which is defined by long-term changes measured over decades. Contrary to popular belief, the Sun has exhibited a slight cooling trend in the past 35 years, even as global temperatures continue to rise. While solar activity does contribute to global temperature increases, its influence is relatively minor. For instance, research published in the journal *Atmospheric Chemistry and Physics* in December 2011 showed that even during an extended period of reduced solar activity, Earth continued to warm (Hansen et al., 2011).

Plants do absorb CO_2 through photosynthesis, but it is important to remember that CO_2 also contributes significantly to the greenhouse effect, leading to global warming. Additionally, it can act as a direct pollutant, causing ocean acidification. A 2007 study demonstrated that while global warming might extend the growing season in areas like Greenland, it also exacerbates water scarcity, intensifies wildfires, and expands desertification in other regions. Climate models have accurately replicated global temperatures since 1900 across various environments—terrestrial, atmospheric, and marine. As stated by the climate scientist Michael Mann of the Pennsylvania State University, these models formalize our current understanding of processes governing the atmosphere, oceans, ice sheets, and more (Carrington, 2019). A consensus among 97% of climate scientists confirms the existence of anthropogenic global warming (Skeptical Science: https://skepticalscience.com/). These scientists assert that human activities that harm the environment are driving climate change and significantly degrading the planetary ecosystem (biosphere). Their primary concern is to mitigate these detrimental effects.

7.2.7 Pessimistic Scenario

If we contemplate the most extreme possible outcome for Earth's climate future, and given the seeming indifference of political and global corporate elites to heed scientific advice or public sentiment toward decarbonizing the economy, then we are potentially looking at an increase of global temperatures by 5.4 °F (3 °C) by 2050. Scientists predict that at this threshold, the world's ice sheets would disappear, severe droughts would decimate vast sections of the Amazon forest (eradicating one of the world's largest carbon offsets), and Earth would enter a feedback loop of increasingly intense and deadly conditions (Carleton et al., 2021; Hansen et al., 2011; Hsiang et al., 2019; Mulholland, 2019). In this bleak scenario, 35% of the global land area and 55% of the global population would be exposed to lethal heat conditions for more than 20 days a year, pushing the limits of human survivability. Concurrently, droughts, floods, and wildfires would regularly devastate the landscape. About one-third of the world's land surface would transform into deserts. Complete ecosystems, starting with the planet's coral reefs, rainforests, Arctic ice sheets, and the total deglaciation of the Hindu Kush Himalayan region, would collapse. The agricultural productivity of the tropics would be devastated, displacing more than one billion people. This mass migration, coupled with diminishing coastlines and acute shortages of food and water, would strain the societal fabric of the world's most populous nations. Conflicts over resources could escalate, potentially even culminating in nuclear war. If not urgently addressed with comprehensive mitigation strategies and restoration efforts, this scenario could mark the end of global human civilization as we currently know it. It is like sailing in a leaky boat on a stormy sea.

7.3 Breaking the Back of Fossil Fuel Nexus

The most immediate and urgent task is to break the back of the fossil fuel nexus. The global corporate industrial nexus of fossil fuels is perhaps the strongest organized force resisting the climate change policies in the world today. Their investment in political establishment and lobbying activities is huge, which apparently makes it impossible to stop their tentacles from either directly or indirectly reaching their political benefactors who resist policy changes toward climate change and, moreover, environmental legislature and regulations. It is clear that the global corporate enterprises of fossil fuels are the major culprits responsible for greenhouse gas emissions, and climate change. Scientists had warned a long time ago that keeping global warming below the Paris Climate Accord's 2 °C limit requires stopping the burning of the majority of the existing fossil fuel reserves (coal, oil, and gas). On the contrary, fossil fuel companies continue to spend billions of dollars each year searching for new forms of fossil fuels and influencing the countries in the world to support fossil fuel production to boost their economic growth. It is paradoxical that many countries continue to support fossil fuel production despite their pledges and stated ambitions to cut greenhouse gas emissions as per the goals of the Paris Climate Agreement. Scholars, activists, and policymakers have recognized that the biggest challenge is the fossil fuel industries' stranglehold on politics to resist serious climate action policies. The fossil fuel industries spend hundreds of millions of dollars every year to deliberately mislead the public and stifle policies through lobbying and spreading disinformation. It is also equally important to remember, remind, and inform the public about how the extraction of coal, gas, and oil is causally interlinked with water and air pollution, habitat destruction, and environmental injustices, besides global warming and climate change.

It is interesting that over the last 25 years of international climate negotiation, including the Paris Climate Accord, there has not been a single proposal, debate, or even a position paper on limiting fossil fuel production. This was perhaps the largest and most damaging mis-framing in climate policy. It makes no sense to talk about climate change policy by continuing fossil fuel extraction, production, and consumption as usual. Politicians in the United States have rarely talked about limiting the use of fossil fuels until very recently. However, the political landscape seems to be gradually changing even in the United States, with most democratic candidates for the presidential race putting forward climate plans that consist of a multitude of supply-focused policies. It was inconceivable even 5–6 years ago to see US mainstream politicians' commitments to phase out fossil fuel production and hold the industry legally, morally, and financially accountable with Bernie Sanders being a notable exception. A number of actors have played pivotal roles in shifting and shaping the political momentum. The climate champion Jay Inslee, climate activism, and grassroots movements demanding that fossil fuel companies comply with the Paris Agreement have reframed the climate narratives as not just the technocratic problem about greenhouse gases but also a moral and political problem about fossil fuels.

Another force that converged and complimented the activism and grassroots voices is the economic warnings and divestment decisions from many of the world's foremost investors and finance experts. Market analysts have reported that in the past year, all major oil and gas companies have approved projects that were consistent with the Paris Climate Accord's goals. It is projected that more than 90% of ExxonMobil's potential spending through 2030 would be stranded in low-carbon projects. This certainly sounds good but raises valid skepticism, given the dubious and deceitful character of the fossil fuel industries. Science and policy debates have generated a powerful stimulus to shape public opinion in favor of the climate change policy formulation. Climate scientists have clearly shown that existing fossil fuel reserves already exceed the global carbon budget and have appealed to political decision-makers to pursue smart climate actions that require seriously addressing both demand and supply. Academicians, researchers, and nongovernmental organizations (NGOs) have also echoed this message about the climate crisis to the decision-makers and public. Investigative journalists and scholars play a crucial role in uncovering the environmental damage caused by the fossil fuel industries. Journalists have done a remarkable job of uncovering the skeletons of the fossil fuel industries' closet, holding these companies accountable for environmental damage and also for denying and delaying compliance with the existing environmental laws and regulations. If the Green New Deal legislation, advanced by some democratic congressmen, is passed by the Congress, then there will be a significant development toward climate change policymaking that will lead to fossil fuel polluters paying for the damages caused and misconducts. The policy shift toward climate change may consist of a number of measures, including setting climate targets based on the amount of oil, coal, and gas extracted; ending all fossil fuel production subsidies; banning or restricting production; suing the fossil fuel industries for the damages caused and obstructionism; rejecting all fossil fuel campaign contributions; clamping down on the fossil fuel industries' lobbying with the government; and creating just alternative economic opportunities for fossil fuel-dependent communities.

7.3.1 Light at the End of the Tunnel?

Is there a way to prevent this catastrophic vision of the future? It may only be possible if people and governments across the globe accept climate change an as issue of emergency and get to work immediately. Most researchers and scientists believe that the human race has about one decade left to mount a global movement to transition the world economy to a zero-carbon emission system. Achieving zero-carbon emissions requires either not emitting carbon at all or balancing carbon emissions with carbon removal. It is a fact that after nuclear war, human-induced global warming and climate change is the greatest threat to human life on the planet, and human life on Earth may be on the way to extinction, in the most horrible way.

There is hope and optimism that the current environmental and climate crisis can be resolved, but those feelings are diminishing when we see that this issue lies in the hands of the far-right climate denial radicals in the US political landscape (Trump and his allies in the Republican party). The European Union (EU) has always been a subordinate ally of the United States, but time has come for the EU to take a lead role and move away from its subordinate position and lead to resolve the current environmental (climate) crisis. Young people across all nations in the world are the real hope of the future. They can influence and impact the development and environmental policies of their respective countries, particularly in bringing about desirable policy and structural changes in the global corporate capitalist system. Without discussing the policies of global corporate capitalism, the present environmental and climate crisis cannot be resolved. Global capitalism and its financial systems cannot be dismantled without a viable, alternative global financial system in place. What we require is changes in the system for which those who are governing the global capitalist system need to understand the complexity of the problem and change their mindsets. The urgency of the present environmental and climate crisis requires that it must be dealt with from within the framework of the existing system and change or modify the current system so that the system becomes progressively more responsible and friendly to planet Earth.

First, the ruling elites have to realize how the present global capitalist and financial systems are driving the breakdown of planetary ecosystems, which has resulted in the present environmental and climate crisis. It is true that for the resolution of the crisis of such magnitude, it requires a tremendous amount of money and resources; however, money is not a problem and the world financial system is not going to be depleted of money. The problem is how we utilize money and other resources to resolve this crisis. If 1 or 2% of the current global GDP (85 trillion dollars in 2021) is allocated for the protection and management of global environmental commons (Earth's planetary ecosystems) and eradication of poverty and empowerment of marginalized communities in developing countries, then the current environmental, social, and climate crises can be resolved. It cannot happen without some structural changes in the global financial capitalism, but, if this happens, then it will create an enabling environment for the emergence of an ecologically sustainable development system.

The current policies of the world government for climate actions have failed miserably to materialize and achieve the intended results, as a result of which tipping points toward climate and ecological catastrophes are rapidly approaching. Given that the political discourse on climate action has not produced any tangible results and that carbon emissions continue to accelerate, the world is experiencing life-threatening climate change-related events, such as water crises and shrinkage of lands and forests at an alarming rate. Robert MacNeil (2022), in his climate manifesto, points out why these far-reaching consequential changes on planet Earth now require humanity to take drastic steps, not just as individuals but also as whole societies. He poses a difficult question: "*What it means to be humans on a planet that can no longer support our globalized industrial-consumer economy and our predatory way of living?... how we can make the changes we need for humanity to survive and thrive on the planet.*" This "*predatory way of living*" has accelerated climate change that is intimately entangled with the health and functioning of the biosphere. Climate change critically impacts ecosystems through changes in the mean climatic variables, coupled with other associated changes such as increased temperatures, ocean acidification and atmospheric carbon dioxide concentrations, and species extinction. It also exerts pressures on ecosystems, including degradation, defaunation, and fragmentation. It is important to understand how the ecological dynamics of the climate impact hotspots of vulnerability and resilience and management interventions may help in the biosphere's resilience to climate change. It is equally important to know how ecosystems adapt to climate change and how their adaptation may affect climate change. The scientific epistemology must address the important issues of how ecosystems respond to climate change, how ecosystem resilience can be enhanced, and how ecosystems can assist in addressing the challenge of a changing climate. Ecological research scholarship should be directed at developing novel approaches to maximize the potential for maintaining a diverse, resilient, and well-functioning biosphere under the challenging conditions of the twenty-first century.

Disease pandemics, climate, and environmental pollutions do not recognize national borders. The globalization of wealth that benefits only 1% needs to be brought under the control of the 99%. The states depend entirely on the taxes paid by the people; therefore, people have the power to control the states and the states should exert control over reckless capitalism. Humans are capable of overcoming and resolving the climate crisis with their cooperative endeavors, sharing of resources, and knowledge. People empowerment with information, knowledge, social and material resources is the most seminal task to deal with and manage crisis

An Economist's Guide to Climate Change Science

Source: Hsiang, Solomon, and Kopp, Robert E. (2018). An Economist's Guide to Climate Change Science, *Journal of Economic Perspectives*, Volume 32, Number 4, Fall 2018, pp. 3–32.

Sunlight continuously enters our planet's atmosphere from space. For Earth to maintain a stable surface temperature, this flow of incoming energy must be balanced by a flow of energy leaving the atmosphere. Earth reflects back about 30% of incident sunlight immediately out to space from its surface or from clouds. The remaining 70% is absorbed by Earth's surface and atmosphere and must be balanced by the planet's own emission of infrared radiation to space, which intensifies with higher temperatures. Without greenhouse gases, the global mean surface temperature would be −18 °C (−0.4 °F), fully

(continued)

determined by the Sun's temperature, Earth's distance from the Sun, and Earth's reflectivity (also known as "albedo").

If a larger flow of energy were to somehow reach Earth's surface—for example, if the Sun were to grow in brightness or Earth were to decline in albedo—then our planet would heat up until the additional outgoing flow of infrared radiation exactly offsets this new source of energy. Greenhouse gases distort Earth's energy balance because they are transparent to the incoming visible and ultraviolet sunlight but absorb infrared radiation, thus hindering the return flow of this energy from the surface and the lower atmosphere into space. When a greenhouse gas molecule intercepts infrared radiation headed from the surface to space, the absorbed energy is reemitted in all directions, sending some energy that might otherwise have escaped to space back down to the surface of Earth. This causes the surface and lower atmosphere to warm up, thus slightly increasing their emission of infrared radiation.

One part-per-million CO_2 in the atmosphere is equal to about 7.8 Gt CO_2 in physical mass (An Economist's Guide to Climate Change Science, 2018). However, cumulative emissions of CO_2 are nonetheless a useful metric, as CO_2-caused warming is approximately proportional to cumulative emissions (Allen 2009), with every trillion tons of CO_2 causing about 0.2–0.7 °C of warming. According to the estimated cumulative emissions of CO_2 from fossil fuels and cement production during 1751–2014, as well as the flow of emissions in 2014 (Boden et al., 2017), the United States is responsible for more than one-fourth of historical emissions, followed by China (12%) and Russia (11%, including the former Soviet Union); together with Germany (6%) and the United Kingdom (5%), these five countries account for 60% of historical emissions.

However, if one examines flows today rather than the stock of historical emissions, then the picture is changing; China (30%) dominated emissions in 2014, followed by the United States (15%), India (7%), Russia (5%), and Japan (4%). Germany is the largest emitter in the European Union (2.1%), with the EU-28 collectively ranking third in global CO_2 emissions, responsible for about 10% (Janssens-Maenhout et al., 2017). High national emissions reflect high carbon intensity per capita (per capita emissions are 16.2 tones/year in the United States, 3.4 times the global average), high population levels (per capita emissions in India, the third-leading emitter, are about one-third the global average), or a mix of both factors (per capita emissions in China are about 60% more than the global average).

Source: World Wildlife Fund (WWF) Report 2014

When the World Wildlife Fund (WWF) released its *Living Planet Report 2014* on September 30, it was not the usual doom-and-gloom environmental news story that is forgotten the next day. This report—the result of a science-based study using 10,380 populations from 3038 species of amphibians, birds, fish, mammals, and reptiles from around the globe—is garnering worldwide attention for its sit-up-and-take-notice findings: between 1970 and 2010, the planet has lost 52% of its biodiversity. In the same 40-year period, the human population has nearly doubled. These figures take a while to sink in, especially since the previous WWF report that analyzed animal populations, published in 2012, showed a decline of only 28% over a similar time frame.

Specifically, the WWF biennial report found that we have lost 76% of freshwater wildlife, 39% of terrestrial wildlife, and 39% of marine wildlife since 1970. Although some animal species numbers are increasing and some are stable, the declining populations are decreasing so rapidly that the overall trend is down. Latin American biodiversity took the biggest plunge, diminishing by 83%.

Elsewhere in the tropics, populations are down 56%. Temperate zones fared better, with a loss of 36%, whereas terrestrial animal populations in parks and wildlife refuges are down 18%, indicating that protected areas can limit losses. Low-income countries are suffering a disproportionately greater loss of biodiversity: a 58% decline. Although high-income countries actually showed a 10% "increase" in biodiversity, a loss of 18% in middle-income countries and the astounding figure for low-income countries canceled those gains out.

Despite the fact that low-income countries are suffering the greatest ecosystem losses, high-income countries are using five times the ecological resources that they do. Those living in high-income countries are consuming more resources per person than nature can replenish, which means that the per capita ecological footprints in high-income countries are greater than the amount of biocapacity (the ability of an ecosystem to produce useful biological materials for food, fuel, building, and other needs and to absorb carbon dioxide emissions) available per person. People residing in middle- and low-income countries have had little increase in their per capita footprints over the same time period.

Those statistics boil down to the fact that every year, we use 1.5 planet's worth of natural resources. If we all lived the lifestyle of a typical United States resident, then we would need 3.9 planets per year. If we all had the footprint of the average citizen of Qatar, then we would need 4.8 planets. The term "overshoot day" is defined as the date when we have used up our annual supply of renewable resources and start spending down Earth's natural capital. In 2014, that day was August 20.

(continued)

The cause for this staggering demise in biodiversity is human activities. We have degraded natural habitats by clearing forests, plowing grasslands, and polluting waters, and have overhunted the land and overfished the oceans. A single culprit, climate change, is now responsible for 7.1% of the current declines in animal populations, but its toll is on the rise. The declining trend in worldwide biodiversity can be mitigated and reversed. To achieve sustainability again, each country's per capita ecological footprint must be less than the per capita biocapacity available while still maintaining a decent standard of living for its people. The WWF suggests that we can do that by shifting to smarter food and energy production; consuming responsibly at corporate, government, and personal levels; and putting a high value on natural capital when making policy and development decisions. Just two countries account for a third of the world's total ecological footprint: China, at 19%, and the United States, with nearly 14%. Perhaps those who take the most from the world should be the ones working hardest to replenish it.

humanity is facing. *Collective consciousness is the shared beliefs and moral attitude, which operate as a unifying force within a society. The collective consciousness of common people across nation states is the most potent force that can change the trajectory of the current development path.*

7.4 Investment on Nature

UNEP (2021) report on State of Finance for Nature states that world financial investment on environmental management is 133 billion US dollars a year. This amount is 0.16% of the current global GDP (85 trillion US dollars). The report states that a total investment of 8.1 trillion US dollars is required to maintain the biodiversity and natural habitats vital to human civilization, reaching 536 billion US dollars a year by 2050. This amount is 0.64% of the current global GDP. This means that the world needs to quadruple its annual investment in Nature if the climate, biodiversity, and land degradation crises are to be tackled by the middle of the century according to the UNEP report published in 2021. Out of 133 billion dollars/year, 86% (115 billion dollars/year) comes from public funds and the remaining 15% (18 billion dollars/year) from private finance. Nearly two-thirds of 115 billion US dollars is invested in forest restoration, peatland restoration, regenerative agriculture, water conservation, and natural pollution control system, and one-third is invested by national governments into protection of biodiversity and landscapes. The private sector finance of 18 billion US dollars/year spans through biodiversity offsets, sustainable supply chains, private equity impact investment, etc. As we can see, the total volume of finance flowing into environmental management (nature) is so limited that it is like feeding an elephant with just an ounce of food compared to what humanity gets from Nature's contribution, which is estimated to be at least 126 trillion US dollars/year (Costanza et al., 2014). Even UNEP's recommended amount of 536 billion dollars/year (0.64% of the GDP) is a tiny amount to be invested for solving environmental and climate problems given the magnitude of the problems.

The magnitude of environmental and climate problems demands that humanity spend at least 2–4% of the global GDP, which is quite feasible if the major polluters (the United States, China, the EU, Russia, India, Japan, etc.), who are the ones primarily responsible for the current environmental and climate crisis, come to an agreement to allocate 2% of their GDP for nature-based solutions to this crisis. They must understand that the threats to human civilization posed by the environmental and climate crisis is much bigger and real than are those that pose to each other with their ideological and political differences. If they are prudent and wise, then they should be investing far less on their military–industrial complexes and substantially more on the environmental and climate crisis, for which they are responsible. Talking about the threat to their national security from each other is simply a hoax of national psychology of fear and insecurity they have created. It is a necessary illusion of fear and insecurity to amass public support and divert taxpayers' money and resources to military–industrial complexes. This is true here in the United States and in China, Russia, India, and elsewhere.

In the event of a third world war, there will be no winner; all will be losers because all these countries are well-equipped with dreadful nuclear weapons that can destroy the whole world not once but several times. I believe they know this, but what I do not know is why they persistently increase their investment in military hardware. Should not they all jointly concentrate their efforts to fight against the environmental and climate crisis, the common formidable enemy of humanity? Even if just 1% of the global GDP (840 billion dollars/year) is allocated for environmental and climate crisis management, the current environmental and climate crisis and the problem of poverty and social insecurity can be resolved within a decade and planet Earth would become a much better place to live for everyone. Investment on Nature-based ecosystem solutions that consist of drastic reduction in greenhouse gas emissions, stabilization of climate and global warming, restoration of damaged ecosystems, reforestation, sustainable harvest of natural resources, protection of ecosystem processes and services, and design of a circular economy (recycle, refurbish, and reuse) with little waste throughput can restore Earth's lost regenerative capacity, making her productive and hospitable. Should not humanity strive for its own longevity?

7.4.1 *Ethical Imperative*

Climate change, primarily characterized by global warming, presents a severe ecological predicament. This broad-based issue has moral dimensions known as "the ethics of climate change." This ethics concept is grounded in the consensus among a substantial majority of the global scientific community that recognizes global warming as predominantly "anthropogenic" or human-induced. This perspective is prominently articulated in the reports issued by the Intergovernmental Panel on Climate Change (IPCC), which provides a robust basis for an ethical imperative to address climate change urgently and collaboratively (IPCC, 2014a, b). The ethical implications stem from the understanding that human actions have induced drastic changes in the world's climate, causing irreversible harm. This damage not only affects our generation but also has severe repercussions for future generations and other species. We have a moral responsibility to minimize further harm and to adapt to the changes that are already in effect (Gardiner, 2004).

This moral imperative involves developing and implementing robust policies of mitigation, adaptation, and compensation. Mitigation refers to strategies designed to reduce or prevent the emission of greenhouse gases, primarily through decarbonization of energy systems, increased energy efficiency, and shifts to renewable energy (Lenton, 2010). These strategies can be observed in the sincere commitment of all nations to reduce greenhouse gas emissions as part of the Paris Climate Agreement in 2015. This historic agreement sets a global goal to limit global warming to well below 2 °C, with efforts to limit the temperature increase even further to 1.5 °C (UNFCCC, 2015), but, regrettably, unless the biggest polluters, namely, the United States, China, India, the European Union, Brazil, and others truly commit to the Paris Climate Protocol, it is simply an empty rhetoric.

Adaptation, on the other hand, is about making adjustments to social or economic systems in response to actual or expected adverse climatic impacts. This includes changes in the processes, practices, and structures that moderate potential damages, taking advantage of opportunities or coping with consequences (IPCC, 2014a, b). Compensation involves holding accountable those primarily responsible for emissions and supporting vulnerable populations who are disproportionately impacted by climate change. Climate justice calls for more developed nations, which are primarily responsible for the historical accumulation of greenhouse gases, to compensate less-developed countries that bear the brunt of climate change impacts (Shue, 1999).

Another approach to the problem of climate change is climate engineering or geoengineering, which involves deliberate large-scale interventions in Earth's natural systems to counteract climate change (Royal Society, 2009). It is a controversial and potentially risky area, yet some consider it a necessary part of a comprehensive response to climate change. There are proven and far more effective technological systems such as restoration and development of degraded ecosystems, afforestation, renewable energy, agroforestry, agroecology, and ecological farming, which directly contribute to the regenerative biocapacity of planet Earth and decarbonization through CO_2 sequestration. In conclusion, climate change is a complex, multifaceted problem with significant ethical implications. It requires comprehensive, integrated, and ethically grounded strategies for mitigation, adaptation, compensation, and, possibly, climate engineering.

7.5 Conclusions

Even with the allocation of 1% of the current global GDP (85 trillion dollars in 2021) for the protection and management of global environmental commons (Earth's planetary ecosystems) and global poverty, the current environmental and climate crisis and social problems can be resolved. This will be a great milestone in the direction of resolving the environmental and climate crisis and in developing socially and ecologically sustainable development systems. It is possible to make structural changes in global corporate capitalism and make it a more humane, socially, and ecologically sustainable development system that can resolve antagonistic contradictions between humanity and Nature created by its unhealthy and reckless engine of infinite growth and profiteering.

Bill Gates (2023), in his climate book (*How to Avoid a Climate Disaster*), rightly points out that the level of global warming is directly proportional to the volume of carbon dioxide that human activities release into the atmosphere and argues that global carbon emissions need to drop to zero for climate change to stabilize. If drastic mitigating measures are not taken, then there is no way to achieve the 2015 Paris Climate Accord goals of 1.5 °C compared to preindustrial levels. The current yearly greenhouse gas emissions of 50 billion tons must be reduced to net zero by 2050 to avoid climate disasters. We must accelerate the deployment of solar, winds, and other cutting-edge breakthrough technologies with market and policy incentives for the production of green and clean consumer products. If political leaders, governments, and the ruling elites of the major polluters states and the global corporate world, including the United States, China, Russia, Japan, India, the European Union (EU), the United Kingdom, and others are politically determined to resolve today's environmental, social, and climate crises, putting aside their political and ideological differences, then we would have a resilient, healthy, and vibrant Earth system in the next two decades. If they address following recommendations with formulation of appropriate policy instruments and allocation of financial resources (1% of global GDP), then the current environmental and social problems can be resolved; the regenerative capacity of planet Earth can be restored; climate can be stabilized; and an ecologically sustainable global development system can evolve. The following conclusions precisely establish the rationale for the powerful nation states of the world to come together to resolve the environmental, social and climate crisis and make planet Earth a better place to live for humanity and all other living systems:

- Ecological problems arise from the functional degradation of ecological systems, and, therefore, the most important scientific research question is to understand how ecological systems function, how they change, and what interventions approaches are needed to maintain their functional integrity and restore the degraded ecosystems. The functional integrity of an ecosystem can be understood in terms of its health. Therefore, the primary emphasis of research scholarship in ecology should be on the study of ecosystem health and the processes that degrade, enhance, or maintain the health of ecosystems. Such studies build the foundation on which science-based ecosystem management policy instruments, approaches, techniques, and technologies can be invented. There is no integrated unified framework for the need of such study. It is highly imperative that universities, governments, nongovernmental research institutions, and private and public institutions join hands on the global scale for the development of an institutional framework for the meta-study of ecological systems. This is not possible without integrating political economy and political ecology. Until now, it is *political economy* that has been driving the engine of social and economic development, but, now, it is the time for *political ecology* to replace political economy or at least work together to drive the engine of ecologically sustained development. It is not possible to socially and ecologically develop, design, and implement just solutions for today's environmental and climate problems without their integration.

- The current environmental problems, including climate change, must be addressed with ecologically holistic approaches and methodologies. An ecosystem-based system approach (EBSA), which entails the maintenance of ecosystem health, restoration of degraded ecosystems, climate mitigations, and adaptation measures directed to the stewardship of Nature along with the implementation of policy instruments to address fossil fuel emissions, has the greatest potential to rescue humanity from the current unfortunate predicament of the environmental and climate crisis. Ecology-based solutions may offer great potential for sustainable fisheries, agriculture, ecosystem management, ecosystem services, and restoration of degraded ecosystems. Scientific epistemology and research must be directed to develop ecology-based solutions to the current environmental problems.
- Environmental and climate change problems are global problems that require multi-jurisdictional and multinational governance. Solutions to environmental problems can be achieved only through the effective participation of local people in the implementation of environmental management approaches and practices. The global community must invest in the capacity building and empowerment of local communities who are the real custodians of the environmental resources at the local level. Every local environmental problem will have a cumulative effect and impact on the global environmental state; therefore, *act locally think globally must be realized practically everywhere on planet Earth but not just rhetorically.*
- There exists a huge knowledge gap in the study of complex ecological systems, which remains intractable for timescales available to implement science-based solutions to relatively simpler ecosystems. For a better understanding of the behavior of complex ecological systems, it is necessary to identify the key elements of the complexity and how they enhance or degrade resilience and adaptations of the complex system.
- Different ecosystem processes and services can have synergistic effects upon each other, and the identification of synergies and trade-offs will greatly enhance human interventions that can maximize synergies between different ecosystem services, implement and maintain long-term monitoring for a proper understanding of the dynamics and trajectories of a complex system, and evaluate the success of management interventions.

Beyond any doubt, the current environmental and climate crisis has resulted from the emissions of greenhouse gases and the degradation and destruction of the planetary ecosystem from *hyper-anthropocentric* activities. Human civilization cannot perpetuate without bringing about drastic human behavioral changes toward its treatment of planet Earth. Decarbonization and judicious protection of ecosystem processes and implementation of science-based solutions for rebuilding and restoring degraded ecosystems can not only stabilize global temperatures but also help in the subsequent cooling down of planet Earth. It may not be possible to regenerate the lost biodiversity (species and life-forms that became extinct), but it is possible that with the implementation of ecosystem-based system solutions (EBSSs) on a large scale, the ecological processes and functional integrity of the ecosystems can be maintained and restored, ultimately recreating a vibrant and resilient biosphere indispensable to the perpetuation of humanity and all the life-forms on planet Earth. This will be the beginning of an eco-based human civilization more appropriately called an "*ecocivilization*."

Michael Mann (2023) rightly pointed out the fact that today's environmental crisis faces its biggest obstacles in the political arena than in science and technology: "*There is still time for us to avert the worst impacts of climate change if we act now and act boldly, but there is no time left for dead ends, wrong turns, and false solutions. We have technology—in the form of renewable energy, storage technology, and efficiency and conservation measures. The only obstacles at this point are not the laws of physics, but the flaws in our politics.*" There is no doubt that today's environmental crisis is mainly the problem of the political economy deeply rooted in the ecologically hostile consumer culture industry created by modern industrial corporate capitalism. Today's environmental crises cannot be addressed as such by science and technology alone. They require multipronged strategies directed toward changing the currently unsustainable consumer culture industry with ethically guided development along with the application of innovative science and technologies that can maintain and enable the regenerative biocapacity of planet Earth. It is not possible without reforming the inherent flaws of modern consumer capitalism to make it more compatible with the proper functioning of both planet Earth and modern democracy.

In the final analysis, informed politicians, who truly understand the gravity of the environmental and climate crisis, must not shy away from putting these measures in their policy framework. There is hope and optimism that the current environmental and climate crisis can be resolved. Investigative journalists and scholars have done a remarkable job of uncovering the skeletons of the fossil fuel industries' closet, holding these companies accountable for environmental damage and also for denying and delaying compliance with the existing environmental laws and regulations. When scientists, researchers, policymakers, analysts, journalists, social and environmental activists, NGOs, and grassroots movements join hands and converge their forces, their unified voices will resonate with a powerful *wave of awakening*, capable of changing the arrogant and rigid mindsets of mainstream politicians and corporate executives, and then showing light at the end of the dark tunnel. When the old superstructure and its *modus operandi* crumbles and a new superstructure with *ecological wisdom* begins to govern the planet, then only is the future of humanity and planetary Earth systems ensured. I believe that there is no alternative to the integration of ecology and political economy that can end humanity's alienation from Nature. When this happens, I would say a new cosmological civilization will begin, which may very well be known as an *ecological civilization*, one that may be in the process of making.

Chapter 8
Valuation of Biodiversity, Ecosystem Services, and Natural Capital

Living systems are cognitive systems, and living as a process is a process of cognition. This statement is valid for all organisms, with and without a nervous system. Our cognitive process differs from the cognitive processes of other organisms only in the kinds of interactions into which we can enter, such as linguistic interactions, and not in the nature of the cognitive process itself.

Maturana (1980)

8.1 Introduction

The valuation of natural systems, which includes biodiversity and ecosystem processes that generate a wide array of vital ecosystem services and goods, has become a focal point of debate among professionals in development fields such as conservation, environmentalism, and economics. This chapter concentrates on how the prevailing faulty valuation approach toward biodiversity, ecosystem services, and natural capital has accelerated environmental degradation, the extinction of a multitude of diverse species, and the ensuing climate crisis. The discussion will outline how the well-being of human socioeconomic systems—entailing human happiness and well-being—is intricately linked with the health of ecological systems. We will stress the need for our policies and valuation approach to focus on the protection and maintenance of the ecological systems'health. The chapter aims to delve into the complexity and significance of valuing biodiversity, ecosystem services, and natural capital.

G. Upreti, *Ecosociocentrism*, https://doi.org/10.1007/978-3-031-41754-2_8

8.2 Valuation Complexity

Biodiversity and ecosystem services constitute the conditions and processes via which natural ecosystems and the species therein meet human needs and sustain all life on Earth. Mainstream economists' prevailing valuation approach recognizes only tangible benefits, the commodity values of organisms dictated by market forces. This approach overlooks the ecological values of the life support services and material goods provided by diverse biotic communities within ecosystems, despite these ecosystems being the sources of all material inputs and services for man-made goods and utilities. Recently, environmental and ecological economists have developed valuation approaches to integrate ecological facts, values, and economic considerations, addressing the traditional neglect of biodiversity and ecosystem services valuation in policy decisions. However, this approach has yet to influence the policy decisions of nation-states (Costanza et al., 2014; Costanza et al., 1997; Balvanera et al., 2014; Baveye et al., 2013; Upreti, 1994, 1996).

Appropriate application of economic valuation techniques to ecosystem services can offer valuable insights for conceptualizing decision choices and evaluating management alternatives. Despite a growing interest and investment in ecosystem health and services in the global science and policy sphere, it remains ambiguous how ecosystem services, particularly changes in those services, should be quantified. The social and ecological factors, and their interactions, that create and modify ecosystem services are inherently complex. Measuring and managing ecosystem services necessitates a nuanced systems-based approach that accounts for how these services are generated by interconnected social and ecological systems (SES), how different services interact with each other, and how alterations in the overall bundle of services affect human well-being (HWB). Understanding how changes in HWB feedback impact ecosystem health and ecosystem services is crucial. It is of utmost importance to comprehend the dialectical relationship between ecological infrastructures (ecosystems and natural capitals) and human social superstructures, which not only interact with each other but also shape and determine this relationship quantitatively and qualitatively. A scientific understanding of this process is vital, not only to develop appropriate methodological approaches and techniques for the valuation of biodiversity, ecosystem health, and ecosystem services but also to reshape the current development approaches and human ethical paradigm that ultimately guide and reshape human behavior.

In their groundbreaking research in 1997 and 2014, Costanza et al. (1997, 2014) strove to place a monetary value on ecosystems and their services, recognizing the vital roles these natural capital stocks play in Earth's life support systems. They argued that these systems, directly and indirectly, contribute to human welfare, constituting a significant part of Earth's total economic value. They estimated the economic value of 17 ecosystem services across 16 biomes and deduced that the average global value of ecosystem services was about $33 trillion/year in 1997, a figure that exceeded the Global Gross National Product (GGNP) of about $18 trillion/year. This pioneering endeavor marked the first attempt by ecologists and ecological economists to quantify the monetary value of ecosystem services, though the estimates were deemed conservative due to the inherent uncertainties in the process. Their revised 2014 estimate, considering updated values of individual ecosystem services and changes in land use between 1997 and 2011, placed the total global value of ecosystem services at $125–145 trillion/year. From these calculations, they surmised that the loss of ecosystem services due to land use change between 1997 and 2011 ranged from $4.3 to $20.2 trillion/year. The monetary estimates generated underscored the magnitude of ecosystem services, emphasizing that many of these services are best viewed as public goods or common pool resources, which conventional markets effectively ignored. It was posited that sustainable development rhetoric would remain hollow unless the value of ecosystem services was integrated into the current economic system.

The research raised pertinent questions about the irreplaceability of natural capital stocks by man-made assets. Given the dependence of human survival and well-being on these ecosystems and their services, it was argued that human activities should be conducted to preserve the ecosystems' structure and integrity. This moral obligation extends from the belief that both humans and the biotic community should coexist and flourish together. The protection of ecosystem health, biodiversity, and ecosystem services requires professionals from diverse fields to develop a comprehensive integrated valuation method. The valuation of biodiversity and ecosystem services must be a critical part of decision-making, requiring a *paradigm shift* to regard these elements as fundamental *ecological infrastructures* underpinning human *socioeconomic superstructures*. Unless such a shift occurs, the development of suitable techniques to measure and integrate the value of biodiversity, ecosystems, and ecosystem services will remain elusive. This paradigm shift is necessary because our assumptions and worldviews shape and determine our methodological approaches.

8.2.1 Ecological Footprint and Biodiversity

Ecological footprint can be defined as the impact of human activities measured in terms of the area of biologically productive land and water required to produce goods and services consumed and to assimilate the waste throughputs generated (WWF, 1990). The major impacts causing human activities include agriculture, fishing, raising livestock, building, industrial, military, and all infrastructures associated with human socio-economic-cultural life. A high level of Ecological Footprint index is linked to high consumption of natural and man-made resources, which causes negative impact on the environment (Monserrate et al., 2020). With increased populations and consumption, humanity has been placing far greater demands on ecosystems and ecosystem services that are essential for the survival of not only mankind but also the biotic community. Climate change, water shortages, overharvesting, and habitat disruption, and destructions are caused by accelerated human pressure on finite resources of planet Earth, driving the extinction of the fauna and flora species and wildlife populations worldwide. Carpenter et al. (2009) point out that it is rare to find linear causal path of changes in drivers leading to biodiversity changes to ecosystem services and to human well-being and feedback to drivers from humans. The casual patterns are much more complex, and the drivers may jump backward or forward and may affect human well-being without affecting biodiversity or ecosystem services. The leaders of the world, under the auspices of Convention on Biological Diversity (CBD) in 2002, committed to significantly reducing the rate of biodiversity loss by 2010 by adopting a suite of indicators in a program called Biodiversity Indicators Partnership (BIP). The objective of the BIP was to provide information on biodiversity trends and assess progress toward their target (the target of halting biodiversity loss). GFN (2022) informs us that as humanity's Ecological footprint has grown 140% since 1961, the vertebrate species population has declined by 69% from 1970 to 2018. The rapid increase in humanity's ecological footprint has accelerated the rate of the loss of the diverse fauna and flora and the decline in the regenerative capacity of the planetary ecosystem. Ecological Footprint has been officially adopted by CBD and included in its biodiversity indicators. Ecological Footprint provides an indicator of the pressure on ecosystems and biodiversity by measuring the competing level of ecological demands that humans place upon the natural capitals (ecosystems and ecosystem services) of the biosphere.

Carbon footprint is the most significant part of humanity's *Ecological Footprint*. Data clearly show that a rapid increase in humanity's Ecological footprint is due to an increase in our carbon footprint since the 1960s. One cannot stress enough the vital importance of reducing the carbon footprint; it is increasingly becoming an economic necessity for all nations. Global Ecological Footprint data indicate that humanity is using resources (natural capital) and producing CO_2 emissions at a rate 60% greater than what Nature can regenerate and reabsorb. This gap, known as ecological overshoot, causes the depletion of the natural capital that all species, including *Homo sapiens*, depend on for their survival and livelihood. It also causes the accumulation of carbon dioxide that leads to climate change, with profound implications for ecosystems, ecosystem services, and the species they support, as well as for our societies' well-being and economic stability. Ecological Footprint of humanity has grown 80% over the last four decades as is evident from the graph below (Fig. 8.1). The greater the gap between human demand and the regenerative capacity of nature, the more pressure there will be on the resources other species need to survive, accelerating the rate of biodiversity loss and ecosystem services (GFN, 2022).

As we can see at current global rates of consumption and waste production, humanity's collective impacts require the equivalent bio-productivity of 1.6 planet Earth to meet global demand. In other words, it takes the Earth 1 year and 6 months to regenerate what humanity consumes in a year. A 2010 report published in the journal *Science* provided a stark assessment pointing out that the world's governments had not met the target set by the CBD in 2002 and, instead, presided over enormous losses and declines. In October 2010, the parties to the CBD met in Nagoya, Japan, to decide whether to adopt a new biodiversity target and new indicators for the post-2010 era (CBD, 2010). At the conference, the BIP presented a list

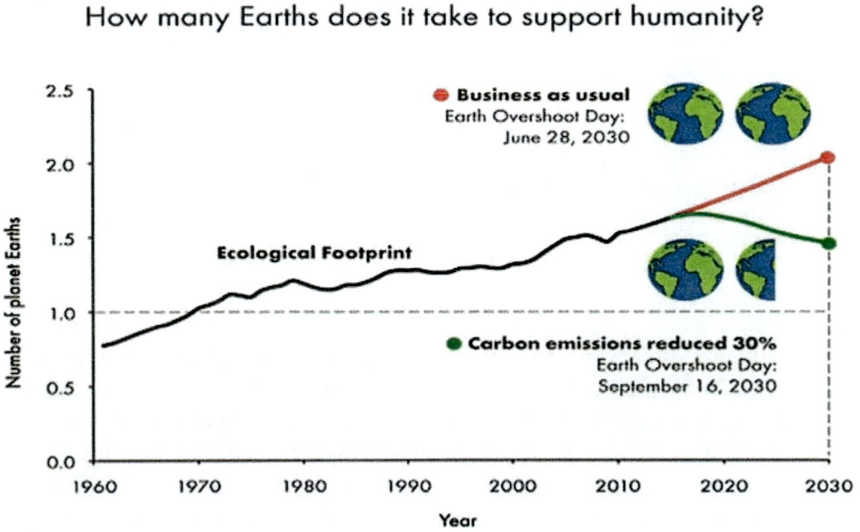

Fig. 8.1 Ecological footprints and the sustainability of the Earth systems. (Source: Adapted from Wahl (2017). Ecological Footprint, Earth Overshoot Day, and Happy Planet Index)

of strategic goals, including means, milestones, and indicators, for achieving the goals set forth in the CBD. Ultimately, halting species loss and enabling biodiversity to thrive will require bringing human consumption demands for ecological services within the carrying capacity of the Earth's systems and what Earth can renewably supply. Hill et al. (2020) argue that attention to the interactions within the global socio-economic-ecological system provides new insights for guiding longer-term actions for the conservation of biodiversity. The destruction and degradation of ecosystems is the major driver of biodiversity loss and, therefore, without ecosystem protection biodiversity conservation efforts are not going to be effective, and biodiversity loss cannot be prevented. It remains to be seen how effectively these recommendations will be implemented by the parties to CBP as some very powerful states like the United States and others have openly criticized and refused to comply with these agreements and recommendations. Nevertheless, the efforts made by leaders of the world's governments are commendable.

It is important to understand the impact of human activities on the environment, and one way to measure this impact is through Ecological footprint. This measures the area of biologically productive land and water required to produce goods and services consumed, as well as the waste generated. Unfortunately, a high Ecological footprint is linked to negative impacts on the environment, like habitat disruption and destruction. This puts the survival of both humans and other forms of life at risk. Carbon footprint is a significant part of ecological footprint, and we need to reduce it urgently. Humanity's Ecological footprint has grown 80% over the last four decades, and we are using resources and producing CO_2 emissions much faster than nature can regenerate and reabsorb them. This leads to ecological overshoot, which causes the depletion of natural resources and climate change. It is crucial to take action to reduce our Ecological footprint and avoid irreversible damage to the environment and our societies.

8.2.2 Millennium Ecosystem Assessment (MA)

Kofi Annan, the late secretary-general of the United Nations, launched the Millennium Ecosystem Assessment (MA) in 2000 with the goal of determining the effects of ecosystem change on human well-being and the scientific basis for actions required to improve their conservation and sustainable use, as well as their contribution to human well-being (MA, 2005). The MA was intended to be an integrated assessment that spans sectors, incorporating perspectives from the natural and social sciences as well as an integrated multi-scale assessment approach that included component assessments conducted at various spatial scales, including global, sub-global, regional, national, basin, and local levels (MEA, 2005). The MEA (2005) combined knowledge from the private sector, practitioners, local communities, and indigenous peoples with material from the scientific literature, data sets, and models. However, the paucity of data and literature did prompt several sub-global evaluations to conduct fresh study and data collecting, notably those at small sizes. The assessment results have, in every instance, proven helpful in identifying information gaps and research objectives. Around the world, more than 1360 specialists have contributed to the MA. The following are the main conclusions of MEA (2005):

More quickly and widely than at any other similar moment in human history, humans have altered ecosystems over the past 50 years. This is partly due to the rapidly rising demands for food, freshwater, lumber, fiber, and fuel. The diversity of life on Earth has suffered a significant and largely irreparable loss as a result.

Significant net gains in human well-being and economic development have resulted from ecosystem changes, but these gains have come at an increasing cost in the form of degraded ecosystem services, increased risks from nonlinear changes, and the aggravation of poverty for some groups of people.

If these issues are not resolved, future generations will no longer gain as much from ecosystems.

A barrier to achieving the Millennium is the loss of ecosystem services, which has the potential to get much worse throughout the first half of this century.

Under these scenarios taken into consideration by the MA, the task of reversing ecological degradation while meeting the rising service demand can be partially accomplished, but it will need considerable adjustments in policies, institutions, and practices that are not now under progress. There are numerous strategies to preserve or improve a particular ecosystem service while minimizing negative trade-offs or maximizing positive interactions with other ecosystem services. The main conclusion of the MA is that human activities are depleting Earth's ecosystems and natural capital, placing such stress on the environment that it can no longer be assumed that the planet's ecosystems will be able to sustain future generations. The MA also demonstrates that, over the next 50 years, it is possible to stop the loss of many ecosystem services with the right steps, but the necessary changes in policy and practice are quite large and not now in place (Carpenter et al., 2009). The MA evaluated current knowledge, scientific literature, data at the most fundamental level, and integrated previously unreleased information with fresh research discoveries, much to the Intergovernmental Panel on Climate Change (IPCC) (IPCC, 2007). Developing policy instruments for their protection, conservation, and sustainable uses and legally recognizing the values of biodiversity and ecosystem services was, in fact, a key accomplishment in the history of the UN system.

8.2.3 Knowledge Gap

There is a significant knowledge gap between the local and the national levels, and even less is known about the economic significance of nonmarketed ecosystem services. Information regarding the state of many ecosystem services is also scarce. Additionally, it is uncommon for national economic accounts to keep track of the expenses associated with the loss of ecological services. Surprisingly little basic information is available on the size and trends of various ecosystem types and land use around the world. The ability of models to predict future environmental and economic situations to consider ecological "feedbacks," such as nonlinear changes in ecosystems or behavioral feedback like learning that may occur through adaptive management of ecosystems, is restricted.

Assessments are helpful in identifying any remaining scientific ambiguities. Uncertainties can be used to support a "wait and see" strategy, but they can also be used to support a preventative strategy. The global findings among the MA findings often have a high level of certainty. Perhaps the biggest question mark around one of the most crucial aspects of ecosystem change on a global scale is how much land degradation there is in drylands. The area and the population concerned are nevertheless considerable even when adopting very cautious estimates of land degradation (10–20% deteriorated) (Steffen et al., 2015). At the municipal or national level, unpredictability is the biggest challenge. For instance, at the local level, decisions are frequently made without sufficient knowledge of the entire economic costs and advantages of alternative uses of ecosystems. This evaluation illustrates the importance of gathering that information as well as how to do so (the MA sub-global assessments serve as a model for one method to carry out that more in-depth local or national assessment).

The MA's overarching goals were to improve decision-making about ecosystem management, human well-being, and the development of scientific capacity for ecosystem service assessments. The degree to which the MA conclusions are used by decision-makers at both the global (e.g., conventions) and sub-global scales will determine the MA's final influence. Through involvement in the MA, significant assessment capacity has already been established globally. Additionally, it is anticipated that the conceptual framework, approaches, and methodologies of the MA will be heavily incorporated into the current efforts and programs of the different institutions that have collaborated on the MA process. A new framework for studying social-ecological systems was developed by the Millennium Ecosystem Assessment (MA), which has had a significant impact on both the scientific and policy communities. To better understand and manage the dynamics of the link between humans and the ecosystems on which they depend, new research is required that considers the entire ensemble of processes and feedback for a variety of biophysical and social systems (Carpenter et al., 2009). The capacity to solve basic issues with complex social-ecological systems will increase as a result of this research, which will also evaluate the presumptions underlying the policies and practices designed to improve human well-being through increased ecosystem services.

Reckless profiteering, together with institutional dysfunction, policy gaps in scientific knowledge, and unanticipated events, is the main cause of the global destruction and degradation of biodiversity, ecosystems, and ecosystem services. The MA demonstrated that, as more natural landscapes were used for human purposes, they became more homogeneous and that, in comparison to the average rate across geologic time, genetic diversity and species diversity are disappearing at rapid rates. According to Carpenter et al. (2009), causes that influence biodiversity also directly affect ecosystem services, and these changes in ecosystem services may subsequently elicit feedback from human responses. Carpenter et al. (2009) claim that it is important to understand how biodiversity effects relate to social-ecological contexts. Causal patterns are far more complicated, and links may move in unexpected directions (e.g., drivers may have an impact on human well-being without also influencing environmental services or biodiversity, or ecological processes may provide direct feedback to drivers without the need for human intervention). The primary factor behind the ecosystem services' rapid decline is the failure to consider their genuine values while making economic decisions. Since all decisions are dependent on market prices, many ecosystem services have no markets and no clear indication of their worth (Constanza et al., 2007; Carpenter et al., 2009). The underlying social value of nonmarketed ecosystem services, which are essential for all stakeholders, is not well understood by mainstream economists. The value of nonmarketed ecosystem services can be estimated using current procedures, such as accounting prices or shadow pricing, although decision-makers seldom adopt these approaches.

Researchers contend that integrated, quantitative models of social-ecological systems fall short of the scope of current conceptual and qualitative models (Daily & Matson, 2008; Schröter et al., 2005; Costanzaa et al., 2014). Existing models are the reductionistic models created to handle specific intersections of concerns (such as biodiversity and land use change) or isolated sectors (such as agriculture, marine fisheries, land use, and water supply). Additionally, in order to evaluate or estimate ecosystem services, models for these sectors must be combined with projections from other models of the climate, demographics, macroeconomic development, and other causes. It is crucial to have models that are equivalent to the conceptual frameworks applied by the MA in terms of breadth and substance. Instead of combining current models with diverse scales and purposes, integrated models should be developed to address the scales and drivers that are directly important for the study of interactions between such factors and the ecosystem services. The vital services individuals receive from ecosystems are known as ecosystem services. These consist of provisioning services, like the supply of food and water; regulating services, like the prevention of floods and the control of diseases; cultural services, like the provision of spiritual, recreational, and cultural benefits; and sustaining services, like the cycling of nutrients that preserve the circumstances necessary for life on Earth. A holistic, integrated approach is necessary for an evaluation of the state of ecosystems, the delivery of services, and their relationship to human well-being. This makes it possible to decide which service or group of services is valued the most and how to create strategies for maintaining services through system-wide sustainable management.

Following the MA initiative, research has focused on fresh issues in the fundamental science required to evaluate, project, and manage ecosystem service flows and their impacts on human well-being. One significant barrier is our inability to reach broad conclusions because of the emphasis on specialized fields that are unable to completely appreciate the social-ecological system. The fact that some policies and practices aimed at enhancing ecosystem services and human well-being are based on unproven hypotheses and scant data is also crucial to comprehend (Carpenter et al., 2006, 2009). Finding creative management solutions to enhance ecosystem services and human well-being is the largest issue. To better understand and manage the dynamics of the link between humans and ecosystems, integrated multidisciplinary research is required that considers the entire ensemble of processes and feedbacks for a variety of biophysical and social systems. We will be better able to answer fundamental questions regarding intricate, interactive social and ecological systems as a result of integrated multidisciplinary research, which will also allow us to assess the presumptions underlying the policies and practices aimed at enhancing human well-being through increased ecosystem services.

8.2.4 Protecting Ecosystem Services

Considerable attention has been given to the human dimension of conservation projects, but no attempt has been made for the conservation of ecosystem services. This may be as Chan et al. (2016) argue because flows of ecosystem services remain poorly characterized at local to regional scales, and their protection has not generally been made a priority. Ecosystem services such as carbon storage, flood control, forage production, outdoor recreation, crop pollination, water provision, nutrient cycling, hydrological cycling, biodiversity, soil regeneration and soil erosion control, watershed, oxygen generation and recreation, etc., constitute the very resource base on which human beings largely depend. Researchers (Chan et al., 2016; Hill et al., 2020) have found that targeting ecosystem services directly can meet the multiple ecosystem services and biodiversity goals more efficiently than for targeting biodiversity protection (biodiversity losses of 44% relative to targeting biodiversity alone) alone. Strategically targeting only biodiversity conservation yields much less benefits compared to integrated conservation of biodiversity and ecosystem services, suggesting that integrating biodiversity and ecosystem services in conservation planning offers scope for identifying valuable synergies. In recent years, however, there have been tremendous advances in the science, economic valuation, institutional design, and social capacity needed for conservation of ecosystem services.

Some examples of ecosystem services which are important locally, regionally, or globally as described by Chan et al. (2016) are as follows: carbon storage locked up in above- and below-ground biomass of primary producers. When natural vegetation cover is converted to agriculture or urban land, carbon is released to the atmosphere as carbon dioxide, exacerbating climate change. Since 1850, more than a third of anthropogenic CO_2 emissions have resulted from land conversion. Accordingly, intact ecosystems provide a service to the global population by storing carbon. Crop pollination by natural pollinators is another important service. Between 15% and 30% of the US food supply depends on insects/animal-mediated pollination, and it is likely that many insect species other than the widely cultivated European honeybee contribute importantly to numerous crops. Accordingly, pollinators from natural habitats provide a service to local food producers which is also critically important to humanity. Flood control is the mitigation of flood risk, which is damaging to areas of agriculture and human settlement, that is mediated by land cover. Land use change is an important contributor to increasing vulnerability to floods, and maintaining the right configuration of natural cover could result in savings of billions of dollars in damages.

Forage production for grazing rangeland livestock is an important economic land use in the grasslands and oak savannahs of this ecoregion. Water provision is the supply of freshwater to meet the demand of the agricultural, industrial, and residential sectors that comprise important element of economy. Maintaining the flow of this service requires both limiting the degradation of water quality from agriculture and urban development and maintaining the active purification of water in wetlands and other habitats. The quantity of this critical ecosystem service is in direct proportion to the amount of useable water that is available for human use. Outdoor recreation is the provision of recreation opportunities by natural and seminatural landscapes. Such outdoor activities are critically important to the economy and to the well-being of people.

As human impacts on the environment expand in intensity and extent, there is a critical need to understand the degree of intersection between conservation priorities for biodiversity and for ecosystem services. This intersection of conservation priorities could achieve a measured and thoughtful balance between previously competing goals, while providing new sources of funding for ecosystem services attracting new conservation partners and funds. Planning for the conservation of ecosystem services would involve a tremendous payoff for both biodiversity conservation and human well-being, promising to sustain critical ecosystem services, open new revenue streams, and make conservation broad-based and commonplace. The goal of simultaneously maximizing biodiversity conservation and ecosystem services critical to general human well-being is one that can be embraced by all. To motivate action, conservationists often mix diverse ethical and practical objectives, hoping they will reinforce each other. But attention given to one goal may instead diminish the prospects for achieving others. If conservation strategies and efforts are devoted to the protection and conservation of ecosystems and ecosystem services, the goal of achieving biodiversity conservation becomes realistically tenable.

8.3 Valuation of Nature

Scientists, scholars, and environmentalists have shown grave concerns that everyday human activities have been pushing the planetary ecosystem or biosphere beyond its capacity to bounce back or regenerate its disrupted component subsystems, undermining the conditions for humankind's own survival (Upreti, 1994; Berry, 1999; Ehrlich, 2002; Steffen, 2004; IPCC, 2014a, b; IPBES, 2015). Humanity has been confronted with severe social and economic impacts of environmental degradation that resulted in ecological conflicts all over the world. The major bottleneck lies in the inherent bias of the dominant development ideology that does not consider the valuation of Nature's services (ecosystem services) and the internalization of the environmental externalities into the economic analysis (Upreti, 1994, 1996; Costanza et al., 2014; Balvanera et al., 2014). From a valuation perspective, environmental problems originating from the destruction and degradation of Nature have not been adequately represented in the decision-making processes of the current development paradigm. Unless the current development paradigm integrates ecology and economics in the decision-making process, it is impossible to pursue sustainable development, which, in fact, has become an *empty rhetoric*, a rhetoric that environmental professionals and politicians repeat every day in their development deliberation as a matter of habit. It has become a *Sisyphean Myth* without political determinism of the super economies (United States, China, Japan, India, and European Union) of the world to resolve the inherent contradictions that exist in the dominant development paradigm pursued by these economies which are responsible for world's 90% of the environmental problems and the destruction of planetary ecosystem and the climate crisis (Upreti, 1994).

The super economies of the world must come to adopt and comply with a common unified approach and methodological standards to integrate Nature's diverse values in the management decision and their development strategies with a goal of slowing down entropy (environmental pollution) and protect ecological processes that can restore the resilience of the planetary ecosystem (use of natural resource bases such as land, groundwater, forest, and seas within the biocapacity of the planet Earth). With increased public concerns and nongovernmental organizations (NGOs) pressures concerning climate changes, global warming, and destruction and degradation of natural landscapes and biodiversity, governments and private companies have begun to recognize the gravity of environmental problems and sustainability challenges; however, the dominant economic interests to maintain *status quo* (unsustainable natural resource use) are so powerful that the voices for environmentally sound and sustainable development strategies have not been adequately heard. There are a plethora of scientific studies and literature about the valuation of our environment, ecosystem services, and the sustainable use of the natural resource base. These valuation methods have emerged from the traditions in ecological as well as environmental economics (Gómez-Baggethun et al., 2010; Baveye et al., 2013), environmental justice (Martinez-Alier, 2002), and ecosystem service assessment practices. Though valuation of Nature and its services has become central to academic scholarship (Fisher et al., 2009; Seppelt et al., 2011) and the policy initiatives such as the European Biodiversity Strategy to 2020, the Sustainable Development Goals, and the Intergovernmental Platform on Biodiversity and Ecosystem Services (IPBES) have substantially contributed to the proliferation of the academic literature, there is not much progress in the valuation of the environment and ecosystem services in the real development practices of the nation-states. This is the most concerning matter when it comes to the implementation of environmentally enabling policies and programs.

Different stakeholders hold different value dimensions (ecological, cultural, economic, self-interest, electoral, or ethical), which implicitly or explicitly enter into the decision-making processes and their justification (Kelemen et al., 2015). It is argued here that not all the value dimensions are equally important. Some value dimensions are critically more important than others. This is particularly important for the value dimensions of renewable natural capital resources (land, water, forest, biodiversity, fisheries, coral reefs, etc.). From the sustainable development perspective, the ecological or environmental value dimensions of the natural capital resources become critically more important than others, since all other values (economic, cultural, and self-interest) are basically derived from ecological/environmental values. Ecological values are the basis for other values that make human existence possible. If the ecological values of Nature disappear due to the destruction or the degradation of Nature itself, the economic, cultural, and self-interest values associated with the ecological values will disappear. The key challenge is how to represent the ecological value dimensions such as ecosystem services, ecosystem processes, evolutionary processes, intrinsic and relational values along with instrumental values in the economic decision-making processes. A growing number of scientists, scholars, and practitioners have been exploring possible ways of combining ecological, sociocultural, and economic valuation tools that can support the principles of sustainable natural capital resource uses and ecosystem services in decision-making processes. The integrated valuation approach that is emerging is a significant development, but it is still far from becoming the dominant tool in terms of valuing the ecosystem services (Gomez-Baggethun et al., 2014, 2016; Kelemen et al., 2015, Barton et al., 2016; IPBES, 2015).

The debate on the valuation of ecosystem services is a never-ending one, and the scattered pieces of empirical studies cannot establish the *raison d'etre* of any valuation method. It requires an understanding of ecological processes that govern natural capital resources (ecosystems and ecosystem services), epistemological and ethical approaches to deal with the ecosystem resources with respect to their uses and preservation. The synthesis from the integration of ecology, economics, and normative philosophy will, perhaps, give rise to the optimal valuation method of ecosystem processes and ecosystem services of Nature.

8.3.1 System Interdependence

Biodiversity as an interacting system implies that the biophysical environment in which we exist is the product of the interaction of diverse life forms with their environment over millions of years in the past. Gee (1992) reminds us that the pollution of the atmosphere, the denudation of soil, and the diminution of supplies of freshwater in many parts of the world could be seen, in part, as consequences of the loss of biodiversity. The plant and the animal species are not only an interrelated parts and components of natural ecosystems but also are an essential component of human economic life, the means of survival. Biodiversity has provided a basic foundation for the evolution of human civilization. Thousands of species of plants and animals supported the development of early societies, providing the basis for evolution from hunting and gathering to subsistence farming to present-day agricultural and industrial levels of organizations. Ever since humans started to live in social groups, right after the hunting and foraging stage, they systematically domesticated the plant and animal species on which their very survival depended. In this process, they discovered and developed agriculture upon which human civilization was founded and flourished. The pyramid of the biotic community and the associated food web in natural ecosystems is so intricate and complex that a quantitative change in one trophic level may cause a qualitative change in another level. The whole system is so interconnected, interdependent, and interactive that synergistic relationships are widespread and feedback mechanisms maintain the system. For this reason, system ecologists do not accept the reductionist's thesis that the total effect is simply the sum of the effects of its components because the components interact synergistically. It is in this context that the importance of biodiversity becomes apparent.

Economists consider only tangible benefits and commodity values determined by market forces and consequently often overlook the ecological value of the life support services and materials provided by the diverse biotic community in the ecosystem, even though it is the source of all the material inputs and services for the production of man-made goods and services. In other words, all material inputs and services for the production of human-made goods are derived from diverse life forms operating in natural ecosystems for which there is no system of economic evaluation. Mankind will enter a new dimension of development when the life support services and material inputs of the natural ecosystem are considered part of the economic production and valuation system and the principle of opportunity cost is applied to maintaining a healthy natural ecosystem.

8.3.2 Biodiversity and Environmental Services

Apart from direct use value of biodiversity for which economic valuation or market valuation exists, however imperfect it may be, the economic valuation of the ecological services rendered by biodiversity is equally important for judging the true significance of the loss of biodiversity. It is only from the perspective of valuing the environmental (ecological) services of biodiversity that we can come to realize the true significance and the problems of the destruction of biodiversity and their habitats, and that only from such realization can we develop an appropriate policy framework for their protection and conservation. In what follows subsequently are the arguments and the rationales in favor of such a valuation approach.

Ecological or environmental services are defined here as the services such as food, oxygen, water, soils, and waste assimilation, which are essential for the very survival of human beings and cannot be substituted at large. Photosynthesis provides the ultimate energetic basis for the activities of biota involved in the generation of such services. The biota removes approximately 200 billion tons of carbon from the atmosphere in the form of carbon dioxide every year and incorporates in the production of complex organic compounds called carbohydrates, the food materials for the entire living organisms (Rabinowitch, 1945). In photosynthesis, light photon energy converts carbon dioxide and water into organic compounds (carbohydrates) and oxygen, and in respiration, the organic compounds are broken down into carbon dioxide and water with the release of energy (ATP).

Most scientists and ecologists believe that gaseous oxygen, the very basis of our survival, currently making up 21% of the atmosphere, is the product of photosynthesis. It is generally believed that as a result of the million years of this process, the oxygen released and accumulated in the atmosphere made it possible for the evolution of higher organisms on planet Earth. The biota not only removes carbon dioxide from the atmosphere to produce food materials but also releases an enormous amount of water into the atmosphere through the process of transpiration. As Salisbury and Ross point out (1985), the leaves of the plants may lose hundreds of kilograms of water for every kilogram of carbon dioxide fixed. Ecologist Paul Ehrlich (1992) also points out that the amount of water a single rainforest tree returns to the atmosphere in its lifetime of 100 years is on the order of 2.5 million gallons. The water vapor, lost from transpiration, upon condensation in the atmosphere, returns to the biota in the form of rainwater, thus completing the cycle. Recently, an Indian scientist quantified the environmental services rendered by a tree over a period of 50 years life span in economic terms which amounts to $40,000. The breakdown for major environmental services is as follows: oxygen production worth $7000, soil conservation and fertility maintenance worth $7000, water recycling and humidity control worth $9000, and air pollution control worth $16,000 (Uniyal, 1993).

Most of the ecological goods and services produced and sustained by the continuous interactions between populations and communities of the species and their environment are clearly indispensable to humanity. These are: maintenance of gaseous quality of the atmosphere, amelioration of climate, maintenance of hydrological cycle including flood control and drinking water supply, waste assimilation, recycling of nutrients, generation of soils, crop pollination, marine food, and maintenance of genetic library service. It is interesting to note that environmental services produced by the natural ecosystems are so obvious and vitally so important for our own existence and biospheric stability on the one hand, and on the other hand, such services do not seem to receive adequate attention even from the scientific community that is committed to a piecemeal approach of preserving and conserving a specific organism here, and there let alone talk about the economists and politicians who hardly acknowledge the fact that the human economic system is dependent on the larger ecological system and that it has to operate within the bound of ecological laws for its own survival and stability.

An ecosystem is usually composed of thousands of interacting species, each species filling a particular niche and contributing in its own way to the health and stability of the system. Ecosystem services are the results of ecosystem processes involving a large number of species and their population constantly interacting among themselves and with their environments. An organism has both direct and ecological values. The market exists only for the direct use values of some of the organisms but not for all.

Economic valuation of ecological services of the species, along with its direct use value, comprises its true value (even though this value concept is imperfect but is at least better than the one that does not consider ecological services). This valuation concept must be extended to incorporate and internalize the environmental life support services of ecosystems into our economic subsystem. Galston (1992) eloquently elaborates on the role of plants in creating life's support system and services: "Within Earth's biosphere, as within enclosed spaces, illuminated green plants are thus able to furnish food, oxygen, and water for animal and human use. Ultimately, all animal, human, and organic wastes are microbially degraded, and most components recycled back to plant growth. This ability of plants to furnish food, oxygen, and pure water from human wastes, including gaseous carbon dioxide, is the rational basis for the use of plants for life support in NASA's Controlled Ecological Life Support system (CELSS) program."

The long-term goal of NASA's CELSS program is to create an integrated, self-sustaining system capable of providing food, potable water, and breathable atmosphere for space crews on an extended mission (Galston, 1992). The research has shown that human life support needs and services can be created by a bio-regenerative life support system based on higher plants, algae, microbes, or a combination of them (MacElroy et al., 1987). Plants, through photosynthesis, convert carbon dioxide and water to carbohydrates and oxygen, and through transpiration, they produce potable water and, they reclaim many components from recycled human and plant wastes. Hence, it can be conceived without much difficulty that the biota, with their photosynthetic ability, play a central role in the generation of life support systems and services.

Ecosystems can suffer from some natural extinction of species without having their functional properties seriously degraded as a part of ecological and evolutionary processes. It is the human-induced unprecedented extinction rate of biodiversity that is alarming because it has severely disrupted the ecological and genetic library services. When the assault on ecosystems is of such magnitude that it changes the basic physicochemical conditions, natural succession becomes unlikely or impossible, or the timescale of restoration of the ecosystem is beyond human interest. When a large area of tropical moist forest ecosystem is cleared, climate change and soil and nutrient loss often dictate a pattern of succession that will not soon restore the forest (Ehrlich & Mooney, 1983).

The Earth and its atmosphere constitute a complex system in which they are linked by many biophysical processes. The living organisms and their physical environments, through mutual interaction, have changed and molded each other and generated today's diverse ecosystems, and the biota with living organisms in it is the major participant in the biogeochemical cycles of the elements such as nitrogen, sulfur, and phosphorous (Orians, 1990). Ecologist Bormann (1976) points out the cost of substituting technologies for the loss of ecological services following deforestation: "We must find replacements for wood products, build erosion control works, enlarge reservoirs, upgrade air pollution control technology, install flood control works, improve water purification plants, increase air conditioning, and provide new recreational facilities. These substitutes represent an enormous tax burden, a drain on the world's supply of natural resources, and increased stress on natural system that remains. Clearly, the diminution of solar powered natural systems and the expansion of fossil-powered human systems are currently locked in a positive feedback cycle. Increased consumption of fossil energy means increased stress on natural systems, which in turn means still more consumption of fossil energy to replace lost natural functions if the quality of life is to be maintained." (Table 8.1).

Entire ecosystems should be valued for the goods and services they produce. A marsh stores water during periods of heavy rains and thereby reduces floods. Forests, particularly on the slopes of watersheds, also reduce flood downstream. Evaporation of water from forest canopies generates rain, thereby maintaining the hydrological cycle. Soil systems detoxify chemicals and carry out valuable biological activities.

Table 8.1 Environmental functions of forest ecosystem

Source of materials and services	Sink for wastes	General and life support services
Timber	Absorption of waste	Genetic pool
Fuelwood	Recycling nutrients	Climate regulation
Business product	Watershed protection	Carbon fixing
Non-wood product	Protecting soil	Habitat for people, flora, and fauna
Genetic resources		
Agricultural production		Aesthetic, cultural, and spiritual source and scientific data
Recreation and tourism		

Source: Adapted from Worldwide Fund (1990)

Environment disperses, processes, and assimilates human wastes and pollution (Orians, 1990). From this perspective, one can argue that the valuation of a forest ecosystem can best be done on the basis of how much oxygen it produces and releases into the atmosphere, how much water it releases into the atmosphere through transpiration, and how much effects it exhorts in regulating humidity and temperature of the environment, how much forest contributes to flood control, and how much it contributes to the control of air pollution, waste assimilation, and soil regeneration.

8.3.3 Valuation of Biodiversity as a System

When it comes to the valuation of biodiversity, economists seem to pay little attention to it. The economic tool they often use to measure the value of biodiversity is the willingness to pay for the preservation of specific endangered species. Economists have assumed that natural environment is insignificant to the functioning and behavior of the economy. Neoclassical economists find it hard to recognize that communities of species with their ecosystems generate life-supporting services on which the survival of human species and its future depends. Folke et al. (1992) argue that the driving force for the loss of biodiversity and natural ecosystems lies within the human economic system itself and warn that the threat to biodiversity is also a threat to our basic life support systems. It is important for human beings to develop an understanding of the role that biodiversity plays in the function of a life support system, especially in terms of how species actively participate in the generation and performance of ecological services that are so fundamentally vital to human societies. It is argued here that species with their ecosystems are not only essential in the production of life support systems but are also essential for sustaining the whole human economy and, therefore, the value of biodiversity must be analyzed primarily in this context and also in the context of informational value, and other use and nonuse values.

The central concern of this section is to show how biodiversity is related to ecological services human beings obtain from the ecosphere/biosphere. Perring et al. (1992) have described the relationship between biodiversity and ecological services as follows: "The most important anthropocentric reason for conserving biological diversity is the role that the mix of micro-organism, plants, and animals play in providing ecological services of value to humanity. A multiplicity of organisms underpins the ecological life-support functions that enable human societies to exist. The values of biological diversity, thus, lies in the values of ecological services supported by the interaction between the organisms, populations and communities of natural environment, and the value of biodiversity loss reflects the sensitivity of ecological services to both the depletion and the deletion of species. There is a threshold of diversity below which most ecosystems cannot function. That is, all self-organizing living systems require a minimum diversity of species to capture the sun's energy and to develop the cyclic relation of fundamental compounds between producers, consumers, and decomposers." To answer the question, of what biodiversity loss means in the context of ecosystem and ecological services, it is important to understand how biodiversity affects the functioning and behavior of the ecosystem and its resilience. The functioning and the structure of an ecosystem are sustained by the synergistic feedback between the population and the communities of the diverse species and their environment. The nature of the interaction between the physical system (physical environment) and the biological subsystem (species and their population) is crucial to understand the overall structure and functioning of the ecosystem.

8.4 Valuation Approaches

Ecological economists have recognized the biophysical foundation of resources, which makes it possible for the human socioeconomic system to operate; however, mainstream economists do not seem to recognize the biophysical resource constraints (Barbier, 1987; Daly, 1990; Repetto, 1992). The mainstream neoclassical economists believe that technological innovations can solve all human and environmental problems and overcome resource constraints. The biggest disconnect between ecological and mainstream economists lies in the valuation approach of ecosystems and ecosystem services. Ecological economists have been advocating to recognize and internalize the valuation of ecosystem services (life-sustaining services) of biophysical systems (planetary ecosystem), while mainstream economists simply dismiss this as untenable because there is no methodology or technique that exists for the internalization of environmental services in the current economic system. It has been long pointed out that the development of methodological approach and technique for the valuation of ecosystem services and environmental externalities requires a multidisciplinary research approach involving economists, ecologists, system engineers, biologists, agriculturists, and social scientists (Upreti, 1994; Daly, 1993; Costanza et al., 2014). Unless there is a change in the neoclassical economic approach that seeks only the technological solution to such a complex problem, ecological degradation and systemic breakdown of the planetary ecosystem cannot be prevented. The current socioeconomic system may be able to operate as usual in the short run at the expense of the gradual systemic breakdown of Earth systems, but it would be too late to bail out humanity from the catastrophic breakdown of the planetary ecosystem. It is a crucial time in history for development thinkers, political economists, development planners, ecologists, system engineers, and Earth scientists to seriously ponder into the valuation of ecosystem processes and ecosystem services that can sustain current systems and also continue to sustain future generations. As was pointed out by visionary Nicholas Polunin (1972) more than four decades ago: "Clearly, we have a grave responsibility, to our descendants and to Nature, to make the future world a fit place to live and work and relax in—which means living in unison with Nature and not the ugly hotch-potch of toil and strife which, even at the best, it is likely to become with increasing degradation of environment if stern action is not taken. It is also clear, indeed axiomatic, that time is not on our side."

It is now clearly established that ecosystems play a critical role in regulating and maintaining ecological services and that the rampant destruction and degradation of these ecosystems have disrupted the flow of these services. Escalating impacts of human activities on forests, wetlands, and other natural ecosystems imperil the delivery of numerous vital services. Human society would cease to exist in the absence of these ecosystem services, and their immense value to humanity is unquestionable. Quantifying the value of ecosystem services and measuring their values against competing uses is a complex task and simply cannot be operationalized without a change in our assumptions of how we should value the nature's vital services. The human economic system depends on the free services of ecosystems. These Nature-supplied services are worth many trillions of dollars, but unfortunately, our current economic system does not have a method for accounting these services. If the ecosystem services are not internalized in the economic system through proper valuation, the social and economic system is bound to collapse. In order to prevent this, there is a need for drastic policy changes that can ensure a balance between sustaining ecosystem services and pursuing the sustained economic development.

8.4.1 Safe Minimum Standard (SMS)

Goodland and Ledec (1987) argue that "Safe Minimum Standard" (SMS) analysis can be used as a methodology to address ecological concerns, which hitherto have not been given any attention in economic cost–benefit analysis (CBA). SMS refers to any noneconomic criterion which a project must meet to be approved. It is time-tested standard operating procedure widespread throughout engineering design, health planning, and industrial worker safety. For example, a bridge is commonly designed with a safety factor of three or more to accommodate the unexpected and the unknown. The concept of SMS can be established to deal with specific ecological concerns, which otherwise go unnoticed or undervalued in conventional CBA because of the problems associated with measurements and valuation, discounting, and irreversibility of environmental resources and services. Some economists have been resisting the use of SMS approach because they argue that SMS criteria are established subjectively and arbitrarily without any reference to what is economically efficient or optimal (Upreti, 1996). In response to such criticism, Goodland and Ledec argue that reliance on the marketplace as the yardstick to measure social well-being (as is done in CBA) is no less arbitrary and no more objective or optimal than SMS analysis, and hence, there is no reason why it cannot be used. The World Bank has also pushed the use of the SMS approach considerably in the appraisal of the proposed developmental projects. A number of workers have suggested the use of *the Safe Minimum Standard* as a methodological and policy approach to protect species in natural ecosystems. This approach was first formulated by

Ciriacy-Wantrup (1959) and later developed by Richard Bishop (1978). Norton and Ulanowicz (1992) have advocated the use of this approach in protecting landscape-level ecological processes, thereby protecting as many species as possible. The central premise of this approach is to protect all species by protecting ecosystem-level processes.

It can be argued that the health of the human economic system depends on the health of the natural ecosystem; the best policy instrument that may reflect the actual values of biodiversity appears to be the *Safe Minimum Standard*. The rational response to the irreversibility of the environmental effects, as Perring et al. (1992) point out, is to slow down the rate of environmental exploitation. It is prudent not just to slow down the economic rate of exploitation where such exploitation approaches the critical thresholds but to stop it completely and help its regeneration. The use of SMS to ensure the perpetuation of self-organizing features of ecological systems and habitats capable of perpetuating species for many generations must be emphasized for achieving the social goal of species preservation and sustainable development. Economists have a great role to play in determining the costs of various alternatives for maintaining a functioning habitat or ecosystem to make the protection efforts more efficient. It is interesting to note that, in reality, ecological systems function as an infrastructure upon which human economic, political, and cultural superstructures are built, but the decision processes involved in the creation of these superstructures have not yet sought the participation of ecologists, ecological economists, and biological and natural scientists (Upreti, 1996).

Safe Minimum Standard (SMS), as a valuation methodology, will bring both economists and ecologists together to an interactive process where they would have to determine various parameters as to what constitutes the minimum threshold of the ecological processes required to perpetuate species and ecological (environmental) services, what are the *keystone species* and keystone ecological processes involved in the regulation and maintenance of the ecosystem, the nature of trophic level, and finally what is the cost of the alternative to maintaining a functioning ecosystem. The Safe Minimum Standard (SMS) as a valuation approach provides an opportunity for ecosystem-level management to safeguard landscape-level processes required to maintain the health of the ecosystem. This valuation approach may be developed in greater detail to entail the following three *central principles* postulated by Norton and Ulanowicz (1992) to guide biodiversity policy:

Maintenance of the total diversity of the landscape over multiple generations should be the focus of maintaining biodiversity. Appropriate management must involve both public participation and expert knowledge because the goal of biodiversity protection and scientific understanding results from interactive and experimental process.

Diversity must be understood in terms of dynamically healthy processes rather than as the mere maintenance of the elements of the system. One can employ appropriate scale and perspective to understand and manage larger ecological systems.

A human economic system that complements and enhance, rather than opposes and degrade ecological processes, are to be preferred and encouraged. Given the fact that natural systems react creatively to change, economic incentives should be developed to encourage economic systems that imitate natural disturbances.

From the standpoint of ecological sustainability and intergenerational equity, it stands to reason that our policy measures be directed to protect complex ecosystem processes to ensure the continuation of ecological services and the perpetuation of diverse species and organisms across multiple human generations. The popular catchphrase "sustainable development"' makes no sense unless concerted attempts are made to motivate human cultural behavior toward maintaining ecological sustainability. The Safe Minimum Standard as a policy measure has the potential to maintain the minimum level of ecosystem processes required to protect species and ecosystem services which are vitally important not only for the present generation but also for the future generation of humanity.

How biodiversity contributes to ecosystem services and how such services can be quantified is a complex problem because this involves highly interactive multi-level ecosystem processes. Since ecological phenomena occurring at higher level ecosystem processes cannot be explained at lower-level ecosystem processes and, more specifically, the behavior of the individual organism cannot explain the behavior of a community or the ecosystem. This necessitates the valuation of the ecosystem services as the appropriate methodological approach for the valuation of biological diversity, since each organism and species participates in the generation of ecological services. It can be argued that the valuation of biological diversity as a system of generating ecological services, along with their valuation as individual organisms or species, can be the best approach. If such an approach of valuation is adopted, then it becomes necessary to consider landscape-level ecosystems as a unit of consideration, since the quantification of ecological services generated by landscape ecosystems is practically more feasible and tangible. This may also reflect the value of biological diversity as a system and may have important implications in the protection of ecosystem processes, thereby protecting as many species as possible. Only such a valuation approach can maintain the autopoietic and self-perpetuating features of an ecological system that can achieve the social goal of species preservation and sustainable development (Upreti, 1994, 1996).

An individual's contribution to ecological services cannot be quantified in a reductionistic linear model because the nature of biological interaction is such that it simply cannot be studied in a linear pattern. One has to take a system approach because even if individual organisms produce ecological services, it would not be feasible to measure and quantify their individual contribution. It is the system of dynamic interaction of a population of species among themselves and with their environments that generates quantifiable ecological services. Ecosystem-level processes which generate ecological services and maintain ecosystem health and protect species should be the central values for driving policies to protect the ecosystem and species. As Norton and Ulanowicz (1992) argue, "If we are committed to saving species/biodiversity for future generations and wish to introduce dollar figure into policy debates, we should estimate the total value of the ecosystem dynamic that protect species to be equivalent to costs that would be incurred to maintain individual species in alternative ways."

8.4.2 IPBES Integrated Valuation Approach

The Intergovernmental Platform on Biodiversity and Ecosystem Services (IPBES) is the intergovernmental body which assesses the state of biodiversity and of the ecosystem services (IPBES, 2015). IPBES has been conducting the first global, government-legitimated assessment on values of Nature. IPBES implements a conceptual framework which includes a global diversity of worldviews and considers both the knowledge from occidental science and the indigenous and local knowledge (IPBES, 2015). Further, its valuation framework explicitly recognizes different ways of perceiving the importance of Nature and distinguishes the value of Nature itself, the *intrinsic values*, the importance of Nature to foster desirable relationships between people and Nature, the *relational values,* and the importance of Nature's benefits to humans, the *instrumental values* (Díaz et al., 2015; Chan et al., 2016; IPBES, 2015). For its regional and thematic assessments, IPBES puts forward integrated valuation as the centerpiece of its valuation guidelines (Fig. 8.2).

By providing multiple values, it helps to increase transparency of trade-offs based on values while diminishing the possibility of critique and personal interest

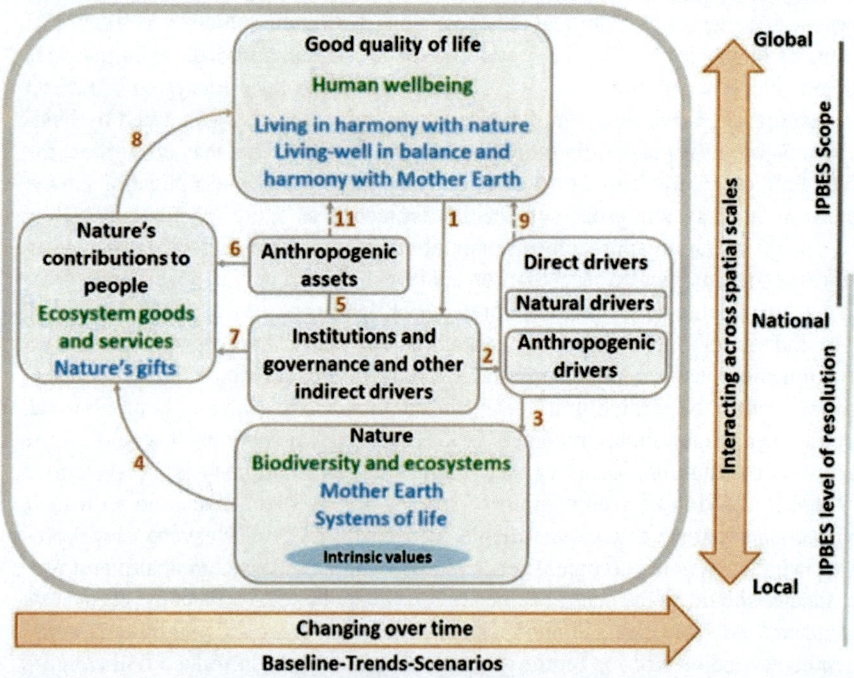

Fig. 8.2 Conceptual framework adopted by IPBES. (Source: Adapted from IPBES conceptual framework 2016 and Díaz et al., 2015) There are six main elements: Nature, Nature's benefits, good quality of life, anthropogenic assets, direct drivers, and institutions and governance. The arrows denote the links between elements, along with temporal and spatial scales (side arrows)

behavior. Consequently, any decision based on integrated valuation is likely to be more fair, sustainable, credible, legitimate, and effective than a decision informed by single-value methods. The integration level of each study depends on the policy question or study context (Schroder et al., 2016). From global policy to local practice, from studies in natural science, social science, and environmental justice research, we observe the rise of a more integrated valuation culture. Taking stock of theories on value pluralism, this new school of valuation explicitly applies a diversity of valuation methods to real-life human–nature issues and aims to account for normative (what should be) and cognitive (what is) complexities and uncertainties of values. It offers a way to articulate between the different value domains (non-anthropocentric, relational, and instrumental) and is inclusive as per definition by involving the broad set of stakeholders concerned with and affected by the outcomes (Jacobs et al., 2016).

Integrated valuation explicitly addresses and highlights the gaps in knowledge, methods, and concepts, especially when these affect the outcomes of applied valuation studies. The current sustainability challenges and the ineffectiveness of single-value approaches to offer relief demonstrate that continuing along a single path is no option. IPBES approach advocates for the adherence of a plural valuation culture, by establishing a common practice, by contesting, rejecting, and complementing ineffective, discriminatory, and counterproductive single-value approaches (IPBES, 2015). The most formidable challenge is to integrate the diversity of values (instrumental, relational, and intrinsic) which require implementing different valuation approaches (Martín-López et al., 2014; IPBES, 2015). Proponents of this approach argue for thorough reflexivity in valuation, as the conceptualization of the research, its fieldwork, its analysis, and communication are political processes. Consciousness on moral assumptions and regular self-reflection should frame the practice of integrated valuation. To achieve this new culture, multilevel communication and education of individuals in the relevant public and private institutions is needed, as is continued comparative research between and within real-life case studies in diverse contexts. In policy contexts with a willingness to improve decision-making, integrated valuation approaches can be blended in existing stakeholder processes, whereas in contexts of power asymmetries and environmental conflicts, integrated valuation can offer sound methodologies to include diverse values in action research, to support the struggle for social and environmental justice.

Most conservation research, emphasis, focus, and funding are oriented toward biodiversity with little tangible effort being directed toward ecosystem services. Although the latter are often used as a justification for the former, little is known about the circumstances under which the two approaches actually contribute to each other (Balvanera et al., 2014). By contrast, neither general priorities nor a methodology have been systematically developed for safeguarding ecosystem services. Mapping the distribution of biodiversity and threats to it is a key tool for turning priorities into plans of action. Analogous maps of ecosystem services priorities, which would locate suppliers, consumers, and threats relevant to each service, are virtually nonexistent. There is not a clear-cut understanding of how biodiversity and the ecosystem services of an ecoregion interact and interrelate with each other and how changes in one affect the other. In other words, how the adverse changes in the ecosystem services would affect the state of the biodiversity in a given ecoregion or vice versa. Not much research has been directed so far to develop an understanding of this complex relationship. Most of the research and thus conservation practices have been singularly directed to the conservation of the biodiversity, and a little or virtually no research has been conducted in the conservation of the ecosystem services of the ecoregion (Upreti, 1994, 1996).

Millennium Ecosystem Assessment (MA) and Intergovernmental Platform on Biodiversity and Ecosystem Services (IPBES) must focus on the development of the integrated methodological approach for the conservation and sustainable uses of both biodiversity and ecosystem services. Biodiversity and ecosystem services can be construed as the products of ecological and evolutionary processes that have been operating in the ecoregions from the time immemorial. The study of their interaction and interrelationship is critically important for laying the foundation for sustainable uses of these resources. It is true that MA and IPBES brought a new paradigm (Fig. 8.3) to bear on conservation issues which has been widely accepted; however, there is much work to be done to build a firmer foundation for the valuation of ecosystem services and to make it more useful to decision-makers at all

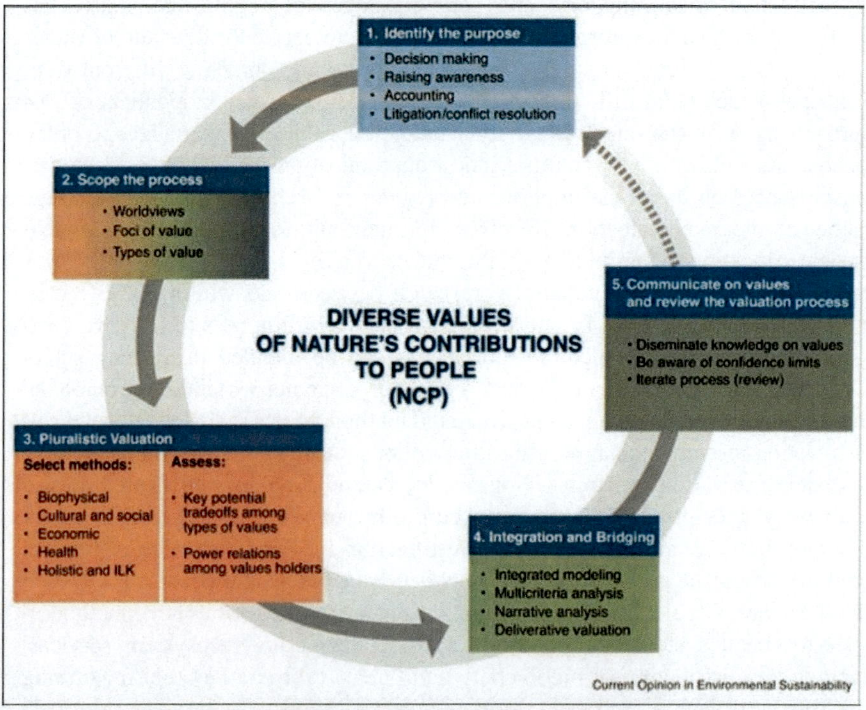

Fig. 8.3 IPBES approach of valuation of Nature's contributions. (Source: Adapted from Pascual et al. (2017). Current Opinion in Environmental Sustainability, Volumes 26–27, 2017, Pages 7–16. https://doi.org/10.1016/j.cosust.2016.12.006)

levels from the household to the globe. As Sander et al. (2015) point out, there is considerable work to be done to develop a better perspective and appreciation of the roles that biodiversity and ecosystem services play not only in the welfare of humankind through their instrumental use values but also the critical roles they play in the continuation of evolutionary processes and the maintenance of the planetary equilibrium. Their roles and contribution in slowing down the entropy (planetary pollution) process should take precedence over other values.

8.5 Conclusions

Biodiversity and functioning ecosystems are the basis of mankind's existence and the foundation of human civilization. Biodiversity and the natural ecosystems provide all the necessary material and energy inputs for the operation and functioning of human economic system. Even more important is their role in generating the ecological services upon which humankind's existence and survival is directly dependent. Biological diversity and ecosystem processes play central role in maintaining the health and proper functioning of the biosphere. One of the biggest problems in the conservation and protection of biodiversity is the lack of appropriate valuation approach that recognizes the appreciation of their roles and contribution in maintaining global planetary health and vital ecological services. The threat to biodiversity and ecosystem destruction is a threat to life-supporting environmental services and thereby constitutes a threat to the survival and well-being of humankind. Unless attempts are made to integrate the contribution and the roles played by biodiversity and ecosystems in the generation of ecological services into economic valuation, effective protection and conservation of biodiversity and natural ecosystem cannot be ensured. It is the problem encountered in the valuation of biodiversity and ecosystem services that is driving the loss and the destruction of biological diversity, ecosystems, and their services.

The most immediate tasks for the ecologists, biologists, agriculturists, and the economists are the development of valuation approach and techniques that can recognize and integrate the *instrumental* along with *relational* and *intrinsic* values rendered by biodiversity and ecosystems in economic valuation. Integration of ecology and economics is fundamentally essential for realizing the social goals of much publicized "sustainable development" paradigm. The use of Safe Minimum Standard (SMS) valuation approach should be advocated to protect landscape-level ecological processes essential for the perpetuation of species and ecosystem services recognizing the Nature's intrinsic values. The SMS valuation approach captures the spirits of the principles embedded in The Earth Charter for building a just, sustainable, and peaceful global society in the twenty-first century. The political decision-making processes that ignore the need of such integrative valuation approach (SMS) and does not recognize ecosystem service and relational and intrinsic values of Nature will lead to greater biodiversity loss and ecosystem destruction, thereby causing more human misery and suffering in the immediate future as we have already witnessed climate change crisis and its adverse impacts all over the globe. It appears that humanity seems to have initiated a process of self-destruction and self-elimination in our evolutionary history.

Can development planning provide policy framework for human actions that encourage a greater human integration and harmony with Nature rather than humanity's alienation from Nature? Modern development planning must pursue an integrated approach based on sound understanding of ecological system and human system knowledge. This calls for increasingly greater integrated roles and approaches of ecological economists, environmentalists, biologists, system ecologists, natural resource managers, social scientists, and the mainstream neoliberal economists in the planning and development processes. Political economy decisions should be made within the epistemological paradigm of ecology if we desire sustainable development and human happiness. The fact that the human socioeconomic system which encompasses human happiness and well-being is intricately interlinked with the health of planetary system makes it clear that our policies must be guided by ecological laws, wisdom, and the values of protecting and maintaining the health of the ecological systems. Evolution of appropriate values and policies for protecting ecosystem health and method of valuation of ecosystem is possible only if economists, ecologists, and natural scientists work through dynamically interactive processes. Planning should not merely be a tool for quick economic-political decision-making, it should rather be an institutional source for correct political decision-making based on the values of Nature's ecosystem functioning and services on which the survival and well-being of humanity and the living system depends on planet Earth.

Chapter 9
Metaphysics of Dominant Development Paradigm and Its Critique

Capitalism is the extraordinary belief that nastiest of men for the nastiest of motives will somehow work for the benefit of all.

John Maynard Keynes

9.1 Introduction

The current environmental crises we confront today—namely climate change, global warming, species extinction on a massive scale, and the desertification of both terrestrial and aquatic ecosystems—primarily originated from the metaphysical foundations of our prevailing development paradigm, a paradigm that has largely defined the *Anthropocene epoch*. The purpose of this chapter is to shed light on the intricate ways in which the underlying *metaphysical assumptions* of the dominant development model (neoliberal corporate capitalism) have led us into the present ecological and social crises. This chapter aims not only to foster a deeper comprehension of the fundamental origins of these environmental and social crises but also to guide us in identifying and conceptualizing alternative development paradigms and trajectories. These innovative alternatives could potentially divert the course of our current highly unsustainable development patterns, steering us toward a future characterized by healthier planet Earth for sustainable living.

9.2 Metaphysical Base of the Mastery of Nature

A paradigm is a description of the worldview, the collections of beliefs, values, norms, and habits which form the frame of reference for the perception and interpretation of the worldly phenomena. A dominant paradigm is the mental image of the social reality that guides expectations and human behavior in a society. The current dominant paradigm of the world is the belief that "*economic growth*," as measured by GDP/GNP, is a measure of progress, the belief that technology can solve

G. Upreti, *Ecosociocentrism*, https://doi.org/10.1007/978-3-031-41754-2_9

our problems (Devall, 2001). This paradigm treats Nature as only the storehouse of resources that need to be exploited to satisfy the ever-increasing demands of humans. This paradigm directs the development of science and technology for the domination, mastery, and control of natural processes. It has been pointed out by scholars and thinkers from ancient times that there exists a faulty human attitude toward treating and understanding Nature. The current development paradigm has been aptly characterized by the metaphysics of dominance and mastery over Nature, as Nature is an external inert object that must be subdued, exploited, manipulated, and controlled for human purposes (Upreti, 1994; Berry, 1999; Raskin, 2002; Tucker, 2007). Michael Bonnet (2017) succinctly postulated the essence of this paradigm: "*Western worldview frames issues in terms that are deeply human centred and manipulative with underlying presumption that everything is to be understood in terms of how it can be brought to serve the human will alone: the purposes that humans give to themselves, increasingly detached from any sensibility of any other source of value and in which ultimately the desire for mastery comes to reign supreme and everything must be brought to order in its service*." As one can see under the aegis of this *hyper anthropocentrism*, all values become *instrumental*, and the world, including the ecosphere/biosphere (planetary ecosystem), becomes nothing more than a *resource* to be exploited and consumed. This motive and its increasing dominance have been spurred by the growth of technological power. This, in effect, led to discard all moral constraints and has expressed itself in the burgeoning of *egocentric consumerism* that has completely ruptured our relationship with Nature.

In the constant interaction with Nature, humans have adopted the unilateral, one-directional modification of Nature and seldom thought of modifying their own behavior to adjust their relationship with the environment or Nature. To gain mastery over Nature, with the use of science and technological power, the multinational fossils fuels and lumber companies have been mining and extracting everything that was there in the ocean, mountains, terrestrial landscapes, tropical rainforests, and arctic region so much so that the regenerative processes of Nature have broken down pushing the planetary ecosystem toward degradation and desertification. Acceleration of drought all over the world, drying of rivers and rapid melting of snow from the mountains and arctic region, global warming, erratic hydrological cycles, and climate change are just the symptoms of an ailing planet. The tempo of the planetary degradation and breakdown is accelerating every day and night with the ever-increasing aggressive anthropocentric idiosyncrasy pushing human civilization closer to the doomsday. It is this *aggressive anthropocentrism* that has led to the current environmental crises and unpleasant human entanglement with Nature.

The biggest problem of this *aggressive anthropocentrism* lies in viewing everything in Nature as a resource to be exploited for profits and material possession, utterly disregarding the creative and regenerative processes of Nature. It can be seen from the lack of understanding of humanity's dependence and interconnectedness with Nature's integrity and stability, consequently resulting in a faulty perception and understanding of the world in which we are embedded. This, in effect, vitiated rational decision-making, inevitably leading to deleterious consequences as manifested in today's environmental, climate, and social crises. It cannot be simply dismissed as unintended outcomes of human interventions in Nature even after having clearly demonstrated by myriad of scientific studies that humanity's ecological footprints have already gone far beyond the regenerative capacity of the planetary Earth's ecosystem. The policies of powerful oligarchic governments like the United States, EU, Brazil, China, India, and powerful corporate conglomerates of the world have not changed, and the business as usual continues. This continued decimation of the natural world (fauna and flora, habitats, and ecosystems), fossil fuels mining, deforestation, and industrial fishing that destroy the sea floor demonstrate very well how (even after when deleterious consequences are well known) narrow short-term interests of material possessions and profit-making prevail over the long-term sustainability and survivability of humanity. As it can be seen from the slaughter of 60 million North American buffalo population to less than 500 in a period of a few decades. How can this be regarded as the right thing to do? Only with the mindset that regards Nature as a disposable resource having no inherent *intrinsic value* could allow such slaughter as a legitimate action.

The most pervasive expression of the underlying attitude of this worldview has exhibited in the strength and the character of the consumerist economic motives that have dominated Western society and have become increasingly globalized. The commodification of everything consisting of natural, social, and human capitals with the sole purpose of profiteering to satisfy human greed is the clear expression of the motive of this mastery (Durning, 1992; Bonnett, 2017; Berry, 1999; Raskin, 2002; Mathews, 1989; Carson, 1962; Nash, 1989). Such egocentric attitude and frame of mind remain immune to the holistic understanding of the natural world, and Nature simply continues to be a pure resource for human consumption. This *aggressive instrumentalism* propelled by *human greed* has not only subverted the subtle natural creative and resilient processes and delicate equilibrium in which human existence is embedded but has also ignored and demeaned the receptive frame of minds that attempt to reason why humanity's existence itself has been entangled with this delicate natural equilibrium.

It is in this sense that the metaphysics of the prevailing paradigm (the paradigm of the mastery of nature) has installed a faulty worldview that inherently and unconsciously works to exclude anything that lies beyond its purview (particularly any sources of *intrinsic values*) that transcend human will and, thus, portraying nature as a quasi-mechanical world that can be controlled and mastered. With such a worldview, it becomes increasingly difficult to address the environmental and social crisis humanity is facing today. When Nature is perceived as purely an object of resource to be exploited and not as a creative and regenerative process, any adverse consequences of exploiting it may appear simply as needing to be fixed by either current or future technologies. This worldview does not recognize the fact that the environmental and social crisis originates from excessive human desires and will to possess but rather sees Nature as an adversary that poses resistance to human will and desires.

Ecology, as a science, has established that things in nature are biophysically interconnected and interdependent; individuals are causally sustained as integral members of local ecosystems, which in turn are nested in overarching regional or global ecosystems. It is argued that from the phenomenological perspective, there is another important sense in which things in nature exist always in relationship in their very occurring and in their being and becoming. Bennet (2017) points out the complexity of the situation in which one must make a certain decision: "*When it comes to deciding how to act in a particular situation* e.g., *whether to preserve or destroy some aspect of nature—the inherent value of one thing alone often cannot determine this; its value needs to be weighed against the values of other things involved, including human-wellbeing. In this sense 'is' does not imply 'ought', but it is a mistake to deploy this as an argument against the idea of intrinsic moral value as it has been taken to do, for in fact it assumes some prior apprehension of intrinsic values.*"

The scientific discovery, inventions, and understanding of the laws of Nature resulted in the emergence of technologies that brought material comfort and progress in human life. The advances in sciences gave rise to a very powerful paradigm, which can rightly be called a *machine paradigm*. This paradigm is characterized by a belief system that the entire universe is material and mechanical. The cosmic and natural laws govern the entire universe in a mechanical way, and even life itself is an expression of the material processes. Critics point out that the machine paradigm brought us powerful technologies such as brain scanners, airplanes, robotics, powerful computation machines, instant communication with people on the other side of the world, the Internet, and information processing technology and Artificial Intelligence (AI). These technologies certainly accelerated the accumulation of the material wealth. Production of the material goods, services, and their consumptions became the primary goal of human beings. Nation states, corporations, and every human individual made an organized effort to compete against each other in the process of producing, consuming, and accumulating material wealth. Gradually people started to believe that the material reality is the only true reality, and the accumulation of as many material things as possible became the fulfillment of life. Machine paradigm made people arrogant and egocentric, which falsely caused them to believe that they had achieved mastery over the laws of Nature, which, in turn, inspired them to build even more powerful destructive devices: explosives, tanks, missiles, nuclear bombs, and chemical and biological weapons, the nuclear arsenals that could destroy the entire planet not once but several times (Matthew, 1992; Bennet, 2017). According to Statista Research Department (2023) (https://www.statista.com/statistics/262742/countries-with-the-highest-military-spending) reports, the global military spending of the nation states for 2022 amounted to $2.25 trillion dollar of which USA and China spent $877 and $292 billion respectively. Some major selected countries' military spending (billion dollar) are shown in parenthesis: Russia (86), India (81), Soudi Arabia (75), UK (68), Germany (56), France (54), South Korea (46), Japan (46) and Ukraine (44). These 11 countries spent 1.73 trillion dollar in 2022 and every year the total military spending in the world is on the rise. This reflects the heightened absurdity of these countries primarily USA, China, Russia, India, EU and others. The disproportionate allocation of resources toward military spending by these countries is utterly irrational considering the pressing global challenges that require collective action to deal with climate change and environmental crisis, and social inequality. This focus on militarization perpetuates the cycle of self-aggrandizement and conflict among nations, undermining the sense of shared responsibility needed to address the world's most urgent crises. By channeling these funds towards collaborative solutions for environmental and social issues, nations could potentially build a more secure, sustainable, and harmonious future for all and unify humanity in the face of existential threats.

9.2.1 Western Worldview and Aggressive Anthropocentrism

The Western worldview has evolved from the scientific revolution of the seventeenth and eighteenth centuries. Galileo, Bacon, Descartes, and Newton's scientific works and thought stimulated changes in thought on a grand scale that eclipsed the earlier worldview of the medieval era. The new scientific worldview was essentially secular and mechanistic, and the methodologies and the principles it advanced were mutually reinforcing. This worldview complemented the growth of capitalism and materialism, forming a cultural nexus that had far-reaching effects on the mainstream thought and action of individuals, societies, and institutions to the present. It became the *dominant consciousness* in the Western industrial nations and affected science, economics, politics, technology, philosophy, art, education, and even conservation and environmental ethics; all are affected by and reflected the beliefs, values, and methodologies associated with this worldview. Though this worldview has been successful in the last 300 years, the major crises we are witnessing now must be viewed as evidence of its fundamental weakness. In other words, this worldview no longer reflects the adequate model of reality now.

Two critical characteristics of this worldview are separation and dissociation. Cartesian logic laid the foundation for the *scientific paradigm* by differentiating mind and body, subject and object, value and fact, and spirit and matter. These distinctions were necessary to the liberation and flowering of scientific inquiry, as some argue, but the contradiction and schism inherent between these opposites are at the heart of the contemporary crises (Capra, 1999, Bennet, 2002). Most importantly, *Cartesian dualism* set human beings apart from and over Nature, thus defining man's relationship with Nature as being primarily exploitative and manipulative. The essence of this *Cartesian worldview* is humans should be the *Masters* and the *possessors* of Nature (Descartes, 1955, 1992). Our worldview has conditioned our perception of the roles of ethics in relation to the environment and its protection or conservation. The positivist influence of Descartes is so powerful that it emasculated the development of ethical thought in Western societies. Positivists hold that because all value judgments are subjective and unreliable, they do not constitute proper knowledge and, therefore, it is not possible to infer "*ought*" from "*is,*" the "*prescriptive*" (value) from the "*descriptive*" (fact). This is the dominant attitude toward the debate on Nature and place of ethics in everyday life, including the domain of nature conservation. The concept of "right" and "wrong" regarding decision-making is largely considered irrelevant. Political, economic, commercial, and technological decisions are backed by an appeal to objective facts, evidence, and probabilities which effectively mask the value of justifying evidence. Where a situation demands that ethical dimensions be more overt, constructive ethical debate can again be undermined by the effects of positivism in the guise of ethical relativism. No ethical position is held to be necessarily better than another because they are all essentially subjective and, therefore, unprovable.

Instrumental values play a predominant role in Western societies. Instrumental values are concerned with the utility of things as opposed to their *intrinsic values*. This means that persons, objects, actions, and all aspects of the natural world tend to be evaluated in terms of their uses. Interestingly, *instrumental values* are often cited in support of an ethical position which should also be ethically untenable. The utility value of a thing is the central premise of the neoclassical economic theory, which is the engine of modern capitalism. The idea of limitless maximization of the profit value of a thing is a key element behind capitalism and much of the environmental destruction and malpractices. Thus, the main argument for conserving a species or ecosystem is its present or future value or economic worth. The softer instrumental values of aesthetic or spiritual worth are a poor second, while conservation of Nature for its own intrinsic worth rarely receives serious consideration. Although many conservationists are aware of these broader values, they feel constrained by the accepted parameters of debate to advance the instrumental ones only. They operate in a society where ethical considerations are regarded as a worthy but separate criterion, essentially unrelated to political or economic criteria. In a trade-off between economics and ethics, the latter is usually dispensable. The conservationists attempt to fight back and reverse this, but the real irony is that they employ arguments based on instrumental values to resist instrumentally advanced forces, which is impossible to fight against.

Rational development thinkers and conservationists display an intuitive feeling that ethics should be a vital part of the decision-making process, including Nature preservation/conservation, but the Western intellectual tradition based on separation and dissociation simply precludes this possibility. This is precisely the reason why we need a worldview and ethos which are holistic and can address the environmental challenges we are facing today. A framework is needed that can provide solution to the complex problems and provide the rational basis for a more caring and sustainable world. This framework already exists in the ecological worldview and has been termed a *holistic, systemic,* and *organic view*.

9.2.2 Basis for Anthropocentrism

From an anthropocentric worldview, it can be argued that it is only human needs, purpose, and interests which matter because human beings have a special place in the scheme of things. Human beings alone are equipped to raise the question of their place in Nature and know how to treat Nature by virtue of the possession of *consciousness (mind).* Possession of consciousness constitutes the metaphysical distinction between human beings and the rest of the Nature. This distinction caused the separation of human beings from Nature, on which Rene Descartes founded his dualism, the dualism of mind and body (Descartes, 1992), that is, the detachment of the human mind from the human body or anything physical or material. The mind was placed on one side of the divide and the rest of the world on the other side of the divide. The dualism of mind and Nature implies that purposes, desires, feelings, and interests are purely mental phenomena that cannot be found in Nature. Nature is devoid of any purpose or aim and cannot have an interest by itself to be protected. Nature can be said to have a purpose only in relation to the interest of mentally conscious human beings. Similarly, religious conception also implied that Nature could have purpose only in relation to the superconscious being, almighty God.

Descartes rejected the concept of God because it is impossible for humans to know what God's purposes would be. According to Descartes, Nature is purposeless, and as human beings, only our own purposes can be known to us, and, therefore, we must regard ourselves as the only source of purpose in the universe (Descartes, 1992). Descartes has been credited to have laid the philosophical and metaphysical foundation of modern science and the current development paradigm which has been aptly characterized by the metaphysics of domination and mastery over Nature, as Nature is an external object that must be subdued, exploited, manipulated, and controlled for human purposes. The physical world (Nature), as conceived by modern science, had been purely mechanical and purposeless system whose laws were to be discovered by the scientists. When scientists discovered the laws of Nature, it became possible for human beings to intervene in Nature and impose their purposes on Nature by manipulating natural processes to achieve their own ends. Values do not exist independently outside the purview of human purposes and interests. Something can be said to have value in relation to realizing a human purpose or satisfying an interest. In other words, *a thing's value is relative to the purpose, interest, or feeling of a conscious human being.* Descartes's dualist conception implies that natural things and processes have no value in themselves but have instrumental values in so far as they further human interests and purposes. Only the human beings themselves have the *intrinsic values.* This dualistic view is the foundation of the human-centered morality or *anthropocentrism.*

Anthropocentrism was pushed further to the extreme by the Christianity. Sarah Oelberg (2002) points out that environmentalists have criticized the Biblical tradition for providing the religious and cultural sanctions for the plundering of the Earth. The notion that humans have the dominion over all other forms of life on Earth has come from Genesis and God's message to man and woman to "*multiply and fill the Earth and subdue it; you are the masters of the fish and birds and all the animals.*" Christianity and Descartes's aggressive anthropocentrism both are responsible for the reckless plundering of Nature because they separated humans from it and regarded it as simply an outside inert object to be exploited and subdued by humans. Descartes's positivist ideology and Christianity provided the main fuel to the development of modern corporate capitalism that proclaims that with the use of science and technology, the free-market economy can solve all problems through continued infinite economic growth. The corporate free-market capitalism's infinite economic growth through the development of science and technology has become today's dominant social paradigm.

9.3 Critique of Dominant Development Paradigm

The critiques of the economic growth model have argued that economic growth cannot continue forever (Meadows et al., 1972; Townsend, 1993; Ehrlich, 1993; Daly, 1993; Jackson, 2017). This has caused serious intellectual headaches and inconvenience for the current dominant neoclassical economic system and its proponents. As Kolasi (2017) points out, Marx argued in the *Grundrisse* that capital cannot tolerate any limits and that the drive for growth and the search for new markets are both necessary for the political and economic survival of corporate capitalism. From this perspective, it seems that the implications of Max's arguments correctly present an existential challenge to the current growth paradigm. Neoclassical economists argue that economic growth could continue without the consumption of additional resources from the environment and that they could do this by doing more with less, investing in clean energy, and developing energy-efficient technologies. In other words, they argue for nothing less than the sustainability of corporate capitalism, ignoring all the science and evidence piling up against the impossibility of infinite growth, which is the existential basis for capitalism.

The general perception that when the GDP of a country/world or economic growth increases, it will reduce poverty and improve distributive justice in society has become a *myth.* Instead, the distribution of costs and benefits from economic growth under neoliberal corporate capitalism is highly unbalanced in almost all the countries in the world, including the United States, except China, which has adopted state-regulated economic system that is distributive in nature. The share of economic benefits reaching those on the lowest incomes has been shrinking. This is paradoxical that under the current system, to generate ever smaller benefits for the poorest, it requires the rich (who are already overconsuming) to consume ever more. As Simms and Johnson (2010) argue: "*The unavoidable result is that, with business as usual in the global economy, long before any general and meaningful reduction in poverty has been won, the very life-support systems we all rely on are likely to have been fundamentally compromised.*" A large number of noted development professionals and economists argue that indefinite global economic growth is unsustainable (Townsend, 1993; Ehrlich, 1993; Daly, 1993; Georgescu-Roegen, 1993; Boulding, 1993; Jackson, 2017). Just as the laws of thermodynamics, as Daly (1993) and Georgescu-Roegen (1993) point out, constrain the maximum efficiency of a heat engine, economic growth is constrained by the finite nature of our planet's natural resources. They propose a new macroeconomic model that allows the human population to thrive without depending on the impossible, endless increases in consumption. They point out that infinite growth in the economy requires that Earth also grow at a commensurate rate which is not the case. The conclusions of the seminal studies by a team of scientists at the Massachusetts Institute of

Technology (MIT) presented below may help us understand why a shift from the growth paradigm is necessary, and we are simply running out of time. Sooner this happens, better for the greater good of humanity and the planet Earth.

9.3.1 Limits to Growth Debate

The 1960s and early 1970s witnessed a vigorous debate on the environmental implications of growth. Scientists at the Massachusetts Institute of Technology (MIT) were commissioned by the Club of Rome to research and publish the then-controversial limits to growth but not controversial anymore now, which came out in 1972. The conclusions of the report by Meadows et al. (1972) have become so crystal-clear today that even the die-hard neoliberal economists cannot ignore them. Meadows et al. (1972), after having gone through the studies and the analysis of the empirical data available at that time, came up with the following three important conclusions:

> If the present growth trends in world population, industrialization, pollution, food production, and resource depletion continue unchanged, the limits to growth on this planet will be reached sometime within the next one hundred years. The most probable result will be a rather sudden and uncontrollable decline in both population and industrial capacity.
>
> It is possible to alter these growth trends and to establish a condition of ecological and economic stability that is sustainable far into the future. The state of global equilibrium could be designed so that the basic material needs of each person on Earth are satisfied and each person has an equal opportunity to realize his individual human potential.
>
> If the world's people decide to strive for this second outcome rather than the first, the sooner they begin working to attain it, the greater will be their chances of success.

In the early 1970s, the environmental effects of industrial activity and the depletion of the natural resource base were not particularly ubiquitous or obvious as they are today. Even then, the magnitude of the work that remained to be done was made clear by how far-reaching and complex their conclusions were. They correctly noted that there must be limitations that can act to prevent exponential growth in any finite system. These restrictions are negative feedback loops. As the expansion gets closer to the carrying capacity of the system's surroundings, the negative feedback loops get stronger and stronger. Growth finally stops when the negative loops balance or outweigh the positive ones. The destruction of productive terrestrial and marine ecosystems, tropical rainforests (Amazonian forests), loss of biodiversity, decreased resilience of the productive systems, and famine are just a few examples of the processes that contribute to negative feedback loops in the Earth's planetary system. According to Meadows et al. (1972), the promise that human equality will result from maintaining our current growth patterns is the most widely believed myth in our culture today. Contrarily, this has demonstrated the exact opposite, showing that current patterns of capital growth have significantly widened the gap between the rich and the poor globally and that the ultimate outcome of continuing to grow in accordance with the current pattern will be a catastrophic collapse. The world's population expansion in emerging nations and the excessive consumerism of the wealthy in industrialized nations are the biggest obstacles to equitable resource allocation.

According to Simms et al. (2010), unrestricted global economic growth is not possible due to the finite nature of the planet Earth's natural resources (biocapacity), which they compare to how the laws of thermodynamics limit the maximum efficiency of a heat engine. Herman Daly, an economist, challenged the idea of sustainable economic growth and asserted that what is sustainable is development without growth, saying that he would accept the possibility of infinite economic growth on the day that his economist colleagues "*could demonstrate that Earth itself could grow at a commensurate rate.*" The economy is a subsystem of the planetary ecosystem, which is finite, nongrowing, and materially closed, so he eloquently states: "*The term 'sustainable development' makes sense for the economy only if it is understood as 'development without growth,'* i.e., *qualitative improvement of a physical economic base that is maintained in a steady state by a throughput of matter energy that is within the regenerative and assimilative capacities of the ecosystem.*" In his opinion, the term of "*sustainable growth*" as used in reference to the economy is an *oxymoron* and self-contradictory. Sustainable development, or development with zero economic growth, can be viewed as a new macroeconomic *steady-state model* that permits the optimum human population to exist on Earth's regenerative biocapacity by ultimately and dramatically lowering the endless rises in consumption.

9.3.2 Transition from Growth to Equilibrium

A growing number of economists and development experts have come to the conclusion that, no matter how desirable it may be, the transition from growth to a sustainable state of global equilibrium is simply not feasible. The transition to a sustainable state is, in my opinion, merely wishful thinking unless the existing growth model is altered and modified to integrate the environmental externalities and the costs associated with the depletion of natural capital and the ecological services in its analytical framework. Although some ecological economists (Costanza, 2014; Daly, 1993; Ehrlich, 1993; Townsend, 1993; Boulding, 1993) have made attempts to create such a model and think it is feasible, they fall well short of influencing the political mentors and economic practices of neoliberal mainstream economists. The principles of thermodynamics govern the physical state of the economy; therefore, it is essential to our existence to know how much energy we need to extract, process, and use the natural resources that power our economy. The limitations of the growth economy have been pointed out explicitly by economist Simon-Kuznets (1973). Growth did not account for the quality of life and left out large and significant portions of the economy where trade was conducted without the use of money. He was referring to the "core economy"—the activities that support the functioning of society and civilization—which includes household duties and volunteer labor in the community.

The unpalatable reality of climate change and environmental degradation can substantially constrain economic progress. However, climate change is not the only natural phenomenon influenced by humans that demands our attention. If we are to preserve Nature's ecological life support system and the perpetuation of human civilization on planet Earth until the advancement of science and technology that enables *Homo sapiens* to safely leave Earth for other livable planets, if such a possibility exists, we must pay attention to the limits of the biosphere's biocapacity that can sustain life on planet Earth. Even if such a possibility exists, it would take place hundreds of years from now, by which time human civilization probably vanish entirely from the planet.

9.3.3 Central Flaws of Neoclassical Growth Model

As noted by Ackerman (2002), Jackson (2017), and Erald-Kolasi (2017), neoclassical theory typically suffers from a number of contradictions, unrealistic assumptions about society, and has no predictive power at all. More output results from increased capital, but some of that output is also reduced by the depreciation of capital assets. There is no longer any growth when the economy eventually enters a stationary condition where growth and depreciation have reached a balance. Neoclassical theory contends that the economy requires a consistent stream of technological advancement, defined as an increase in overall productivity, to produce continuous growth. This gain suggests that when inputs are held constant, output can increase. The idea behind treating production inputs as mainly independent of one another is to allow for substitutions as necessary to maintain or increase the maximum level of production. Neoclassical theory contends that natural resource constraints can be resolved by technical advancement, increased productivity, or other kinds of substitution (Ackerman, 2002; Jackson, 2017). In fact, neoclassical economists frequently hold that long-term capitalism's sustainability is materially conceivable, and all that is required to ensure this sustainability are the appropriate social and institutional frameworks (Ackerman, 2002). This sounds like creating a fairy tale economy where the wheels of capitalism can run on an endless supply of fuel. Solow (1974), the critic of the no limit to growth, entertained the notion that the natural world does not set boundaries on economic growth on the grounds that "*if it is very easy to substitute other factors for natural resources, then there is in principle no 'problem' and exhaustion is just an event, not a catastrophe.*"

His statement captures the overall perspective that many economists have about the inevitable nature of growth under capitalism, even if his model also demonstrated that competition would eventually lead to the exhaustion of natural resources. In other words, regardless of any depletions or instability in the larger natural world produced by such productivity advances, better technologies and higher efficiency would always be available to boost production. Although progress under capitalism can undoubtedly be enabled by better, more efficient technology for a while, it cannot be sustained permanently. Can the fundamentals of quantum physics, gravity, relativity, space-time, and thermodynamics be replaced or altered by technology? Can technology produce new material resources without utilizing those that the planet Earth has so far provided? One could argue that technological advancement will make it possible to exploit the resources of the Moon, Mars, and other planetary systems, but by the time technology reaches that point of development, what will become of the Earth and humanity? *Homo sapiens* and a wide variety of other life-forms are specialized on planet Earth. No organism can live by eliminating the environment that makes up its *niche*. Only if such technological breakthroughs can keep the niche environment in its most productive state under capitalism or socialism would growth under these systems be feasible. Why shouldn't we direct and concentrate our technology advancements on how to build and sustain a healthy, productive niche environment or Nature where humans and other life-forms can flourish? The future of *Homo sapiens* and the living system's existential conditions are the subject of this inquiry.

The consequences of the physical and ecological laws that restrict and direct the socioeconomic system are typically disregarded by neoclassical economists. In other words, they appear to be completely unaware of the interactions between the elements of the ecosphere and sociosphere and how these affect the sociosphere (socioeconomic system). Thermodynamics is the most fundamental limiting factor for the principle of substitution. The greatest efficiency of energy flows through technological systems is constrained by thermodynamic restrictions (Georgescu-Roegen, 1993). *"If global civilization runs out of natural resources, we cannot replace them by investing in commodities through financial markets,"* argues Erald-Kolasi (2017) in his critique of neoclassical growth economics. Money is inedible to humans. At the micro-level of economic activity, substitution may be possible in the long run, but long-term macro-level substitution is just wishful thinking.

This argument emphasizes the crucial fact that, despite their importance, renewable technologies cannot resolve the ecological catastrophe on a global scale under the neoliberal capitalist economic system, which is wholly dependent on the deceptive promise of unending expansion in consumption and production. Even if renewable energy sources were to completely replace fossil fuels while further growth was encouraged, global civilization would nonetheless collapse in a few generations. There is no mechanism at the heart of neoclassical economics that acknowledges the most fundamental things of all, the living, changing, and sustainably functioning Earth systems (ES). The driving forces behind market capitalism are very clearly explained by Speth (2008): *In a capitalist economy, survival requires development, and growth requires profits. This is akin to Charles Darwin's theory of how organisms evolve through natural selection and is known as the "survival of the fittest"* in capitalism. This leads one to the conclusion that current neoliberal market capitalism cannot protect environmental resources and the biocapacity of the planetary ecosystem, and the political system cannot correct the market capitalism. *In the capitalist version, Darwin's idea of fitness—success in producing offspring—becomes success in making profits*. Profit is the fuel that drives market capitalism's motor; the instant it stops producing profit, it vanishes. To address the environmental and social issues that market capitalism brought about on Earth, it is necessary to examine these fundamental beliefs and reformulate them. However, this cannot be done without altering the beliefs that led to the problems in the first place.

9.3.4 Connection Between Energy, Growth, and Emissions

Energy flows are necessary for productivity increases; therefore, all technical advancements are connected with the energetic transformations that enable human existence. Every business activity needs energy. According to Kolasi (2017), by examining the American economy, we can more clearly comprehend the connections between energy, growth, and emissions. An economy can flourish even if primary energy consumption drops if it starts utilizing natural resources with higher energy efficiency and greater power densities. Recent US experience further supports the idea that significant carbon reductions are practically unattainable in an economy that places growth above everything else. For a short while, it is impossible to deny that emissions have decreased somewhat, but over the long term, increased emissions may still occur despite macro-level efficiency improvements and technical advancements due to the unrelenting drive to expand consumption and output.

9.3.5 Decoupling Environmental Impacts

Neoclassical economists contend that human civilization can disentangle the effects of economic growth on the environment (Gross Domestic Product). Although this reasoning may seem persuasive, is it actually possible to decouple economic growth from the environment? The modeling efforts of Ward et al. (2016) showed that, in the end, it was impossible to disentangle GDP growth from increases in material and energy consumption. Not only did they come to the conclusion that decoupling is impossible, but they also made the point that it is completely false to base a growth-oriented strategy on the notion that decoupling is conceivable. They also pointed out that the GDP is a subpar indicator of societal well-being. According to their argument, society cannot sustainably increase its well-being—including the well-being of its natural capital—without abandoning GDP growth in favor of alternative, more all-encompassing indicators of social well-being.

Decoupling would be a rational and natural objective if there were no constraints on the amount of resources (infinite) used to support the growth economy. This would enable GDP to continue growing eternally. Obama cited patterns from his presidency that demonstrated that the economy increased by more than 10% despite a 9.5% decrease in carbon dioxide emissions from the energy sector, according to the Economic Report (Ward et al., 2016) that was presented to the president. The idea that fighting climate change necessitates accepting slower growth or a lower standard of life should thus be put to rest, according to President Obama, who excitedly but incorrectly stated: *"Decoupling" of energy sector emissions and economic growth. The International Energy Agency similarly declared triumphantly but incorrectly in 2016 that the "decoupling of global emissions and economic growth confirmed."* In opposition to this assertion, Hausfather (2017) noted that global greenhouse gas emissions dramatically increased in 2017. Emissions increased once more in 2018 at a quicker rate than in 2017 despite an increase in frightening scientific findings about the risks of global warming (Carrington, 2018).

However, many economists contend that the perceived decoupling was only partially a result of actual efficiency improvements. The remaining consequences include a combination of cost-shifting, financialization, and substitution (Auckerman, 2020; Kolasi, 2017). While increasing economic activity per unit of energy and material cannot be the solution to the problem of perpetual growth, efficiency gains can. It can only delay economic progress for a certain period of time before finally exceeding its capacity. The appearance of decoupling is produced when resource-intensive forms of production are moved away from the point of consumption (Kolasi, 2017). For instance, many products used in Western countries are produced in poor countries. Consuming those commodities thereby increases GDP in the country of consumption, while the environmental effects occur elsewhere (typically in a developing economy where they may not even be measured). A clear example of cost shifting is this.

Critics point out that the existing growth paradigm equates value with systematic exploitation of social and natural systems, which makes it impossible to decouple GDP and its growth from environmental deterioration. For instance, cutting down old-growth forests and selling them result in higher GDP than protecting or regenerating them. The GDP does not fully account for all elements that impact societal well-being. These include the distribution of income and wealth, the state of local and global ecosystems (including the climate), the level of social contact and trust at various sizes, the importance of parenting, household labor, and volunteer work, and the distribution of wealth and income. It is unfortunate and regrettable that these components' contribution to social well-being is not included in GDP.

As a result, we must evaluate human progress using metrics other than the GDP and its rate of rise. According to ecological economists (Townsend, 1993; Ehrlich, 1993; Daly, 1993; Georgescu-Roegen, 1993; Boulding, 1993; Costanza et al., 2011; Jackson, 2017), rise in GDP cannot be separated from growth in material and energy use. In that regard, Ward et al. (2016) correctly noted that *"our model reveals that increase in GDP ultimately cannot feasibly be divorced from growth in material and energy use, unequivocally establishing that GDP growth cannot be sustained permanently. Therefore, basing growth-oriented strategy on the supposition that decoupling is feasible is deceptive. We further point out that the gross domestic product (GDP) has been proved to be a poor proxy for social wellbeing, which is something it was never intended to measure, and that GDP growth is thus, in any event, a dubious long-term societal goal. We contend that it is now necessary to acknowledge the biophysical bounds and to start the long overdue process of reorienting society toward a more realistic and fulfilling set of objectives than perpetual growth.*"

It is crucial to note that technology advancements could result in decoupling for some kinds of consequences. It is feasible, for instance, to replace a polluting activity with a nonpolluting one. Notable examples include the removal of CFCs from propellants and refrigerants, as well as tetraethyl lead from vehicle fuel. By transitioning to 100% renewable energy, it is also conceivable to imagine a future in which GDP growth is uncoupled from the consumption of fossil fuels and associated CO_2 emissions, although this is not the same as doing so (Auckerman, 2020; Kolasi, 2017; Ward et al., 2016).

9.3.6 Kuznets Curve, Growth, and Inequality

The majority of mainstream economists support the idea that economic growth is crucial for reducing inequality and enhancing environmental quality. They contend that while economic growth may at first exacerbate inequality and environmental damage, it ultimately promotes more equitable development and higher environmental standards. A curve known as the Kuznets curve, which has gained enormous popularity, has been used to describe this relationship between economic growth, inequality, and environmental quality.

The national income accounting was developed by American economist Simon Kuznets, who examined the income statistics gathered from the United States, the United Kingdom, and Germany in 1955. He came to the conclusion that as the economy expanded, income disparity initially increased, then leveled off, and then decreased. What best illustrates the relationship between income inequality and economic growth is an inverted U-shaped curve. In his initial analysis, Kuznets believed he had found a trend whereby economic expansion in wealthy nations reduced the gap between rich and poor people's incomes, while in poorer nations, it widened it. He proposed that income inequality increased first and subsequently declined as countries experienced economic growth by observing trends of income inequality in industrialized and developing nations. According to Kuznets, for economies to grow, they needed to switch from the agrarian to the industrial sectors. The income from agriculture varied little, whereas industrialization caused substantial income disparities. Furthermore, as economies grew, mass education increased opportunities, which reduced inequality, and the populace with lesser incomes gained political power to alter governmental policy. All growth-oriented economists and the financial and development agencies, including the World Bank, IMF, and others, adopted the assumption that economic expansion initially widens income inequality before leveling it off and finally narrowing it. Today's majority of economists still have this viewpoint.

As Raworth (2017) and Jackson (2017) note, economists rigorously evaluated Kuznets curve in the 1990s using adequate time series data but discovered no such trends. They found that when countries advanced from low to medium to high income, inequality increased in some, decreased in others, and then increased again. In some other countries, it only increased or decreased, with no trends as indicated by the Kuznets curve. The prosperity of the four East Asian nations—Japan, South Korea, Indonesia, and Malaysia—also refutes the incorrect theory of the Kuznets curve. Due in large part to rural land reforms that increased smallholder farmers' incomes as well as significant public investments in health, education, and industrial policies that increased workers' wages while containing food prices, these countries experienced rapid economic growth, low inequality, and declining poverty rates. It demonstrates how avoiding the Kuznets curve and achieving relative growth with fairness and environmental quality are both achievable with the right policy measures in place.

The ramifications of Thomas Piketty's influential findings must be discussed. Piketty (2014) examined vast historical data on the dynamics of income distribution under capitalism from Europe and the United States and derived certain findings. The distinction between households that own capital, such as real estate, homes, and financial assets, which produce rent, profits, and interest, and households that own only their labor, which produces only wages, is critical, according to Piketty. By comparing the growth patterns of these various income sources in Europe and the United States, he came to the conclusion that the Western economies and other similar countries are headed toward dangerously high levels of inequality. This is because the return on capital typically grows faster ($r > G$) than the economy, which leads to an increase in the concentration of wealth in the hands of capital owners. The interests of the already wealthy are then further promoted through political influence, including corporate lobbying and campaign finance. Piketty comes to a conclusion that "*capitalism automatically generates arbitrary and unsustainable inequalities that radically undermine the meritocratic values on which democratic societies are based.*"

Prior to recently, policies did not focus on inequality. There was a broad belief that tolerance for inequality was necessary to ensure better wealth and opportunity for all people. Even now, this is still very much a neoliberal policy script. From studies on inequality conducted by economists and social scientists (Piketty, 2014; Raworth, 2017; Goerner et al., 2009; Torras & Boyce, 1998; Webster, 2015), the following conclusions can be made that:

If you invest more in the rich, the economy will trickle down income to the poor is a neoliberal myth or hoax created to trickle up income from the poor to the rich.

National inequality, not national wealth, mostly influences national social welfare.

Inequality does not make the economy grow faster; it rather slows it down.

Inequality damages the social fabric of the whole society.

More equal societies, whether they be rich or poor, turn out to be healthier and happier.

A higher level of national inequality tends to increase ecological degradation.

With the implementation of appropriate policies, developing countries can achieve economic growth with equity and environmental sustainability. Given these perceptive findings from the empirical studies, Kate Raworth's advice to act now to reduce inequality is well worth considering: "*Don't wait for economic development to alleviate inequality because it won't. Make an economy that is distributive by design, instead. The capitalist economies of industrialized nations, with its widespread corporate dominance and concentration over all industrial sectors, from agribusiness to the health sector to the media and banks, cannot and should not be copied by developing nations.*" According to Goerner et al. (2009), revitalizing the small-scale, varied *cooperative enterprises* that make up the majority of the economy's network is the best course of action for developing countries. Since such economic growth should place more of an emphasis on fostering small business capital and human community development, which in the long run preserve cross-scale vitality, equity, and environmental quality.

9.3.7 Environmental Kuznets Curve and Growth

When analyzing the correlation between GDP, air, and water pollution in 40 nations, American economists Gene Grossman and Alan Krueger (1995) observed a trend that remarkably resembled the Kuznets inequality Curve. The environmental Kuznets curve was quickly named to capture this phenomenon. Once more, most economists were unable to refrain from coming to the conclusion that economic growth causes the environment to first deteriorate and then improve. It became the catchphrase of growth economists who argued that while environmental quality declines during the early stages of economic growth and development, it eventually improves as the economy or GDP grows. Growth in the economy benefits the environment. Therefore, growth-promoting policies (such as trade liberalization, economic restructuring, and price reform) ought to be environmentally friendly. This is yet another misconception produced by neoliberal mainstream economists. Environmentalists have advanced three reasons in response to their grave criticism that economic growth has significantly harmed Earth's soils, lands, oceans, and ecosystems and created unfavorable climate change:

As economies develop, people have the means to start protecting the environment and demanding higher standards, and as a result, enterprises can begin utilizing cleaner technologies and shifting from manufacturing to services. This explanation could appear plausible at first glance, but closer examination reveals that it is false. To begin with, people do not need to wait for the GDP to rise before demanding clean air and water. Researchers have discovered, through studies in a larger range of nations, particularly in low-income ones, that environments are better in places where income is divided more fairly, where people are more literate, and where civil and political rights are more highly upheld. Environmental quality is protected by people, not by economic progress in and of itself (Torras & Boyce, 1998; Raworth, 2017; Jackson, 2017). The claims put out by Kuznets curve proponents have been entirely false and have caused substantially more environmental damage.

9.3.8 Ecotax and Environmental Management

Governments in Europe, the United States, and other countries have implemented various ecotax laws, and these initiatives have had a real influence on reducing carbon emissions, fossil fuel consumption, and the degradation of air, water, and land quality. Economists preferred market-based techniques to internalize those externalities, since they recognized the detrimental consequences of using fossil fuels and industry in creating harmful environmental externalities. As a result, numerous ecotax policies, sometimes referred to as polluters pay policies, were implemented in the form of property rights quotas, trading and swapping, trading and swapping caps, and so forth. The plan was to impose an eco-tax that was equal to the societal cost of pollution, after which the market would decide how much pollution is acceptable or allowable. The ecotax policies have undoubtedly assisted in reducing emissions and maintaining environmental quality to some level, but it is a pipe dream to think that they will resolve the environmental issues that are currently plaguing the world on the scale and magnitude they are. The existing ecotax policies put in place by governments are simply the *tip of the iceberg* when it comes to solving environmental issues of the current scope and magnitude. The current policies must be completely revised and overhauled in order to achieve this, and new strategic policies and programs must be developed under the direction of a greater level of ecological awareness.

9.3.9 Reforming Modern Capitalism

Political and institutional reforms in the current neoliberal capitalism are necessary to make it compatible with the basic functioning of democracy. Such reforms entail substantial public investment in education, healthcare, and infrastructure—areas often overlooked in a capitalist system, where profit drives resource allocation. Progressive social policies, including universal healthcare and free higher education, can be instrumental in leveling the playing field and ensuring equal opportunities for all (Bernie, 2020). Famous French economist Thomas Piketty is best known for his in-depth empirical research on the evolution of wealth and income inequality (Piketty, 2014). His work has helped to expose important patterns and factors that contribute to economic disparity under neoliberal capitalism, weakening the foundation of an inclusive democracy. The book "Capital in the Twenty-First Century," one of Piketty's landmark works, depicts the inequality relationship as an eq. ($R > G$), where the rate of return on capital (R) outpaces the rate of economic growth (G) over a protracted period of time. According to Piketty's theory, these dynamics—which causes wealth to concentrate in the hands of a small number of individuals and has the potential to widen the wealth gap, causing social unrest and economic instability—is not an aberration but rather a fundamental feature of capitalist economies.

The enormous archive of historical data Piketty's study is based on was created in partnership with other economists, including Emmanuel Saez. This database exposed trends in income inequality, including a sharp decline in top marginal tax rates over several decades that, according to Piketty and Saez (2003), has a substantial impact on rising income disparity. Piketty (2014) and other well-known economists like Emmanuel Saez and Zucman (2019) have recommended a *progressive worldwide tax* on wealth accumulation as a solution to the problems caused by rising wealth and income inequality under neoliberal capitalism. Mitigating economic disparity is a key technique for bringing capitalism into line and making it more democratic. *Progressive taxation*, in which wealthier taxpayers pay higher rates of taxation, can aid in the more equitable distribution of income within society (Saez & Zucman, 2019). Additionally, the implementation of a wealth tax can promote economic equality by thwarting the concentration of wealth accumulation (Piketty, 2014). They claim that such a solution could address a substantial aspect of economic inequality by preventing the excessively quick accumulation of wealth by the ultra-rich (Piketty, 2014). Additionally, Piketty foresees the emergence of *"patrimonial capitalism,"* a socioeconomic system in which wealth is mostly acquired through inheritance rather than through effort of labor or invention. Such a system, according to Piketty (2014), runs the risk of escalating socioeconomic inequalities and eroding the fundamental principles of democracy.

9.4 Alternative Economic Worldviews and Models

It is crucial to understand that the human socioeconomic system (the "sociosphere") functions as a component of the planetary ecosystem (the "ecosphere"). The most fundamental requirement for designers and development thinkers who wish to create an ecologically sustainable economic development model is perhaps an understanding of how various elements of the planetary ecosystem interact to produce natural capital and ecological services on which the human socioeconomic subsystem (sociosphere) operates. It is hard to comprehend the limitations of the existing growth model without a thorough grasp of the system structures and their interactions with the planetary ecosystem (ecosphere/biosphere). The network (web) of physical, biological, and ecological relationships that connect any human population, its natural environment, and its economic activities essentially governs the behavior of the socioeconomic subsystem or *sociosphere*. Our socioeconomic subsystem's architecture and underlying network relationship cannot be properly handled unless they are fully understood and studied. A car system analogy can be used to illustrate it. Ecologically sustainable development cannot be achieved without an understanding of how the various parts of the planetary ecosystem interact with one another and influence one another, just as an automobile cannot be kept in good working order without understanding how its many parts are connected to one another, interact and influence one another's functionality. Since the sustainability and viability of the *sociosphere* (socioeconomic subsystem) in the *ecosphere/biosphere* depends on the health and functional state of our planet Earth's systems and their biocapacity, the ecological footprints that humanity has left behind will send out some worrying signals about those conditions.

9.4.1 The Ecological Footprints

The ecological footprint is a well-known method for assessing the *biocapacity* of the planet Earth to create resources and digest waste products produced by human economic activity. It was first devised by the Canadian geographer William Rees in the early 1980s. The *ecological footprint* concept gauges how much biologically productive land and water are needed to create the commodities we use and to ingest the throughput we generate. It contrasts the pace of resource consumption and waste production by humans, measured in terms of greenhouse gas emissions, with the biocapacity that is available to support agriculture, fishery, and forestry as well as to absorb waste from human economic activity. Humans today consume as much ecological resources as 1.7 Earths were to regenerate for us. Accordingly, the Earth needs 19 months to generate the ecological services that humanity requires during a 12-month period. Earth Overshoot Day, which occurred on July 29, according to Lin et al. (2021), meant that people had used the equivalent of 1.7 heaps of earth's resources. The primary factor was an increase in carbon footprint of 6.6% from 2020 to 2021. Global forest biocapacity has decreased by 0.5% since 2020, largely as a result of an increase in Amazonian deforestation. In Brazil alone, 1.1 million hectares of forest were lost in 2020, and predictions for 2021 show a rise in deforestation of up to 43% from the previous year (Lin et al., 2021). By 2030, if current trends hold true, two planet Earths will be required to provide all of humanity's demands in terms of products and services. Unfortunately, Earth is the only planet we have. The ecological footprint is the sole indicator that weighs human, governmental, and commercial resource consumption against the planet's capacity for biological regeneration.

Humanity would require the equivalent of at least 3.4 planets like Earth if all people consumed and generated throughputs at the same rate as the average individual in the United Kingdom (Raworth, 2017). Most concerningly, there are indications that the biocapacity that is still available is being depleted by the current levels of abuse, creating a vicious cycle of overconsumption, and weakened ability to create and provide. Data on human usage of biophysical resources (biocapacity) send several frightening messages to everyone, but especially to those who think unrestricted eternal economic growth is possible. Our global ecological footprint is expanding quickly, far exceeding the capacity of the biosphere to provide for us and to absorb it. This is hastening the depletion of the biocapacity, or the base of biophysical resources, on which we rely. According to Simms and Johnson (2010), we are using resources and emitting carbon emissions 44% more quickly than nature can replenish and reabsorb what we use and the waste we produce.

9.4.2 Planetary Boundaries

The "*planetary borders*" report was initially published in Nature (Nature 461, 472–475; 2009), a prestigious scientific magazine. A team of eminent academics led by Johan Rockström of the Stockholm Resilience Centre presented a new framework in 2009 for evaluating stress to Earth systems (ES) and defining a safe operating space for human habitation on this planet. "*Planetary Boundaries*" (PB) was the name they suggested for this system. They claimed that in order to prevent a catastrophic environmental change, humans must respect set boundaries for a variety of crucial Earth systems (ES) processes (Nature 461, 472–475; 2009). Climate Change, Rate of Biodiversity Loss (Terrestrial and Marine), Interference with the Nitrogen and Phosphorus Cycles, Stratospheric Ozone Depletion, Ocean Acidification, Global Freshwater Use, Change in Land Uses, Chemical Pollution, and Atmospheric Aerosol Loading are the nine biosphere processes they identified for which they felt it necessary to define them as "*planetary boundaries*." Three of these nine planetary boundaries—*climate change, interference with the nitrogen cycle*, and *biodiversity loss*—have already been crossed, according to the scientists (see Chart 9.1). Setting limits is difficult. Earth systems adapt and respond frequently and nonlinearly. The resilience and behavior of one system can be impacted by the erosion or overburdening of another system. According to the findings, once one barrier is crossed, other borders are seriously put at risk as well. For instance, large changes in land use in the Amazon could have an impact on water resources in Tibet.

The second "Planetary Boundaries" report from the Stockholm Resilience Center was released in 2015. Five years later, Steffen et al. (2015) reviewed PB (Fig. 9.1). Based on the available scientific evidence, their investigation unequivocally demonstrated that boundaries are interconnected and function as an interdependent component of the Earth as a single, complex, integrated system. These processes and their interactions can provide either stabilizing or destabilizing feedbacks in the Earth system, which functions in clearly defined states. This underlines the necessity of addressing numerous interrelated environmental processes at once (e.g., stabilizing the climate system necessitates sustainable forest management and stable ocean ecosystems), which has significant implications for global sustainability. It is interesting to consider how they came to the following conclusion: "*PBs are scientifically supported levels of human disruption of the ES beyond which ES functioning may be significantly affected. Thus, breaking the PBs significantly increases the chance that the Holocene condition of the ES, where modern societies have developed, may become unstable. The PB framework makes no recommendations about how society should advance. These are political choices that necessitate taking into account human factors, like as equity, which are not taken into account by the PB*

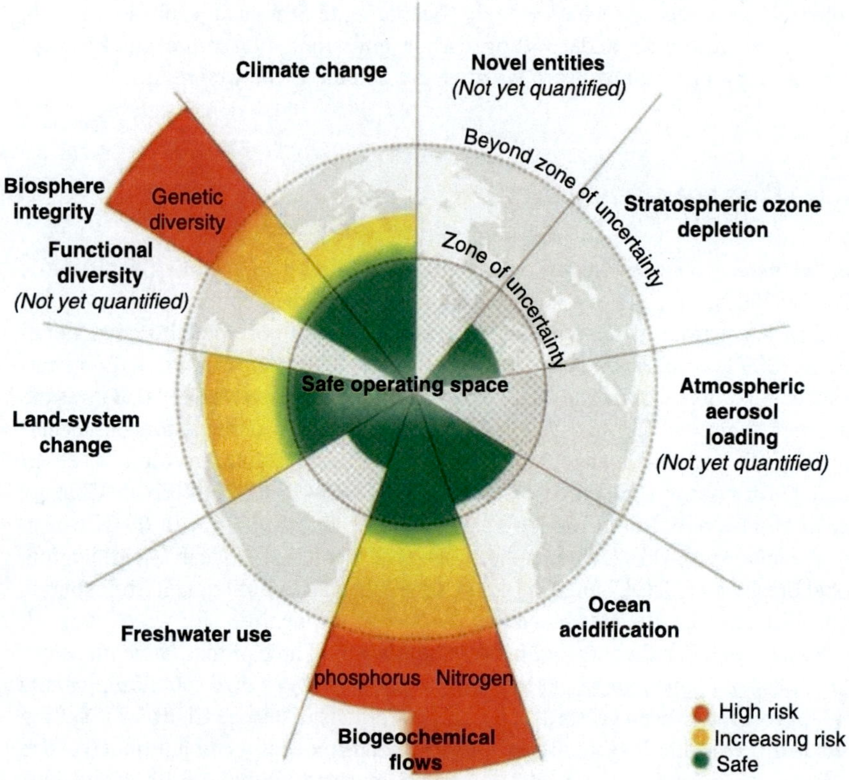

Fig. 9.1 Planetary boundaries (PB): guiding human development on a changing plane. (Source: Adapted from Steffen et al. (2015), Science Vol. 347(622))

framework." However, the PB framework can help decision-makers in mapping out desirable paths for social progress by establishing a safe operating environment for humans on Earth. The planetary boundaries research supports and enhances the ecological footprint approach. Assessments of the portion of the Earth's resources and services that are available for safe human economic use are expected to become more realistic and, almost unavoidably, lower as more nuanced planetary boundaries are attempted to be defined with regard to various Earth systems.

Neoclassical economic theories suggest that the economy is a closed system in which labor and capital are used to produce things. These models do not take into account wastes, ecosystem services, and natural resources, which are still, at best, the secondary concerns. They are typically excluded from economic calculations. Increased labor and capital inputs, improvements in the quality of those inputs (better-skilled people), and technological advancements are all seen to contribute to economic growth. Energy is believed to play a relatively small role in increasing productivity and economic growth, as Sorell and Ockwell (2020) pointed out, and it is further believed that labor and capital may replace energy should it become more expensive.

Environmental economists, on the other hand, contend that neoliberal models fail to take into account how the economy is intertwined with the planetary ecosystem and its biocapacity (Ward, 2016; Sutton, 2017; Werner, 2016; Costanza, 2016; Mohr et al. 2016; Simmons, 2016). They believe that flows of high-quality energy and resources, which are later recycled into the environment in the form of unusable low-quality heat energy or *entropy*, are what maintain economic activity. Since energy cannot be created or recycled, it serves as the main input for economic production. The system is powered by solar energy, both directly and as it is stored in fossil fuels. Contrarily, ecological economics view labor and capital as intermediate inputs because they require energy to develop and maintain them. Instead of being a subsidiary, energy is now the main issue on which industries and their applications are centered. They further contend that while the advancement of industrial methods reduces energy use per unit of economic output, they also enhance overall productivity and outputs to the point where total energy use rises. This could help to explain why economies are seeing increased energy consumption even as their energy intensity is decreasing.

Political and business leaders who have been brainwashed by the current economic paradigm (neoliberal capitalism) think that economic growth is the cure-all for all ills. The majority of economists in the field today are not even willing to consider the possibility of a model other than economic growth model. In actuality, the only way to save human civilization from impending destruction may very well be to imagine and realize alternatives to the current economic development model. Kolasi (2018) makes the following remark, which is well taken: "*Instead of building our society and economy around the idea of growth, we should organize them around the premise of sustainable human development, which necessitates the metabolic stability of the broader ecosphere.*" We may prevent the recurring booms and crises of capitalism while simultaneously extending the time of human civilization by strictly limiting the levels of production and consumption around some dynamic equilibrium and emphasizing qualitative human–social relations rather than the currency nexus.

The "*circular economy*" concept, which is gaining momentum among environmental and development economists and thinkers, was promoted by the Ellen McArthur Foundation in 2012. Its applicability is crucial given the escalating environmental issues and the unsustainable use of natural and synthetic technical resources brought about by the existing linear market economy's focus on fast development. We can better understand how resources should be used with zero, if possible, or very little waste generation to sustain and maintain the economy and environmental quality by having a brief discussion of this alternative approach, which has some potential for resolving today's economic and environmental problems. The existing dominant economic paradigm without alteration or modification cannot resolve the ecological or environmental and social problems, especially the planetary ecological crisis and the negative effects of climate change. Therefore, a shift in consciousness toward adopting a steady-state circular economy that produces a very low level of entropy and steers humanity to the development of a sustainable society, becomes necessary.

The current development paradigm and its *modus operandi* run counter to the ecological systems of the planet Earth. While the planet relies on a cycle of regeneration and balance to function, the corporate market capitalism relies on a linear model of infinite growth and consumption to thrive. This inherent contradiction in the neoliberal paradigm is the principal cause of today's environmental destruction and climate crisis, which endanger our ability to survive. To change this, we must reform this neoliberal market capitalism and push it in the direction of circular regenerative economy which remains within the bounds of the Earth's biocapacity mimicking its natural processes. Furthermore, capitalism, as it exists now, violates the democratic value of equity and inclusiveness. It encourages social injustice, deepens the income divide, and concentrates wealth and power in the hands of a select few. We need to humanize neoliberal market capitalism with distributive social justice and equal opportunities for all to bring it into line with modern democracy. This entails putting a priority on empowering vulnerable communities, distributing resources fairly, and fostering social inclusion.

9.4.3 The Circular or Cyclical Economy

The circular economy refers to economic, technological, and environmental systems that optimize resource reuse and zero waste. To build a closed-loop system based on the principle of preserving "*virgin*" materials, minimizing waste creation, and increasing waste reuse, circular economy systems employ design of *recycling, reuse, remanufacturing*, and *refurbishment* (Raworth, 2017). The need to complete the circle of waste generation is greater than ever as human population and consumptions continue to put pressure on our landfill, recycling facilities, and natural resources. It is obvious that the overuse of natural resources necessary for economic development and growth has had a negative influence on the environment as well as their availability and cost. This makes the *circular economy* a workable substitute that presents fresh approaches to developing a more sustainable form of economic growth. According to the statement made by the Ellen McArthur Foundation (2012), "*We must reform all aspect of our take-make-waste system, including resource management, product creation and usage, and material disposal. Then and only then, within the bounds of our planet, can we establish a healthy circular economy. To ensure there is no or very little waste, a circular economy envisions separate components of goods that may be recycled and replaced. The elements that need to be replaced are either recycled back into the biosphere of the Earth or made into other goods or services.*" The Ellen MacArthur Foundation estimates that a circular economy could result in $630 billion per year in cost savings for the medium-lived products sector in the European Union. The Foundation makes the audacious claim that "*there is no waste and no pollution, and the planet's natural systems are regenerated in the process.*" According to estimates, it might result in annual savings of $700 billion for fast-moving consumer items globally.

The circular economy's main goal and task is to create technological design innovations that can replace the existing waste economy with one in which waste is significantly reduced, if not entirely eliminated, resources are circulated, and Nature is rejuvenated. The circular economy, according to its proponents (McArthur Foundation, 2012; Webster, 2015; Raworth, 2017; Jackson, 2017), aims to address critical societal needs, reduce greenhouse gas emissions, waste, and pollution while addressing climate change, biodiversity loss, and planetary ecological degradation. This is undoubtedly a step toward a sustainable society that adheres to the maxim that less is more desirable.

The term "*circular economy*" is widely used when discussing how to rethink and restructure the current neoliberal market economy in an effort to solve or lessen the environmental crises it has caused. The ability to redesign or restructure industrial manufacturing and material consumption processes from their current *degenerative processes* to *regenerative processes*, allowing materials inputs to be recycled and reused with very little waste throughput, is a key point made by advocates of a circular economy (Webster, 2015; Raworth, 2017). There is no waste in the traditional sense, or if there is, it is little. This is what has come to be known as the *"circular economy."* The wastes generated enter as fresh material inputs again in the manufacturing or production processes. The current corporate industrial economy, according to its proponents, is a degenerative linear caterpillar economy that can be transformed into a regenerative butterfly economy. According to the *butterfly economy*, one wing of the butterfly stands for the process of regeneration, which captures the value at each stage of decomposition, and the other wing for the process of restoration, which captures the mending, reusing, refurbishing, and recycling of the material inputs (Ellen McArthur Foundation, 2012; Raworth, 2017). Whether they are biological, synthetic, or metallic, the waste products from one industrial process serve as the raw materials for the subsequent one in a circular economy. Biological (soil, microbes, plants, animals, ecosystems, etc.) and synthetic–metallic (plastics, synthetics, metals, etc.) cycles can be considered of as the two groups that all materials fall under. According to Raworth (2017), the reuse and renewal cycles, which are shown in the diagram of the following figure, turn the two cycles into the butterfly's two wings since they allow materials to be utilized repeatedly (Fig. 9.2).

The circular economy diagram designed by Ellen MacArthur Foundation captures the essence of the circular economy in the above diagram, which is nicknamed as the "butterfly diagram." The diagram captures the flow of materials, nutrients, components, and products, while adding an element of financial value. The diagram consists of biological and technical components representing the flows of biological and technical materials.

From bio-ecological perspective, the best example of circular economy is the *regenerative agriculture* and *forestry* in which all material inputs and nutrients are eventually consumed and regenerated through the living Earth system. For example, we can use forests in an endless cycle by harvesting forest products at a rate no

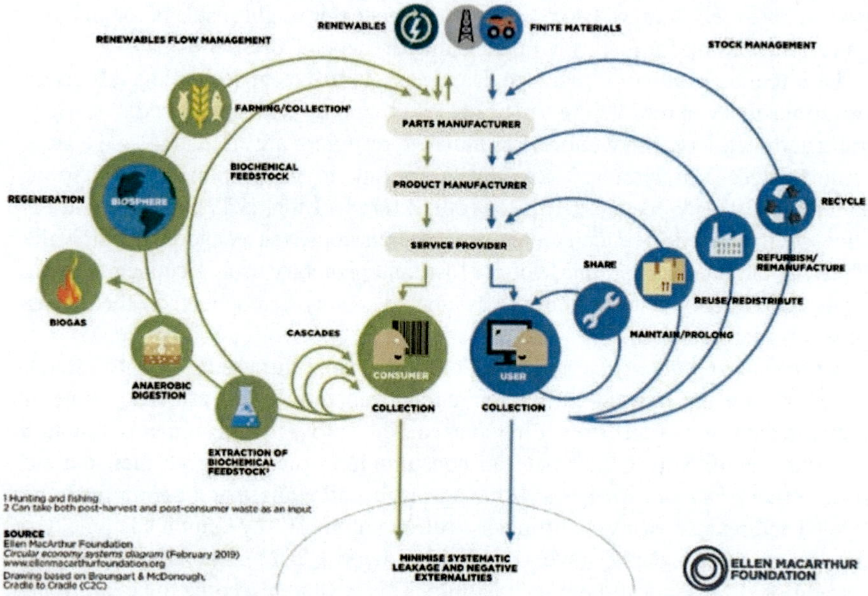

Fig. 9.2 The circular/cyclical economy system diagram. (Adapted from Ellen MacArthur Foundation, 2012). (Source: Adapted from Ellen MacArthur Foundation, Circular Economy Systems diagram, 2019. https://www.circularity.com.au/what-is-circularity.html)

faster than forests regenerate them. The rate and the amount of forest harvests must be less than the rate and the amount of forest regeneration. Similarly, take the example of coffee beans: less than 2% of every bean ends up in a cup of coffee. The leftover coffee materials/grounds are rich in cellulose, lignin, nitrogen, and sugars, and they can be the inputs of another production process, for example, coffee grounds make an ideal medium for growing mushrooms, and can be used to feed cattle, chickens, and pigs and eventually returned to the soil as organic manure for crop and food production. This is just an example of coffee to illustrate the nature's *regenerative process*, but this principle can be scaled up and applied to all foods, crops and every home, farm, firm, and institution. *Agriculture, forestry*, and *food industries* can be transformed from the current *degenerative* production and consumption processes into *regenerative* processes which can maintain the living systems on which they depend.

Similarly, the other wing of butterfly consists of synthetic and metallic products such as metals, plastics, and synthetic fibers, which do not naturally decompose but can be designed to be restored through *repairs, reuse, refurbishment*, and *recycling*. As Ellen McArthur Foundation (2012) reports, in EU 160 million mobile phones were sold in 2010 but only 6% of the used phones were reused, and 9% disassembled for recycling and the remaining 85% ended up in the landfills. This is what happens in a degenerative use and throw away economy as opposed to circular economy where a system of collections and disassembly can be designed leading to their refurbishment and resale or the reuse of all their parts. If the design principle of collection, disassembly, refurbishment, and resale or reuse is scaled up and applied to all manufacturing industries, it is possible to turn twentieth-century manufacturing industrial waste into twenty-first century manufacturing industrial inputs. Raworth warns that it is impossible to create truly *circular economy* and a more accurate name for it would be *cyclical economy* because no industrial loop can recapture and reuse 100% of its materials. For example, Japan impressively recycles 98% of metal used domestically, but there is still 2% leaking from the *cyclical loop* (Raworth, 2017). The entropy law states that there is no 100% transformation of energy or matter from one state to another, and some percentage of loss is inevitable in the process of transformation.

Whatever name one gives to it *circular* or *cyclical* should not matter even if it can reuse 75% materials. If the present degenerative economic design can be restructured or replaced by regenerative economic design that can recapture and reuse even 75–80% materials in the production process, it will not only drastically reduce the impacts of negative environmental externalities on planet Earth but also restore the lost *regenerative* and *resilient* capacity of the planet Earth. Raworth (2017) rightly points out "*Regenerative industrial design can be fully realized if it is underpinned by regenerative economic design.*" It calls for a *gestalt approach* of redefining the role of markets, state, the purpose of business, and the function of finance. Such industrial redesign process emerges only from the technological innovations, research, and experiments. This is the biggest challenge of twenty-first century, a century of transition from *degenerative industrial economy* to *regenerative industrial economy* that can function within the bounds of regenerative biocapacity of planet Earth.

The most important question that needs to be embedded into any development model or framework is the question of values. What kinds of values human society should pursue and implement to guide its own behavior in its relation to the planet Earth and planetary ecosystem? The choice of the values has become more imperative today than ever before. Given the fact that society cannot maximize everything for everybody, it must make choices. Should there be more wealth or more people; should there be more healthy and productive ecosystems and wilderness or more automobiles and jungles of concretes; should there be more food and basic need resources for the poor or more ecologically hostile consumer goods for the riches? These are all value-related questions and the societal answers to questions like these and translating those answers into policy is the essence of the political economy toward transitioning to a more sustainable society from its current unsustainable growth society.

9.5 Conclusion

The current neoliberal market capitalism's assaults on planet Earth have crossed the redline and pushed the Earth's many component systems beyond thresholds point from which it makes almost impossible to reverse. We have clearly seen the breakdown in the resilience of these systems and are spiraling into intense degradation. The most disappointing fact is that proponents and adherents of neoliberal market capitalism (the mainstream economists) and politicians still fail to see or recognize the need to change or reformulate the neoclassical tenets and assumptions to make it compatible with planet Earth systems' functioning. This is clearly a path of self-annihilation and destruction. We cannot solve environmental and climate crisis and social problems with the same mode of thinking and doing that created these very problems. The current market capitalism's economic and political *modus operandi* cannot free humanity from the impasse and challenges posed by Anthropocene *era*. We need new development paradigm rooted in the integration of the ecology of planet Earth and human political economy that can offer a way out from this unfortunate predicament. The assumption of the neoclassical "growth economics" that economy can grow forever is the biggest fallacy in the history of economic thought. Mainstream economists who promote and implement this "growth model" do not realize and recognize the fact that nothing can grow forever in the planet Earth which is a limited container by size and volume which cannot contain such exponential growth. The following statement by Kenneth Boulding (1993) succinctly reflects this absurdity: "Anyone who believes that exponential growth can go on forever in a finite world is either a madman or an economist." This is true, h*owever, not all economists but only those economists trained and falsely indoctrinated by neoliberal market capitalism.*

Hyper anthropocentrism resulted in the inflated sense of human self-importance and quest to dominate Nature. This overemphasis on human self-importance and sovereignty with neoliberal capitalism led to the instrumental manipulation of Nature. The *anthropocentric view* that humans are the measure of all things and masters of Nature inevitably gave rise to the objectification and manipulation of Nature by humans (Eckersley, 1999). What we need today is the transformation of our relation to and knowledge of Nature such that Nature would once again be regarded as possessing functional and survival values essential for the perpetuation of humans and all living system. Such transformation becomes possible with *ecological wisdom awareness* that recognizes the necessity of developing science and technology grounded in the experience of Nature as a totality of life to be protected and cultivated, and applying them to the reconstruction and healing of the ruptured Nature.

Chapter 10
Environmental Ethics, Nature Conservation, and Sustainable Development

We don't have the right to ask whether we are going to succeed or not. The only question we have a right to ask is what is the right thing to do? What does this Earth require of us if we want to continue to live on it?

Wendell Berry (2017)

10.1 Introduction

This chapter is devoted to the discourse on environmental ethics for nature conservation and sustainable development. The chapter argues why environmental ethics is necessary in the context of the development in general and dealing with environmental challenges in particular. The first part provides the critical review on why current development approach needs to be restructured and guided by pragmatic development ethics that lays the foundation of Nature protection, conservation, and sustainable development. The second part attempts to provide the metaphysical basis for environmental ethics and its implications in achieving nature protection, conservation, and sustainable development.

10.2 Development Ideologies and Ethics

Development can be identified with two interrelated but separate concepts. The dominant concept of development can be conceived as the increased production of material goods and services. This notion of development focuses on "*economic growth*" which is characterized by quantitative relation of output/input and the use of resources. The economic growth is good and must be pursued by whatever means that can make it possible. This notion of development, however, inadequate it may

G. Upreti, *Ecosociocentrism*, https://doi.org/10.1007/978-3-031-41754-2_10

be, is perhaps the most pervasive and dominating in modern society today. The measuring indices of this kind of development are purely economic growth matrix (GDP and GNP) characterized by increased production and consumption of consumer goods and services, and quantitative scale of the physical infrastructure development such as roads, building, bridges, dams, and high-rise building infrastructures. Development, on the other hand, with equal or even more emphasis can be conceived as a qualitative change in the life of people and the nature of their social relations. This notion of development focuses on "*social development*," where development is emphatically based on qualitative and distributive changes in the structures of societies. Development programs and policies initiated from this framework emphasize the elimination of discrimination and the structurally determined exploitation, creation of equal opportunities, and the more equitable distribution of the benefits of economic growth and empowerment of the local community. This concept of development is based on distributive justice and entails socioeconomic policies and programs whose ultimate objectives is to achieve a qualitative improvement in people's lives. In developing countries, a balanced economic growth with equitable distribution of economic benefits is *sin-qua-non* for bringing desirable changes in the lives of people. For this to happen, it is highly imperative that our political, social, and economic programs and policies be tailored to both wheels of the development, the economic growth, and the social development.

10.2.1 Ideology and Strategy

It has become increasingly difficult to understand what we mean by the term "development." There is plethora of literature on development ideologies, issues, and strategies, and yet, these treatises have not resulted in one unified theory of development. Perhaps, the evolution of divergent views on development ideologies and strategies has something to do with the cultural and technological evolution of human societies in ecological or environmental setting at historical time. It is logical to assume that a society develops its *ideological superstructures* on the *material infrastructures* it has built. It follows that the development ideologies and the strategies (superstructures) in society emerge from the material and the environmental conditions prevailing in the society. Development ideologies and strategies must evolve from the life conditions of people so that such development strategies can be tailored to the improvement of those life conditions of people. Improvement in the life conditions of people in a society should be the overriding goal of all development ideologies and strategies. Often, it is difficult to distinguish between economic growth and development. Traditional economists (neoliberal economists) have a strong tendency to equate economic growth with development, whereas anthropologists and social and development scientists may have different views and not necessarily equate economic growth with the development, because economic growth may not necessarily bring a qualitative improvement in the life conditions of most people. In this respect, development must be defined in terms of those human activities and human endeavors designed to bring a qualitative improvement in the life of people so that they can actualize their inherent human potentials. I argue that development must be understood in terms of the development of human potentials, and the human capacities, but not in terms of the mere growth of physical structures such as high-rise buildings, highways, and accumulation of wealth. The physical growth must not be considered as *development per se* but must be considered as means to achieve human development potential. *This means, development is not about things, but it is about the fulfillment of human potential and capacity.* Development is about the realization of people's potentialities. It may involve economic growth with distributive justice as a means for the realization of such human potentialities. Development may occur with or without economic growth, whereas economic growth may not necessarily mean development. Development ultimately reflects personal values conditioned by social paradigm in which we live. The values and the ideologies grounded in social paradigm become the basis or norms by which development is judged.

10.2.2 Bottom-Up Versus Top-Down Approach

There are some basic values associated with development approach from below. It is argued that it is a development determined from within by the people themselves based on their own physical, institutional, and human resources. This means that the development strategy adopted is unique to the society in which it evolves. Such development strategy by nature becomes egalitarian and self-reliant which emphasizes meeting the basic needs of all the members of society. The ultimate aim of any development strategy is to bring qualitative improvement in the lives of people in any society. Such development strategy entails value-based growth, distribution, employment creation, self-reliance, community empowerment, good governance, and respect for human dignity. Such development strategy facilitates and initiates the processes for greater participation of people in development activities. Any development ideology or strategy that does not seek or encourage peoples' active and direct participation in development process cannot succeed in achieving its stated goals. The fundamental objective of any development strategy or ideology should be to ensure and achieve active participation of people in development process. The failure and the success of such strategy must be judged by the degree of people's active participation in the process. In this respect, participatory inclusive development strategy becomes at the same time both a development strategy and development ideology. Participatory inclusive development strategy or ideology is a process-oriented rather than the quick fix result-oriented approach.

Bottom-up development approaches are bound to be participatory and inclusive in nature. One of the basic features of bottom-up development ideology is its inherent flexibility. It does not prescribe any rigid set of policies and ideas. Rather, it advocates and allows for the emergence of development policies and strategies from the interaction of people with their resource base, integration of local knowledge and resources. Bottom-up development ideology facilitates the emergence of appropriate policies and development strategies through the participatory process of peoples in a local context. It is natural to assume that over the time and space, such approach may receive many different responses. It regards development as a process by which the participating agents undergo a positive qualitative change that eventually results in the actualization of the potentials inherent in the participating agents. Bottom-up development approach must be based on integrated local and regional resources utilization at different spatial scales. Resources integration must be emphasized at the lowest possible scale so that the communities become organically linked with each other. Bottom-up development approaches are essential for achieving regional self-reliance without many changes in external relationships.

Centralization and decentralization in the past had been thought of being contradictory rather than complementary. Development experts argue that a synthetic approach (combining centralization and decentralization) may be essential in bringing the desired success depending on the situation. One obvious example is China which has achieved an impressive success through such synthetic approach. Bottom-up development strategies may need considerable external inputs. Central planning may be crucial in facilitating the equal access of all population to the production and consumption of goods, services, and welfare at the local or regional level. Hence, development policies must entail some distributionary assistances form central decision-making units. Though development professionals seem to advocate for bottom-up and inclusive participatory approach, the dominant mode of development is still very much experts driven top-down approach with some semblance of participation. Government's central development planning agency drives the process of formulating development policies and strategies, however, this scenario is changing.

10.2.3 Development Ethics

Development thinkers and experts have now become increasingly concerned about the meaning and the ethical values of development. Development discourses, symposium, workshops, and conferences have been organized in the national and international development arenas to get a grip on the shared meaning of development and underdevelopment or faulty development. What seems to be the development from one perspective does not seem or seem at least different form another perspective. It has also raised a serious question that the development for one class or section of people in society takes place at the expense of other class or section of people and development of one nation causes underdevelopment of other nation(s). Many may argue that underdevelopment of developing world is basically the consequence of the development of Western developed countries at the cost of developing world. The direct colonialism of the past has been replaced by the neocolonialism of the present. The developed countries want to keep developing world's development within the sphere of their influence and control so that they could direct the development in developing world to serve their interests ultimately.

It becomes quite apparent when one examines the roles played by the international financial and trade agencies such as World Bank (WB), International Monetary Fund (IMF), and World Trade Organization (WTO) in developing world. It is interesting to note that these agencies under the guise of development assistance compel developing countries to take loans, invest the loan money in a manner that directly serve the interest of the Western countries, and open up the market for the finished products of developed world. Many disinterested development experts believe that WTO, IMF, and WB exclusively serve the interest of the Western world and their roles lie in keeping underdeveloped and developing countries dependent on the developed world. The ultimate goal of these international development agencies is to create a consumer class in developing world for their manufactured products, because Western corporate capitalism can only survive on the ever-increasing consumption which requires the ever-increasing markets for such consumption. Hence, naturally it follows that for the Western countries, the notion of development in developing countries is, by and large, the creation of consumer classes and the markets for Western manufactured products with some negligible trickle-down effects to the people. Western countries do not want to see developing countries to produce and market their own manufactured goods not because they could compete the Western products in the international markets but because they severely limit the access of Western goods to their domestic markets in developing world. A considerable number of developing countries like China, India, Malaysia, and Vietnam have rapidly developed their economies and their products substantially captured the international markets outcompeting Western products. The seizure of the international and domestic markets by their products have posed serious threats to the Western world particularly United States. They have done it by the play book of World Trade Organization (WTO) and IMF, and the trade wars between United States and the countries like China, India, and Malaysia can be seen as the result of the conflict arising from the control of the international and domestic markets for their products. Once these institutions cannot function effectively to protect the economic interest of the United States and Western world by containing the emerging rival economies like China, India, and others in the international arena, they would probably be dysfunctional and eliminated. Their continued existence depends on inventing new tools and techniques in expanding the financial interest and market control of US led conglomerates.

This international and national development predicament poses a serious question on the ethical aspect of the development. What kind of development is to be understood by the term development? Within what kind of ethical paradigm, the development of a nation be carried out? What is development ethics if there is such a thing called development ethics? Development thinkers believe that *development ethics is a normative assessment of the ends and means of development. Should the concept of development be ethically positive, negative, or neutral and should the benefits and burdens of development be equitably distributed? Should developed nations or regions take a moral responsibility for impoverished nations, regions, and classes of people?* These questions have ethical implications and certainly deserve the attention of political leaders, development thinkers of the world, and those who advocate the necessity of sustainable development.

10.2.4 Reductionism and Environmentalism

It is important to realize how the *reductionistic* mode of thinking in science and development practices affects our understanding of the environmental problems, in general, and the development practices, in particular. We need a conceptual framework that clarifies the distinction between reductionistic and system thinking (integrative thinking) and provides guidelines for developing system approach or holistic thinking. The reductionistic approach in development promoted unhealthy competition and desire to dominate Nature and control resources instead of promoting sustainable use of resources, conservation and protection of ecosystem processes, cooperation, equity, and social justice. This approach has been, both, ultimately negatively anthropocentric and negatively ecosystemic. It is negatively anthropocentric because the consequences of this approach have the potentials to do far greater harms than goods to humankind itself. It is negatively ecosystemic, since it does not recognize the fact that everything is interconnected with everything in Nature, existence of all lifeforms including humankind depend directly on health and the integrity of the Earth's planetary ecosystem (Upreti, 1994). The prevailing worldview is a major source of today's human and environmental problems, and it must be replaced by a new ethical paradigm that nurtures the notion of *interconnectedness, interdependence, diversity, health*, and the *integrity* of *ecosphere* in which each species including *Homo sapiens* is a member of a highly interactive community of interdependent systems in the *biosphere* and each species has its own role and value. For development professionals, the primary goal and the major challenge is to develop an appropriate framework of development ethics from which to design and pursue development strategies with the objectives of achieving *environmental* and *social sustainability. Environmental ethics, that embraces the ecological principles of diversity, autopoiesis (resilience), interconnectedness and interdependence, coevolution, and symbiosis, provides a basis for sustainable development.* From such an understanding, an analytical and explanatory framework for appropriate values, beliefs, worldviews, and development strategies can be worked out. It may certainly sound absurd to the mechanistic and reductionistic mode of thinking that is dominant in the modern development arena, but holistic or integrative thinking that derives its sustenance from system thinking and social ecology is the hope for mankind's future. The task is to reestablish mankind's relationship with Nature, and this cannot be done without rupturing the *Cartesian mechanistic reductionism* and replacing it with a paradigm grounded in the normative philosophy that values environment as a very basis (the source and the sustenance) for the survival of all living things including humankind (Capra, 1996). If the development professionals and the politicians tinker with *environmental ethics* and derive some wisdom on the subject that has great implications for Earth's planetary ecosystem, they would significantly contribute to the making of a better world, a world in which both *sociosphere* (human sphere) and ecosphere/biosphere can coexist, coevolve, and flourish together. This is what the wisdom tradition of Buddhism (Eco-Dharma), the Gaia principle, and System Theory teach us. I believe, this is the most crucial time in human history to revisit these ideas and concepts and derive the wisdoms from them to guide and reorient our *consciousness* and *modus operandi* to avoid humanity's existential threats.

10.2.5 Critique of Deep Ecology

The main difference in the position of deep and shallow ecology revolves around their value theory particularly in the treatment of Nature. The "*sole value assumption*," the assumption that only human beings have moral standing and that only the good or the interests of humans constitutes the *intrinsic value*, is the bone of contention between deep and shallow ecology view. Deep ecology rejects this, while shallow ecology accepts this. Holders of shallow ecology view advocate the conservation of resources, habitats, and ecosystems solely for the sake of human interests and uses, whether that is for present or future generation of human beings. This can be called anthropocentric view in the sense that it is concerned with the welfare and goods of human beings from a long-term perspective of human welfare and interest at the expense of Nature and other beings in it. Contrary to this view, proponents of deep ecology argue that human and nonhuman life-forms have the same natural rights to exist in Nature and both have *intrinsic values* (Naess, 1973; Fox, 2006). The proponents of shallow ecology or anthropocentrism have been well placed in the position of governments and public institutions from where they design and execute instrumental value policy to conserve Nature and other things in it.

Fox (2006) professes that deep ecology is neither ethical nor axiological but metaphysical in its emphasis, and that view of deep ecology about the value of nonhuman Nature expresses the characteristic attitude of those who go along with the metaphysical teaching. His concept of "*transpersonal ecology*" is about identification with wider world that goes well beyond our normal range of personally based identifications. Fox (1990) describes three basic forms of identification, which he referred to as "*personally based identification*," "*ontologically based identification*," and "*cosmologically based identification*." These forms of identification originate from the felt commonality with other entities that are brought about through personal involvement with them; through a deep-seated realization that we and all other entities are aspects of a single unfolding reality. According to Fox, the central tenets of deep ecology include a *total-field* or *relational view of reality*, facilitating in ourselves the *ecological consciousness* and identification with nonhuman Nature, an enlarged conception of self and emphasis on self-realization. This enlarged conception of self requires that any diminution of natural entities amounts to the diminution of oneself. Fox (2006) has summarized the main tenets of deep ecology in the statement that "*The ideal state of being is one that sustains the widest and deepest possible identification and, hence, sense of self.*"

One of the fundamental tenets in Arne Naess's (1973) original statement of deep ecology is *biospheric egalitarianism* which advocates a belief in the equal right of everything alive to blossom and flourish. This tenet has become highly vulnerable in the sense that even if each life-form has intrinsic value, it does not follow that all have equal intrinsic value. One of the main weaknesses of deep ecology is the assumption of *absolute value system* which gives equal treatment to the well-being of human beings and to that of nonhuman life-forms. This position may be theoretically possible but practically impossible to carry out. Attfield (1993) presents an example to illustrate this fact: "*Thus, granted this principle, and other things being equal, when water is scarce, and the small quantity available can be given either to a dying human or to a dying plant, it is indifferent to which it should be given, as there is no further reason to take account. Again, if the death of a human would result in a greater number of flourishing lives (e.g., maggots) than the number of lives which the same human being sustains when alive (e.g., intestinal flora), then it would be better for the human being to die. But no system, I maintain, which can yield these judgements can constitute an operative and defensible morality.*" I think value absolutism is not only difficult but also impossible to realize in real human behavioral context; hence, *value relativism* in place of *value absolutism* would be a more tenable moral endeavor because it is a more saleable ethical hypothesis to say that all life-forms are intrinsically valuable, but human beings are more intrinsically valuable than other life-forms. The implication of *value relativism* would be preservation of all life-forms whenever and wherever is possible and at least conservation of diverse nonhuman life-forms in a situation where human needs override the needs of nonhuman life-forms.

The environmental and deep ecology movement that originated in response to the environmental destruction caused by egocentric anthropocentrism gave rise to radical ecocentric or biocentric views known as *ecocentrism*. The biggest challenge of ecocentric/biocentric argument is to come up with a rational explanation that can confer intrinsic values to other life-forms (biotic community), wilderness, natural processes, and Earth's ecosystems. To establish a rational foundation for the intrinsic values of other living creatures, natural processes, and planetary ecosystems is by no means a Herculean task. One can argue that every living creature has an innate character attribute evolved to protect itself from the adverse environment. There is some limited awareness we can find in other animals like monkeys and higher primates like chimpanzees. They have limited ability to learn and imitate. We cannot deny that higher primates and dolphins and other creatures are not completely unaware of their environment. Griffin (1976) along the line of Darwin provides evidence for the idea that there might be a continuum between what we call consciousness and various manifestations of behavior in animals. The question of what motivates other animals to love, learn, or socialize compared to human is certainly difficult to answer, but one can ask a reasonable question, that is, is it possible to do so without having some level of mental awareness or consciousness? This gives us the basis to attribute intrinsic values to them because in some limited sense, they are aware of their existence. John Bonner (1980), the socio-biologist, argues that there are clearly some attributes in the animals that are anthropomorphic resembling human attributes and those animals having those attributes can have intrinsic values. The following excerpts from his book "The Evolution of Culture in Animals" worth pondering: "*The existence of anthropomorphism is a problem to which there is no solution. Those interested in similarities between man and animal have no fear of anthropomorphism, while those who see man as special in some way feel that our whole man-oriented language is dangerous and misleading when applied to animals.*"

Similarly, the natural creative process known as *autopoietic process* that makes it possible for the evolution of such attributes in living beings can also be thought to have intrinsic values. The adherents of the ecocentrism/biocentrism have put forward the following arguments: It is argued that the relativity theory and quantum physics has undermined the Cartesian and Newtonian paradigm. As Fritjof Capra (1999) articulates to that effect: "*The first three decades of our century changed the whole situation in physics radically. Two separate developments—that of relativity theory and of atomic physics—shattered all the principal concept of the Newtonian worldview: the notion of space and time, the elementary solid particles, the strictly casual nature of physical phenomena, and the ideal of an objective description of nature, none of these concepts could be extended to the new domains into which physics was now penetrating.*" Quantum theory implies that while observing the phenomena at subatomic level, the very process we observe the phenomena affects the character of what is being observed. Capra (1999) further explains: "*The subatomic units of matter are very abstract entities which have a dual aspect. Depending on how we look at them, they appear sometimes as particles, sometimes as waves; and this dual nature is also exhibited by light which can take the form of electromagnetic waves or of particles.*" This apparent contradiction raised the question about the nature of reality. Capra argues that the subatomic particles are not solid objects existing at a particular space at a particular time but a wave-like pattern of probabilities which can only be understood as manifestation of interconnected phenomena between experiment and the subsequent measurements. The observer plays an essential role in determining the state of their reality.

Matthews (1989) argues that even if the human observation interferes with determining the properties of what is observed, it does not follow that what is observed does not exist independently of the human observer. Though Cartesian dualism (separation of mind and body) makes it difficult to account for the interaction of mind and matter (brain and mind) for which it has been criticized in the contemporary philosophy of science, the anthropocentric values readily embrace the Cartesian metaphysics. A certain level of dualism between mental and physical may be unavoidable. One should not conclude that anthropocentrism stands or falls with a dualistic ontology. Even if we reject the Cartesian dualism, we cannot deny the fact that humans are the only creatures on this planet who are capable of self-awareness, reflection, and consciously pursuing purposes and values which make them uniquely distinct in Nature as much as they are the part of Nature. Matthews (1989) succinctly articulates this line of thought: "*We can accept all the philosophical criticism of Cartesianism but still regard human beings as occupying a special place in natural order in virtue of their self-consciousness. It is this self-consciousness, whatever the kind of substance in which it is to be located, which provides the philosophical basis, as I shall argue, for the claim that human beings, uniquely amongst all creatures we know of, possess intrinsic value, and that it is by reference to their interests and purposes alone that other things come to have value.*" It is true that human beings are part of the Nature and are subject to its laws. One can argue that human beings are not simply a part of Nature in the same sense in which other animals, plants, and inanimate objects are. Humans are different in the sense that they can consciously reflect upon their environment which is a precondition for the understanding of that environment. From the understanding of their environment, they can discover various alternatives from which they can choose the one that is good and desirable for them. Therefore, only humans can pursue the conscious purpose to achieve what is good and desirable for them. By the same token, it can be said that only human beings can have an ethics because of their capacity of the *conscious reflection* on what is good and desirable for them and others.

10.3 Environmental Ethics

The interface between environment and normative philosophy is an area that needs rational consideration from the perspective of maintaining balance between nature conservation and human conduct. Environmental ethics works in the interface between environmental space and the normative philosophy of existential life phenomenon. The environmental problems that have resulted from destruction of biological diversity and the vital life support system of the ecosphere/biosphere cannot be overemphasized. Environmental problems have not only generated tremendous awareness and affected the outlooks and worldviews of large number of people in society but also affected scientists and philosophers and their worldviews likewise in academia. *Environmental ethics* is essentially a normative response by humans to a range of environmental problems which collectively make up today's environmental crisis. Humans are largely responsible for the damage to the environment, but what makes it wrong is not only too much of *anthropocentric interest*, deeply rooted in human self-aggrandizement which sequesters natural resources, but also the wanton destruction of Nature itself, the living things such as species, ecosystems, and wilderness which not only have instrumental values but also have existential and some intrinsic values. It is important to realize that environmental destruction is an immoral act because it undermines the security and survival of all life-forms including human beings. It is like putting our own house on fire.

Moral values and cognitive beliefs of a culture play a significant role in human adaptation to the natural environment. The constellation of such belief systems also determines what kind of political and economic systems societies want to create and maintain. The root causes of today's environmental problems lie in contemporary dominant belief structures which organize the way people perceive and interpret life phenomenon and the way the world functions. Perhaps, the first task ethics should perform is to understand and evaluate the moral codes, dominant assumption, and value systems woven into the culture and worldview of the society particularly what they are and how they function to enhance or distort the relationships of human beings to one another and to Nature.

Perhaps, it is in this context that the role and significance of normative philosophy becomes very crucial in providing coherent, integrated view of human knowledge systems that form the basis not only for the perceptual and interpretative framework for worldly phenomena but also for the moral framework for development practices and human behavioral conduct. It does not content itself with the appearance of a thing, it goes much deeper into the matter and examines the origin, nature, meaning, the unity, interconnectedness and relationships among things, their inherent contradictions, and finally, it seeks humankind's role and place in the scheme of things. It takes into accounts of scientific facts and provides value judgment for interpretation of those facts.

10.3.1 Why Environmental Ethics

The theistic religions, particularly Western forms of religion, have been criticized for fostering an egocentric *anthropocentric* attitude to Nature with despotic and domineering approach. It is widely accepted that the roots of current environmental problems lie in the Western Judeo-Christian fundamental belief system, whereas Eastern religious and wisdom traditions including Confucianism, Daoism, and Buddhism have significantly contributed to the care of Nature and environmental system. White (1967) thinks the root cause of the current environmental crisis lies in the Christian dogma of creation in which God created men in his own image and has commanded men to rule over and be the master of his creation. The following excerpts from his article published in Science (1967) worth pondering: "*What we do about ecology depends on our ideas of the man-nature relationship. Our science and technology have grown out of Christian attitudes toward man's relation to nature which are almost universally held not only by Christians and neo-Christians but also by those who fondly regard themselves as post-Christians. Christian dogma of man's transcendence of, and rightful master over, Nature. Man named all the animals, thus establishing his dominance over them. God planned all of this explicitly for man's benefit and rule: no item in the physical creation had any purpose except to serve man's purposes.*" There is no doubt that today's *egocentric anthropocentrism* originated from Western Judeo-Christian tradition and neoclassical capitalism emerged as the faithful carrier of this worldview. There is no other religious tradition that has done more harm to Nature than Judeo-Christianity; however, the current attempt to reform Judeo-Christianity's classical views of Nature should be appreciated and welcomed.

Present human activities can significantly change the future which undeniably may have important consequences and implications to future generation; therefore, present generation has *moral obligation* to use Nature's resources in a way that does not discount future and compromise the ability of future generation to meet their needs. Notwithstanding the fact that future generation and their interests matter, it is widely accepted fact that their interests count far less than the current interests. In other words, the needs and the interests of the present generation always prevail over the interests and the needs of the future generation. There is an insurmountable challenge in striking a balance between needs of present and future generations. Environmental ethics considers what ought or ought not to be done and how that to be done or decided should be based on the principles of right action that emerges from our understanding of ecological values and moral principles. It draws the lesions from different theories of ethics that fail to ensure equity between generations and between species and attempts to come up with a *pragmatic ethical system* that is more consistent, most fruitful, and best serves the needs of future generations and other living entities in Nature.

Climate change (global warming) is existential challenging ecological/environmental phenomenon. Vast majority of scientists have warned that global warming is "*anthropogenic*" as reflected in the reports of Intergovernmental Panel on Climate Change (IPCC 2007, 2015). From the perspective of *environmental ethics*, this presents a strong case for vigorous and concerted action to mitigate the impacts of climate change and to adapt to its effects. Environmental ethics can shape the policies of mitigation, adaptation, and compensation along with the different defensible approaches to these issues, including the 2015 Paris Climate Agreement, reducing greenhouse gas (and other) emissions, and climate engineering. Sustainable development, as defined by Brundtland Report (1987), refers to the development that "*meets the needs of the present without compromising the ability of future generations to meet their own needs.*" The concept envisaged the social, ecological, and economic dimensions of sustainable development and recognizes the needs of future generations, but its recommended policies and strategies based on neoclassical/neoliberal growth economic model cannot achieve the goals of "sustainable development." The Millennium Development Goals set in 2000 and the Sustainable Development Goals set in 2015 added significant policy imperatives and programmatic action plans toward achieving those sustainable development goals.

The impacts of human activities on environment became very visible since twentieth century. Aldo Leopold's *A Sand County Almanac* (1949) and Rachael Carson's *Silent Spring* (1962) raised the environmental awareness which provided the impetus in the rise of ecological science and the emergence of environmental ethics, particularly with the work of philosophers Richard Routley, Arne Naess, and Holmes Rolston III (Attfield, 2018). These philosophers rejected the notion of only anthropocentric instrumental value approach to ethics and argued that there is ample reason to consider the non-anthropocentric (intrinsic) approach to ethics. There are some key concepts that are pivotal to environmental ethics. The most important one being the concept of Nature and the ways in which human beings relate their behavior and attitudes to Nature that surround us and how they relate Nature to their inner nature. Are human beings separate from Nature or simply part of Nature? How the concept of environment relates to Nature? As Attfield (2018) points out "*Only through the concept of the environment as an objective natural system can we make sense of environmental problems in the first place.*" Perhaps, the most important key concept considered in environmental ethics is the moral standing and value of the other living entities and system in Nature which raised the ethics of biocentrism or ecocentrism.

Environmental ethics aims to navigate the relationship between the natural world and human existence. The pressing issue of climate change resulting from the destruction of planetary ecosystems and increased greenhouse gas emissions cannot be overlooked. Despite these clearly visible dangerous impacts, environmental problems have not received ample attention from politicians, policy-makers, economists, executives, philosophers, and scientists. Environmental ethics is a response to the range of environmental problems, including climate change, that collectively make up today's environmental crisis. It is widely acknowledged and recognized that human impacts are largely responsible for environmental destruction due to excessive anthropocentric interests rooted in greed rather than the satisfaction of essential human needs. The wanton destruction of planetary ecosystems, which generate life-sustaining environmental services, undermines the security and survival of all life-forms, including humans. *It is important to realize that human-induced environmental destruction is an immoral act because it undermines the security and survival of all life-forms including human beings.* The question then becomes how long will it take for those who control the political, economic, financial, and corporate power centers and institutions to realize and change their mindsets?

The ecosphere and the sociosphere are in constant interaction that mutually and reciprocally causes changes in each other. This interaction evolves into a new state. Human beings are both subject (actors) and object (part) of this *dialectical* interactive process of social and natural dynamism in which they have to constantly recreate, redefine, and reestablish themselves with the changing social and natural systems (contexts). This *dialectic* gives rise to a new relational state where *dominant social paradigm* becomes no longer functionally relevant and lack explanatory power. Thus, the *dominant paradigm* needs to be replaced by a new and relevant, powerful, and adaptive *paradigm* that provides better perceptive and interpretative framework. When this happens, a *paradigm shift* is said to have occurred (Khun, 1967). Consequently, a worldview evolves with new values, beliefs to guide human understanding, actions, and behavior.

Evolutionary biology, ecology, political economy, and normative philosophy have always attracted mankind's intellectual motivation and curiosity, but the difficulty they have faced is in the development of practical strategies to effect changes in human behavior in the real-life circumstances of people. The biggest challenge for political economists, social scientists, philosophers, system ecologists, environmentalists, and development practitioners lies in the design of appropriate development ethical framework in the interface area between planetary Earth system (ecosphere) and human society (sociosphere). Such development ethical framework cannot evolve until the scientific epistemology is not integrated with normative philosophy to resolve environmental and social problems. This provides the axiological basis for *environmental ethics* on which development strategies and policies can be construed and constructed, and human behavioral conduct toward Nature can be guided and changed.

It is well recognized that there is something radically wrong with the planning and the decision-making processes in the development endeavor of nation-states. This is generally attributed to conceptual inadequacies among participating professional groups and politicians who control the development processes, political and economic institutions, and corporate power centers. The worldview within which the professional groups (scientists and development practitioners) are trained is said to be largely responsible for this. The *paradigm* involved has been variously labeled as *positivist, reductionist, compartmentalized, mechanistic*, and so on. The fundamental characteristic of this paradigm is very narrow vision that reflects inability and unwillingness to look beyond its narrow specialization and sphere of responsibility. Critics of this paradigm argue that professionals of all kinds need to be trained in a new *cognitive paradigm* which puts more emphasis on *synthesis, participation, interdependence*, and *interconnectedness* rather than on traditional reductionistic approach and analysis (Upreti, 1996; Capra, 1996; Attfield, 2018). This new paradigm is often called *holistic* or *integrative thinking* or *system thinking* (Capra, 1999). The development of *system approach* and multidisciplinary research programs has propelled the changes toward integrative thinking; however, this is still far from becoming the dominant mode of thought in sciences and development practices.

It is important to look into how the *reductionistic* mode of thinking in science and development practices affects our understanding of the environmental problems in general and development practices in particular. We need a conceptual framework that clarifies the distinction between reductionistic and integrative thinking and provides guidelines for developing holistic thinking. The reductionistic or mechanistic approach has been criticized for promoting unhealthy competition and desire to dominate Nature, extract, and control resources instead of promoting sustainable use of resources, cooperation, equity, and social justice. This approach has been, both, *negatively anthropocentric* and *negatively ecosystemic*. It is negatively anthropocentric because the consequences of this approach have done more harm than good. It is negatively ecosystemic because it does not recognize the fact that existence of all life-forms, including humankind, depend directly upon the health and integrity of planetary ecosystem. This prevailing worldview propelled by corporate market capitalism, is the major cause of current social and environmental problems, and it must be replaced by a new paradigm with ethical view that nurtures the notion of diversity, integrity and resilience of ecosystem, interconnectedness, autopoiesis, and ecosystem processes in which each species including *Homo sapiens* is an integral member of the community of interdependent systems in the biosphere with each species having its own *existential* and *intrinsic value*.

Perhaps, it is in this context that the role and significance of scientific epistemology and normative philosophy become very crucial to provide a coherent, integrated view of human knowledge systems and values that form the basis for not only the perceptual and interpretative framework for social and natural phenomena but also the moral framework for development practices and changing human behavioral conduct. Such interpretative and ethical framework should not content itself with the existential forms of things, and it should go much deeper into the matter and examine the origin, nature, meaning, the unity, and interrelationships and interconnectedness among things and their inherent contradictions, and finally, it must seek humankind's role and place in the scheme of things. It is based on the assumption that scientific facts and values cannot be treated as entirely separate entities but rather complimentary to each other and must be integrated for the interpretation of social and natural phenomena (Capra, 1996). *Environmental ethics*, the growing normative discipline within philosophy, not only offers a great opportunity to inspire and sustain environmental movements growing all over the world but also has a great potential to provide *pragmatic* human behavioral norms to guide human activities and conducts in the process of mankind's struggle for survival in their daily existential life situation.

Scholars and philosophers (Katz, 1992; Rolston, 1986; Soule, 1985; Sterba, 2001; Johnson, 1992; Callicott, 2006; Weston, 1985) think species and ecosystems have subjective or objective intrinsic values (e.g., for their organizational complexity, diversity, spiritual significance, wilderness, beauty, or wondrousness). Katz (1992) and Rolston (1986) are of the view that natural entities, including species and some ecosystems, have *intrinsic values* by virtue of their independence from human design and control and their connection to human-independent evolutionary processes (Rolston, 1986). This concept of intrinsic value finds its full expression in Soule's following postulate: "*Species have value in themselves, a value neither conferred nor revocable, but springing from a species' long evolutionary heritage and potential.*" Bryan Norton's axiom of ecosystem health concept provides an integrated view of instrumental and intrinsic values when he says "*An ecological system is healthy only when its creative processes, represented by the free flow of energy and active competition to utilize it, remain intact. Unhealthy ecological systems will be characterized by a tendency to undergo rapid change, change such as the rapid disintegration of complexity and integrity.*" The self-creative autopoietic process of living organisms including that of ecosystem can be considered an intrinsic value for itself because this is the process that determines health, its proper functioning, and the existential vale (being) of the ecosystem. The existential and evolutionary processes of living entities in nature need to be recognized as intrinsic values and their destruction by anthropogenic activities ought to be prevented and minimized to the maximum extent possible.

All living organisms have a good of their own but one can argue that which organisms' good or interests human beings ought to care about. Some environmental ethicists (Sterba, 2001; Johnson, 1992) have argued that species and ecosystems have a good of their own (their inherent worth) and that their good needs to be considered. The difficulty with this view is that it is not clear what is there that could be considered the species' or ecosystem's good above and beyond (or distinct from) the good of the individual organisms that comprise them? If we consider nonhuman organisms, species, and ecosystems possess only instrumental values, their values and by extension the conservation and management goals they justify are highly contingent, defeasible, and unstable. They can be treated as comparable to, and substitutable by, other instrumental values, but on the contrary, intrinsic value of an entity is not substitutable or replaceable (Callicott, 2006). If nonhuman organisms, species, or ecosystems have (subjective or objective) intrinsic values, their values are not dependent on other alternative means (e.g., economic or medicinal), and they cannot be traded or substituted for without loss. *For example, can we substitute or replace an elephant by any other animal? Likewise, can we substitute or replace Amazon or tropical ecosystems by anything else?*

For development thinkers, professionals, and the scientists, the major challenge is to develop an appropriate framework of development that can integrate both *science-based facts* and *ethics* from which it becomes possible to design and pursue human development strategies with the objectives of achieving *environmental* and *social sustainability*. This requires a profound knowledge and understanding of how human social and natural systems interact with each other and how such interactions change them and determine the nature of their relationship over time. The scientific epistemology that integrates environmental ethics embraces ecological principles of *diversity, ecosystem stability, resilience* and *health, interconnectedness, self-organizing complexity,* and *life-supporting environmental services* provides basis for social and environmental sustainability. It may sound absurd to the mechanistic or reductionistic mode of thinking which is the dominant mode of thinking in the current development practices, but this holistic epistemological system thinking that derives its sustenance from system ecology and social ecology is the hope for mankind's future on planet Earth. The task is to reestablish mankind's ruptured relationship with Nature, and this cannot be done without rupturing the mechanistic reductionism and replacing it with a paradigm grounded in holistic epistemology (system theory) that can integrate facts and values that treat planetary ecosystem as the foundational basis for the survival and flourishing of all the living things including humankind. If politicians, political economists, global corporate elites, and development professionals do little thoughtful tinkering with *environmental ethics* and recognize its implications for planet Earth systems, they would contribute to the making of a better world, a world in which humankind, Nature, and the biotic community can coexist and flourish together. *What we need today is a pragmatic environmental ethics embedded in the development strategy that envisages an environmentally and socially sustainable global society with the recognition of intergenerational equity and if not full, limited intrinsic value of living system on planet Earth.*

10.4 Metaphysical Basis for Intrinsic Values

The term biodiversity has been defined and used in a variety of ways by the scientific community and by a multitude of decision-makers. The Convention on Biological Diversity (CBD) states that (CBD, Article 2 Convention Text, 1992): "*Biological diversity means the variability among living organisms from all sources including, inter alia, terrestrial, marine and other aquatic ecosystems and the ecological complexes of which they are part; this includes diversity within species, between species and of ecosystems. Biodiversity is the variety of life at all levels from genes, through populations, communities and species to ecosystems and biomes. It includes not just the diversity of the biological components, but also their structures, functions, and the interactions between them.*" This definition is broad and inclusive. MA and IPBES conceptual framework should fully embody this definition. This definition captures the essence of the famous "land ethics" of Aldo Leopold (1949) who, in his famous essays, posthumously published as A Sand County Almanac, argued for a holistic approach: *A thing is right when it tends to preserve the integrity, stability, and of the biotic community. It is wrong when it tends otherwise.*

It is very difficult to argue for a non-anthropocentric value system to appreciate and respect natural world (Planetary Ecosystem) from the perspective of human subjectivity. It is logical to assume that humans assign values to natural world depending on how they perceive and interpret natural phenomena. Human knowledge and understanding about Nature and natural phenomena will certainly impact, modify, and change human perception and interpretation of Nature. Human knowledge and understanding about natural and human cultural world are evolving and so are the scientific and social paradigms of perception and interpretations, value system, and the moral and ethical precepts. Everything, right from material (physical and biological) world to our thought process (consciousness) is evolving and emerging. Everything, including humans, is in the process of evolving and becoming. If we look into the historical process of our own becoming, we, humans, find to our own surprise, how amazingly connected we were to the most primitive unicellular life-form, from which multicellular life evolved and as a result of billions of years of this process, we became what we are today and still in the process of becoming (evolving), never knowing exactly what to become except the fact that we are ever evolving and becoming. What is the nature of this *evolutionary becoming* which is so pervasive and encompassing? Can this process of becoming be viewed as having a value in-itself (intrinsic value)? Can this process exist for its own shake? Can this exist independently of the subjectivity of human consciousness? If we can answer the first question, perhaps we can at least try to answer the subsequent questions.

One of the most troubling questions that is often raised in the criticism of environmental ethics is the question of *intrinsic value of biotic community* or the *natural system*. Intrinsic values are those values that exist for their own fulfillment and actualization, whereas instrumental values are those values that can be used as a means or instrument for the fulfillment and actualization of human values. Thompson argues that the basic requirements for an environmental ethics are of two (Thompson, 1990): *first, establishing reasons that natural entities and/or states or system are intrinsically valuable.* Once we establish that things and states (system) have values for what they are in themselves, then they become the object of our moral consideration. Second, the principles or guidelines for human actions with respect to what is to be done to promote, protect, or bring into existence that which is of intrinsic value. If these requirements are not met by environmental ethics, as Thompson argues, then environmental ethics cannot be called as ethics at all.

I think establishing some intrinsic value of nonhuman biotic community or ecosystem is the most fundamental requirement for an ethics to be called environmental ethics. Once this principle is postulated, the second part becomes theoretically tenable but practically difficult because it deals with the means to achieve ends. What follows subsequently, then, is an attempt to derive metaphysical basis for environmental ethics.

10.4.1 Nature of Being

All life-forms are evolving from their previous mode of being and are in the process of becoming. Becoming is the process of *self-actualization* of intrinsically inherent qualities. In the process of evolving (becoming), a thing acquires new properties that make it different from its previous being. Becoming is a natural historical phenomenon which cannot be reversed. Becoming is not only irreversible but also directional, meaning that life, the first primitive unicellular organism, evolved from nonliving precursors, unicellular organism evolved into multicellular and simple multicellular organism evolved into more complex multicellular organisms and finally evolved the most complex of being, the *Homo Sapiens* from the primate world. In other words, *chemogenesis* (evolution of the precursors of life) set the stage for *biogenesis* (evolution of life from nonliving precursors) and biogenesis in turn set the stage for *sociogenesis* (evolution of human consciousness and culture). Human social evolution can be seen simply as consequence of human consciousness. Hence, the process is directional proceeding from a simple to a more complex being or from a state of simplicity to a state of complexity. Perhaps, complexity can be regarded as the nature of the process of becoming. Complexity, though it is again a human subjective judgment, indeed, appears to be the nature of becoming. We must explain evolution of complexity or the evolution of biological system and provide mechanism that drives a system from simplicity to complexity in order to be able to arrive at some conclusion that might be helpful to answer the question posed above.

Biological systems have remarkable self-organizational properties called (autopoiesis). This can be seen right from cell to organisms to community to ecosystem. Schrodinger postulated that living systems create and maintain exquisite order from disordered elements dominated by carbon (Schrodinger, 1994). He introduced the principle of "*non-equilibrium thermodynamics*" which states that organisms create the state of order by importing free energy from outside and process it to generate a lower entropy state within. Hence, they thrive on *negentropy*. Living systems, cells, organisms, ecosystems, and human civilization all evolved and achieved the present state of complexity through this process of the utilization of free energy imported from outside. Strictly speaking, evolution of living system and ecosystem can be regarded as the phenomenology of bioenergetics.

Hence, succession or ecosystem evolution can be seen as an orderly process, resulting from the modification of the previous forms and environment, and culminating in a more complex stabilized system with maximum biomass per unit of energy flow. Ecosystems are autonomous energy dissipative structures, just as organisms are. Many system ecologists argue that ecosystems, organisms, as well as planetary ecosystem (Earth) itself show the same phenomenological behavior when viewing them thermodynamically as developing systems.

Ecosystem evolution or the evolution of Nature is actually an organizational ordering process, proceeding from simplicity to complexity, but unlike the organization of a single plant or animal which persists for only a generation, ecosystem organization extends over many generations, and the final ecosystem may persist thousands of years.

Ecologically fundamental and the most important variable in this *organizational process* is the diversity. As defined above, diversity refers to a system of highly interacting diverse species, organisms, and their population living in a given environment. This is the basic material basis for the evolution of complex organization. The creative capacity of the ecosystem is tightly coupled with the biological diversity available in the system. The higher the diversity within an ecosystem, the greater is the creative and organizational capacity of the system and consequently greater is the complexity and the stability of the system with their emergent properties. Stability is synonymous with homeostasis, which is maintained by synergistic interaction and the negative feedback loops in the biological system. Hence, it can be said that diversity increases the complexity and the stability of the ecosystem, though many ecologists trained in reductionistic paradigm do not hold this view.

Ecosystem evolution normally proceeds from a simpler to a more complex system; nevertheless, stability does not mean ecosystems are static. Ecosystem stability is a relative concept with respect to spatial and temporal scale. With considerable certainty, we can say that there is no system that can remain stable without bringing certain changes in the system. From an evolutionary perspective, ecosystems, however complex they may be, are in constant flux. They change, organize, and reorganize in a stable way maintaining their structural and functional integrity. Ecosystems are dynamic. But if there is an abrupt change of great magnitude that system cannot handle such change, then it becomes chaotic because structural components cannot function smoothly and maintain the matter and the energy recycling within the system. This is comparable to the anarchy that results from the collapse of social system during civil war or revolution in which magnitude of change is so much that the system cannot maintain its structural and functional integrity. For some time, there is no system of coordinated behavior, and the system is bound to be *chaotic* and *unpredictable*; nevertheless, finally from such chaos is expected to emerge a stable, and smoothly functioning social system, though there is no guarantee that it will be the case.

Does this have any relevance to the evolution of human biological and cultural system? There is a parallel pattern in human biological and cultural evolution. Eugenics and physical anthropology have established that enormous development of the cerebrum led to the evolution of mankind and through cultural evolution to a new quality of life. Perhaps, human is the only organism capable of *self-consciousness*; however, possibility of some level of awareness cannot be ruled out in other member of hominid group and primates. The biological evolution of human brain may be one of the most irrefutable and convincing case of the process of becoming, marveling on the diversity of neuronal network leading toward complexity. Human brain is reported to weigh only three to four pounds comprising a trillion cells, 100 billion of them being neurons linked in the diverse networks that give rise to intelligence, creativity, emotion, memory, and consciousness (Fischbach, 1992). Human brain has been hailed as the most complex object in the universe, and the part of this complexity, as Fischbach argues, lies in the diversity of the nerve cells. Shantz (1992) postulated that the neurons are specifically and intricately connected with one another in ways that make possible memory, vision, learning, thought, consciousness, and other properties of mind. He argues that perception and comprehension of the visual world occur as a result of multistage integration of diverse neuronal activities. His theory hypothesizes that the integration of visual information is a process in which perception and comprehension of the visual world occur simultaneously; indeed, consciousness is the emergent property of a complex neural apparatus that the brain has developed to acquire knowledge (Shantz, 1992). He further argues that mind is an emergent property of the brain's electrical and metabolic activity.

An emergent property is one that cannot be accounted for solely by considering the component parts one at a time. It occurs from the interaction of the interconnected components within the system. Crick and Koch (1992) believe that all aspects of mind including its puzzling attribute, the consciousness, or awareness are likely to be explainable in a more materialistic ways as the behaviors of large sites of interacting neurons. Humphrey (1992) argues that an absolute requirement for consciousness is a recurrent feedback loops to the brain "*with the result that the outgoing signal and the return message meld into a large, longer-lasting, process.*" This supports William James's assertion that "*consciousness is not a thing, but a process.*" Boris Kotchoubey (2018), on the other hand, thinks that consciousness is not a process but a kind of behavior that is controlled by the brain like any behavior. It emerges on the interface between three components of animal behavior: *communication, play*, and *the use of tools*. It is anti-reductionist and anti-eliminativist, and yet, human consciousness is considered as a purely natural biological phenomenon.

The neurophysiological findings put an end to the Cartesian dualism which regards mind and matter as separate entities. Descartes often regarded mind as something immaterial, separate from the brain. This was clearly a mistaken view of the mental phenomenon originated from his reductionistic paradigm, an approach that does not take into account of diversity and the complexity of highly interacting interconnected components and simply attempts to characterize the system, the whole (patterns and process) on the basis of isolated parts of the system. Consciousness is a complex natural biological phenomenon, a phenomenon historically poised to make us human. This is a mental phenomenon that results from the interactions of diverse networks of highly integrated neurophysiological processes characterized by recurrent feedback loops in the brain. In fact, consciousness is a process, an emergent property of the complexity of neurophysiological synergism operating in our brain produced by million years of evolutionary process.

I cannot help quoting Holmes Rolston III (1992) who strongly argues that *diversity, unity*, and *spontaneity* in Nature have values in themselves and human intervention should not disrupt these remarkable properties of organizational process of Nature. The emergence of human consciousness (mind) in human brain can be seen as mimicking these complex organizational processes of Nature. As he states: *Notice how both this diversity and unity feed the human mind. Mind cannot be formed under the homogeneity of a blank wall nor before the heterogeneity of a bewildering jungle. A complex mind evolves in order to deal with a diverse world, yet one through which unifying relationships run. Do we then say that these features are of no value until thickened by the addition of human interest? Or do we wander that just this system, evolving so, did thicken human interest to form the mind prehistorically and that it continues to do so now? The mind is a mirror of these properties in Nature, and there is even a sense in which the mind, founded on the cerebral complexity and integrating capacity, is a product of Nature's inclination both to diversify and to unify* (Rolston, 1992). Perhaps this Nature's inclination to diversify and unify can be better conceived and understood by the Lovelock's Gaia theory which states that the *Earth can be regarded as a single living organism consisting of a system of vast biotic community characterized by synergistic and mutualistic relationship in an interconnected web of life* (Lovelock, 1979).

Let us take a brief look into human cultural evolution. Cultural evolution, here, has been used to denote all human social and technological superstructures (institutions, ideologies, and innovations). Man evolved from culturally simple to a complex being. Human cultural evolution has proceeded from a simpler to a high energy complex system. The creation of today's technologically complex, high-energy ecologically unsustainable consumer cultural society is itself a case in point. This society must culturally transform itself from this high energy, ecologically unsustainable phase, into a low energy, low entropy, and ecologically sustainable phase; otherwise, its structural components will be unable to function and maintain the material resource and energy recycling within the system, ultimately leading to the collapse of the system itself. Regardless of whether it will be able to do so or not, it cannot escape inevitable evolutionary process, the process of becoming. Does this mean that the probable ecologically sustainable society will be a less complex society? On the contrary, such society will be characterized by even more diversity of interconnected networks of mutualism, symbiosis, cooperation, negative and positive feedback, and synergistic interactions not only among humans but also between humans and the biotic community as a whole. Diversity and the creative organizational complexity will inevitably be increased in both "*biotic* and *social web of life*," in the ecologically sustainable society and that is possible only through the balanced interaction between *sociosphere* and *ecosphere.* Human activities within sociosphere must remain within the resilience and regenerative biocapacity of the ecosphere.

There are three undeniable recurring patterns inherent in the process of becoming, be it in natural system (ecosystem) or human social system or the human brain and consciousness. These are *diversity, autopoiesis*, and *complexity* characterized by interconnected and interdependent mutualism, symbiosis, and synergism. Biotic community and ecosystems are intrinsically characterized by diversity, autopoiesis, and interconnected complexity and so are the human cultural systems and human brain (mind) and consciousness. These patterns or properties have nothing to do with our liking or disliking or preferences, but they are all encompassing essential emergent properties of the systems and are valuable in themselves and for themselves because in the absence of these properties, the systems collapse and the ordering process of Nature of which humankind itself is the product of that process becomes inconceivable. Hence, diversity, autopoiesis, and complexity characterized by interconnected symbiotic and mutualistic synergism in Nature, in human cultural system, and in the development of human brain (mind) and consciousness have values in themselves and for themselves. *From this, it follows that these qualities must be preserved and promoted in Nature as well as in human cultural system, because they are values in themselves.* The preservation, protection, and promotion of these qualities both in *ecosphere (biosphere)* and *sociosphere* is the biggest challenge faced by humanity in Anthropocene era. This analytical framework establishes the minimum basis for *environmental ethics*. Social diversity, spontaneity, and interactive complexity embody humanistic values in the sense they all, in effect, create social environments and conditions conducive to the actualization of human potential just as diversity, autopoesis, and interactive complexity are the basis for the continuation and actualization of biotic community in Nature.

Irrespective of whether or not, the above deliberations can satisfy the minimum conditions for something called *intrinsic values,* one of the major pillars of environmental ethics, there is another perspective from which it can be argued that biodiversity, autopoiesis, and complexity in the ecological systems are valuable and therefore ought to be preserved and promoted in the systems because human lives and the conception of ourselves as human beings will be enhanced in a spiritual sense, when we learn that our survival and well-being are inextricably connected with the health and stability of diverse and complex ecological systems. Therefore, human must learn how to live in harmony with Nature's systems characterized by autopoiesis, diversity, stability, and complexity (intrinsic values).

10.4.2 Species, Ecosystem, and Moral Standing

The term "*Autopoiesis*" was first coined by Humberto Maturana and Francisco Varela (1980b) to describe the nature of living systems as opposed to nonliving. Thus, this term has been used to describe the nature of life and its characteristics. In a simple term, *autopoiesis* is the capacity of an entity to reproduce itself. Autopoietic systems are characterized by self-creating processes that entail networks of relationships between components that create the complexity of living organisms. Moreover, these processes serve the same function in the body (i.e., a cell) as in the mind (i.e., cognition). Components are part of self-creating circular networks because they create other components in order to maintain themselves and the structure in its entirety. Maturana and Varela (1980a) formulated a theory of "*Autopoiesis*" as the organizing network of the living system. Cognition is the activity involved in the self-generation and self-perpetuation of living networks (autopoiesis), and therefore cognition is the very process of life. The organizing activity of living system at all levels of life is cognitive activity. As Capra and Luisi (2014) points out the interactions of living organisms such as plant, animal, or human with its environment are cognitive interactions. When the complexity of the living organisms increased over a long history of evolution, their cognitive processes also increased over time. Maturana and Valera advanced a radical concept of cognition that involves the entire process of life—including perception, emotion, and behavior and may not even necessarily require a brain and a nervous system. Maturana puts this idea eloquently in his paper "Biology of Cognition": "*Living systems are cognitive systems, and living as a process is a process of cognition. This statement is valid for all organisms, with and without a nervous system. Our cognitive process differs from the cognitive processes of other organisms only in the kinds of interactions into which we can enter, such as linguistic interactions, and not in the nature of the cognitive process itself.*"

Maturana and Varela's "*autopoiesis theory*" described the characteristic processes that were fundamental to self-organized living organisms (Mingers, 1980, 1981). Fox (1995) argues that the concept of *autopoiesis* provides the best possible support for the claim that non-sentient living entities have interests and contends that this concept can be extended to ecosystems. According to Fox, this self-directed character of living entities implies that, in a sense, they "matter to themselves" as opposed to nonliving objects. Callicott (1999) along the line of Fox proposes that the concept of *autopoiesis* provides a plausible interpretation of Aldo Leopold's notion of "*land health*" understood as a land's "*capacity for self-renewal.*" Johnson (1992) considers that species and ecosystems are living entities with morally significant interests in their own right and has defended against the leading objection that species and ecosystems cannot have interests in their own right because evolution does not proceed on that level. On the contrary, he argues "*Although evolution proceeds on the level of the genotype, those selected are able to cooperate in entities of various higher orders including species and ecosystems. Having their own nature and interests, species and ecosystems can meaningfully be said to have moral standing*." In general systems theory, the following definition of Maturana (1981) is widely accepted: autopoietic systems are defined as "*unities and as networks of production of components that recursively, through their interactions, generate and realize the network that produces them and constitute, in the space in which they exist, the boundaries of the network as components that participate in the realization of the network.*" Species and ecosystems have evolved in such a way that their life processes tend to maintain the viability of the wholeness. Living things are special entities in that they go against the universal trend toward increased entropy through systematic means of self-creation and self-maintenance. As Kenneth Sayre (1976) puts it "*the typifying mark of a living system .... appears to be its persistence state of low entropy, sustained by metabolic processes for accumulating energy and maintained in equilibrium with its environment by homeostatic feedback processes.*"

Lawrence (1992) argues that living beings have self-identity in that what it is and what it serves to maintain it are determined by its own nature. A living being defines its favored states (homeostasis) and the life processes of the being as a whole are integrated toward maintaining them. Accordingly, that which serves to maintain its viability is in the interest of a living entity. Lawrence further elaborates not only living things meet Sayre's characterization, but they also have organic unity and self-identity. When an ecosystem is perturbed in any one of various ways, it bounces back. The members of the ecosystem do just what is necessary to restore the system to equilibrium. Species and ecosystem exhibit stability and resilience that contribute to their ongoing viability and maintaining their homeostasis through intricate feedback mechanism. The genetic diversity that exists within species and the ecosystem is important for the long-term welfare of the species and the ecosystem itself. Species and ecosystems are loaded with genes which manifest themselves in such a way as to fit in with other genes in other beings of various types. Genotypes tend to be selected for when they tend toward manifestation in viable individuals, viable species, and viable ecosystems. Ecosystems are not just the aggregations of plants and animals occurring in a space. Rather, they are living systems with their own organic unity and self-identity, having, and acting so as to maintain their own character.

As to the object of our moral consideration, I would like to emphasize on species and ecosystems or community as a whole because they can be regarded as an entity with a value of their own, and thereby the individual components such as organisms, and the populations of organisms, the animals and plants and their interrelationships with each other are all included in biotic community (ecosystems). This also solves the problem of dichotomy that exists between the individual parts and their whole in connection with which individual organisms or the system constitute the object of our moral consideration. Once we are committed to take care of the whole, we automatically take care of the individual components that make up the whole. What exactly should we preserve and promote in the biotic community (ecosystem) that we consider having *intrinsic values*?

As I have explained above, it is the *diversity*, autopoiesis (creative self-organization), and the *complexity* prevalent in ecosystems or biotic community that constitute the very essence of the system. These values exist for their own actualization and should not be regarded as only *instrumental* for the actualization of other values. Complexity characterizes the actualized or realized state of these values in Nature. Ecosystem as an entity is the actualization of diversity and autopoiesis inherent in it. Human consciousness is also a process of actualizing the diversity and autopoiesis intrinsically inherent in human brain. Diversity and autopoiesis are also prevalent in social system as they can be seen manifested in *cultural diversity* and *social autopoiesis* (complex social network). *Diversity, autopoiesis,* and the *complexity* (biotic web of life) are *intrinsic values* not only in the ecological systems (ecosystems) but also in social system and, hence, they must be preserved and promoted in both *ecosphere* (biosphere) and *sociosphere* (social systems). The ecosystem or biotic community as an object of our moral consideration can encompass and embody all these values (*diversity, autopoiesis,* and *complexity*) in their full actualization.

10.5 Sustainable Development

From a theoretical perspective, sustainable development can hardly be considered an alternative development model. It is essentially the continuation of the same economic growth model of the market capitalism with some cosmetic changes in the language. The practice of sustainable development since Brundtland Commission has been associated with limited goal of nature conservation and accelerated growth economics. It cannot fully comprehend the ecological and social dimensions of sustainability. In my view, neither the World Conservation Strategy (WCS) nor the World Commission on Environment and Development (WCED) went far enough to disassociate the term from dominant growth-driven market capitalism that thrives on excessive extraction from Nature and excessive consumerism. As long as the concept of sustainable development entails only the instrumental view of Nature and does not incorporate moral dimension in the development that recognizes the need to protect *ecosystem health* and *services, resilience, self-organizing attributes* of ecosystem processes, and *interconnectedness* in Nature, it is incomplete and self-defeating concept.

10.5.1 Concept of Sustainable Development

The phrase "sustainable development" has been used wastefully in development literature ever since the World Commission on Environment and Development (WCED) defined it for the first time as the development that "*meets the needs of the present without compromising the ability of future generations to meet their own needs*." The definition tried to vaguely capture the concern and the needs for environmental protection on the one hand and the necessity of economic growth to meet the needs of growing human population on the other hand. A large number of development thinkers and scholars (Barbier, 1987; Mathews, 1989; Milbrath, 1989; Panayotou, 1990; Upreti, 1994) have profoundly elaborated, supplemented, and enriched the discourse on sustainable development with emerging social and ecological dimensions of sustainability that were not explicitly espoused in the original concept of Brundtland Commission's report. Because of the multidimensional nature of the sustainable development, the concept must entail the management and conservation of natural resource base, social, institutional, technological, and cultural changes and, more importantly, changes in the ethical dimension of the development itself.

It is extremely difficult to come up with the ideal definition of "sustainable development"; nevertheless, such definition must be sufficiently broad to capture various dimensions of sustainability. As Gao (1990) and Upreti (1994) point out the definition of sustainable development by FAO (1989) broadly captures the multidimensional aspects of sustainability with respect to the conservation and management of natural resource base: "*Sustainable development is the management and conservation of the natural resource base, and the orientation of technological and institutional changes in such a manner as to ensure the attainment and continued satisfaction of human needs for present and future generations. Such sustainable development (in agriculture, forestry, and fishery sectors) conserves land, water, plant, and animal genetic resources is environmentally non-degrading, technically appropriate, economically viable and socially acceptable.*"

The above definition by FAO captures the multidimensional features of sustainable development and emphasizes the conservation of natural resource capitals (land, water, forests, and biodiversity) and its sustainable uses. This is certainly an encouraging development, but the biggest challenge lies in the design of innovative technology, ecologically and socially effective policy instruments, and strategies and the development ethics that can motivate human behavioral changes for the realization of the underlying objectives and the goals of sustainable development. The sustainable development approach must address certain principles and strategies for improving the basic needs of the poor, but such strategies formulated and implemented must also be environmentally sustainable over time and encourage the grass root participation in the development process. Such principles and strategies demand not only the environmentally sustainable development policy programs but also the socioeconomic restructuring of the society so that the benefits of the development could be shared by a larger section of society. Myrdal (1968) recognized the need for a sustainable development approach that must directly address the "*basic needs*" of the poor and thereby arrest the process of "cumulative causation" of poverty, environmental degradation, and underdevelopment.

One central premise of sustainable development as pointed out by Bartelmus (2013) and articulated by Barbier (1987) is that "*many environmental problems in developing countries originate from the lack of development, that is from the struggle to overcome extreme conditions of poverty.*" Barbier (1987) put forward a basic analytical approach that views that sustainable development results from the interaction of three system goals, namely *the biological natural system goal, the economic system goal*, and *the social system goal*. The biological natural system goal entails *biodiversity, resilience*, and *biological productivity*. The economic system goal entails *satisfying basic human needs,* and *increasing useful goods* and *services*. The social system goal entails *cultural diversity, institutional sustainability*, and *social justice*. Integrating these system goals, sustainable development can be defined as a *development that strives for achieving these goals namely biodiversity, resilience,* and *biological productivity* in natural system; *basic human needs,* and *useful goods* and *services in economic system*; and *cultural diversity, institutional sustainability*, and *social justice in social system.* Development, that can achieve these system goals, can be called sustainable development; otherwise, sustainable development remains elusive and untenable as it has been the case ever since the phrase "*sustainable development*" began to appear wastefully in the development discourse.

The objective of sustainable development is to maximize the goals across all these three systems through the adaptive process of trade-offs. The neoliberal free market capitalism aims at maximizing only the economic system goals and does not pay attention to the other equally important system goals. It assumes that the resources are not supplied by Nature but by human ingenuity, and the only index of scarcity is the price per unit resource and the market will automatically take care of both the natural and social system goals. The neoliberal corporate capitalism that strives for short-term profit maximization by any means at the expense of long-term natural capital resource degradation has been largely responsible for most of today's environmental problems, pollutions, climate crisis, and social problems. The other alternative, the Marxist centralized economic system, unlike the capitalist economic system, advocates maximizing both the economic and social system goals, but it also exploits and regards natural capital resources more or less in the same way as the neoliberal corporate capitalism. The environmental pollutions in the former Soviet Union and the Eastern Europe have been even more terrifying than the ones in the Western world. It is important to understand that the economic and social system goals cannot be achieved without achieving Natural biological (resource) system goals and, in fact, Natural biological system is the very infrastructural basis upon which economic and social superstructures have been built. When natural infrastructures of Earth's systems collapse, how can thc superstructures built upon them survive? The advocates of neoliberal corporate capitalism and Marxist central planners do not get it. Do they?

Overexploitation of natural resources is embedded in economies and governance systems, and the resulting degradation is undermining hard-won development gains and threatening the well-being of future generations. The transformation of degraded and unsustainable systems into the sustainable productive system will promote sustainable development. Ecosystem restoration is one of the most important ways of delivering nature-based solutions for food insecurity, climate change mitigation and adaptation, and biodiversity loss. UNEP (2019) Decade on Ecosystem Restoration 2021–2030 project declared that UN General Assembly aims to massively scale up the restoration of degraded and destroyed ecosystems as a proven measure to fight climate change and enhance food security, water supply, and biodiversity. The degradation of land and marine ecosystems undermines the well-being of 3.2 billion people and costs about 10% of the annual global gross domestic product in loss of species and ecosystems services. Key ecosystems that deliver numerous services essential to food and agriculture, including supply of freshwater, protection against hazards, and provision of habitat for species such as fish and pollinators, are declining rapidly. The project estimated that restoration of 350 million hectares of degraded land between 2021 and 2030 could generate US $9 trillion in ecosystem services and take an additional 13 to 26 gigatons of greenhouse gases out of the atmosphere. This is certainly a welcoming act of UNEP, and such project should be designed and implemented in all countries where possibilities of the restoration of degraded ecosystems exist.

10.5.2 Consumerism and Sustainable Development

The discourse on sustainable development leads us nowhere unless we try to understand it in the context of modern consumerism. As critiques have pointed out the lifestyle and the behavior of modern consumer society have been characterized by ecologically unsustainable production and consumption patterns (Durning, 1989; Upreti, 1994). The consumer culture promoted by market capitalism is not only ecologically hostile but also dangerously motivating human behavior toward individual self-aggrandizement through excessive consumerism of unnecessary goods and services as an end in itself. This trend of excessive consumerism has given rise to dangerous patterns that show very little concerns with economic, social, technological, and political commitments to the concept of sustainable development. It is impossible to construct and maintain infrastructural base of sustainable development without challenging, and changing the very lifestyles of modern consumer society particularly in the West and rapidly emerging consumer elite class in developing countries. Ecologically hostile and excessive consumerism is the very antithesis of sustainable development. To realize the goals and objectives of sustainable development and make it a real "development paradigm," there must be a cultural change in our lifestyles and value systems with equitable access to resources and optimum human population size. And most importantly greed and the accumulation of wealth by any means (corporate greed) are the root causes of social and ecological unsustainability. As Gopi Upreti (1994) rightly articulated to that effect: "*If the current dominant paradigm (greedy and egocentric), that seeks only profit motivated unhealthy competition, is not replaced by a paradigm that seeks ecological and social justice based on the ecological principles of diversity, interconnectedness, and interdependence, symbiosis and coevolution in nature and pluralism, equity, social synergism in society, sustainable development will be a Sisyphus's Myth.*" If we ponder into what Mahatma Gandhi said long time ago that the planet Earth has "*just enough resources to satisfy every body's needs but not every body's greed*". It becomes clear that without such realization, we are unable to change our perceptions, values and assumptions, lifestyles, and behaviors, and cannot restructure our economic, social, political, and cultural institutions. Longer it takes to realize this truth, harder the misery and catastrophes we inflict on ourselves (humanity) and the planet Earth systems.

It is important to point out that strategies and policy instruments designed to achieve sustainable development cannot be the same for developing countries as deemed appropriate for developed Western countries because of the differences in the nature of economic-social and environmental problems prevalent in the Western and developing worlds. The strategies and policy instruments in developing countries need to be developed to address poverty, quality of life, and capacity empowerment of the people to influence the development process. Policy strategies require creation of inclusive social and political institutions that can be entrusted with the judicious allocation of development resources and their sustainable uses. Equitable access to natural and social resources and the capacity building of the people excluded from development process constitutes the essential condition to ensure sustainable development in developing countries. As Horowitz (1988) eloquently articulates to that effect: "*Environmental degradation is not a problem of the relationship between people and their habitats, but of the relationship among people competing for access to productive resources.*"

The strategies and policy instruments in developed countries must be centered around the scale of consumption and production patterns and how that relate to the question of sustainable development. As we know, the per capita consumptions of the material goods and services in developed countries put many times greater stress and impacts on the resources of planet Earth than the per capita consumption in developing countries. The resource extraction and production patterns motivated by ever-increasing consumerism and profiteering in the West and the imitation of the same patterns by developing countries are most dangerous threats to the destruction of the planet Earth's systems. The control of the productive resources by a few both in local and global contexts through unjust social and economic system has forced a vast majority of the people to compete for limited productive resources in developing countries. A similar pattern of the control and concentration of the productive resources is apparent at a global context. As we can see, the Western world with 15% of the world population control over 80% of the world's resources and the rest of the world with 85% world population has to live with 20% the world's resources.

If we look into the basic factsheet about how the capital flights from Global South takes place to the Global North in the form of debt servicing, it raises our hairs with perplexity. Global South suffers miserably from debt pandemic that is preventing them from meeting their basic needs and achieving a meaningful recovery. As Daniel Munevar (2021) reports, between 2000 and 2020, public debt of developing countries has increased from an average of 40.2 to 62.3% of their GDP. This increase spiked in 2020 alone to more than one-third equivalent to $1.9 trillion dollar. Most developing countries now allocate more than 20% of government revenues to debt services. It is pathetic that developing countries have struggled to procure the Covid-19 vaccines, but they continued to pay their external creditors more than US$ 372 billion in debt service in 2020. If we look at the patterns of concentration of resources and wealth, it appears that resources and wealth have moved from vast disorganized masses to a few well-organized and powerful elites in developing countries, and from poor Southern countries to the rich and powerful Western countries. This trend must be stopped and the poor people's access to the productive resources and capital must be increased so that human pressures on natural environments can be minimized. Sustainable development becomes possible only when environmental sustainability and socioeconomic sustainability are maintained in a dynamic equilibrium state. The biggest challenge of development professionals and political decision-makers is to design the development strategies and pathways rooted in development ethics and ecological wisdom consciousness.

10.5.3 Natural capital and Sustainable Development

The most important prerequisite for sustainable development is that a nation's basic stock of ecological capital remains constant or does not decrease over time. A constant if not increasing stock of natural capital is not only needed to meet the needs and the aspiration of present generation but also to ensure a minimum degree of fairness and equity for future generation. It is important that the governments design public policies that are tailored to minimize deforestation, desertification, destruction of habitat and species, and decline of air and water quality. Especially in developing countries, government policies must entail the problems of landless people and create viable alternatives for these people so that the cultivation of marginal land and deforestation can be checked. The planning must address the root causes of environmental deterioration and possible solution for it. Planners must understand the fact that ecological imbalance create conditions for man's physical and economic imbalance. It is particularly true for developing countries of mountainous geomophology when viewed in the context of loss of fertile top soil, deforestation, increasing river floods and landslides—all contributing to negative agricultural productivity both in the hills and the plains on the one hand, and acute shortage of fodder, fuelwood, and gradual desertification of the mountains that sustain considerable portion of the population on the other hand.

There is a tendency on the part of politicians and the planners to regard environmental problems and concerns as something of less importance and inevitable outcome of development process, and those who cry out for environment are influenced by Western environmentalism which is incapable of understanding the problems of developing countries, especially the need to meet the basic needs of poor people. Politicians of developing countries have gone to the extent of choosing either people or Nature in the name of development. This is highly misguided and dangerous attitude that must be corrected. It is important to show to our politicians that people and Nature can live together in harmony without destroying each other's productive capacity. It is important to show to them that environment constitutes the resource base infrastructures upon which human superstructures and socioeconomic system are built and if something goes wrong with the resource base infrastructures, then socioeconomic superstructures cannot remain unaffected and functional. Those who fail to see this dialectical relationship between infrastructures and the superstructures also fail to see the mutually enforcing relationship between people and Nature. If we look into the pattern of natural resource uses in a historical perspective, we find that many human civilizations became extinct because of human-induced irreversible changes in their landscape ecology driven by rapid depletion and unsustainable use of their resource base (Van Dyne, 1969).

10.5.4 Reconceptualizing Sustainable Development

It is absurd to talk about sustainability without trying to understand the foundation of sustainability which is environmental sustainability. We cannot have a meaningful discourse on sustainable development apparently with no substantive discussion on the foundational principles and components of the sustainability. When we do not have the proper understanding of what sustains human existence, the use of sustainable development phrase and sustainability talk simply becomes an empty rhetoric. The magnitude of environmental problems, the disruption in the planetary ecosystem, global warming, and the climate change and their catastrophic consequences warrant a global gestalt approach to deal with this existential threat. Time is running out much more quickly than the realization by the ruling elites, political, corporates, and institutional power centers. A recent article in the New York Times by Paul Krugman (2022) alarmed us: "*The fact that the planet is changing faster than even pessimists expected: ice caps are shrinking, arid zones spreading at a terrifying rate. And according to number of recent studies, catastrophe—a rise in temperature so large as to be almost unthinkable—can no longer be considered a mere possibility. It is, instead, the most likely outcome if we continue along our present course.*"

The policy measures undertaken by the world governments, United Nations, and international development agencies in response to the global warming, climate change, and reduction in carbon emission have proven to be cosmetic and ineffective, and their net effects are questionable. The global protocols they have proposed can only delude us into thinking and making us believe that they are taking responsible action, but on the contrary (to the business as usual), there is even more competition to extract more resources for wealth accumulation from already degraded planetary ecosystems. The biggest predicament of the current development approach is even a deeper delusion and the false notion that humanity can avoid environmental catastrophe without understanding its root cause, the *egocentric consumerism.* The change in human consumption patterns cannot come by itself without change in human behavioral patterns for which change in human consciousness is necessary to effect the real change.

We all know that the underlying cause of the present ecological crisis has resulted from the destruction of the planetary ecosystem by industrial pollution, unsustainable mining, and extraction of resources from ocean, terrestrial ecosystems, and biosphere. This happened because of human desire to accumulate wealth and excessive material possession propelled by human greed and the ecologically hostile consumerism. We know that at the root of this paradox is the human treatment of the Nature (environment) with utter disregard and disrespect as if it is an object only to be exploited and humanity has no existential relational connection with it. This dangerous perception is the product of a human consciousness planted by neoliberal market capitalism that is totally divorced from the humanity's organic interconnectedness with natural world and has originated from the false notion that humans are separate from Nature, and they can do anything with it as they want. This attitude is pervasive among human population that live in big urban environment metropolitan cities, whereas such attitude is unthinkable to indigenous people living in the proximity of the natural world. Their culture is enriched with values and respects for the natural world. Following Industrial Revolution, Western consciousness evolved through mechanistic scientific reasoning that treated natural world as a mere object, and the purpose of human enterprise was to invent tools, discover and master the natural laws, and ultimately conquer the Nature and subject it to the service of humanity. This *egocentric attitude* coupled with the mercantile capitalism propelled by reckless profiteering and modern consumerism became the *dominant social consciousness* that guided the scientific research and development of destructive technologies. The reductionistic approach in science not only undermined the interdependence and interconnectedness of human social system with biophysical natural system but also caused the degradation of natural system.

While defining "sustainable development," two fundamental components that sustainable development must integrate are the satisfaction of basic human needs which are same in all cultures and the maintenance of the *resilience* and the *biocapacity* of the planetary ecosystem (Earth's system). Human socioeconomic and technological throughputs cannot exceed the regenerative biocapacity of the planet Earth in order for it to be sustainable. We should be mindful of the fact that sustainability is a relative concept, and truly speaking, there is eternally nothing sustainable. The totality of human activities (throughputs) in the *sociosphere* must remain within the regenerative biocapacity of *ecosphere* (biosphere). The complexity of sustainability can be reduced to a simple but fundamental question: how can human socioeconomic system be sustainable when it consumes more than what Earth can regenerate? *Sustainable living can be possible if a value-based development is pursued, and science and technological innovations are turned around toward maintaining and increasing the biocapacity of the planet Earth.*

10.5.5 Principles of Sustainable Development

It is imperative that sustainable development goals entail two very similar goals in both ecological and social systems. The ecological system goal entails *biodiversity, autopoiesis,* and the *symbiosis* (interdependence and interconnectedness) and the social system goal entails *cultural diversity, social autopoiesis*, and the *social symbiosis*. These system goals need to be promoted and maximized both in ecological and social systems for operationalization of sustainable development. Following principles grounded in intrinsic and instrumental values, as postulated by Upreti (1994), are considered indispensable for achieving sustainable development:

1. The first principle is maintaining ecosystem health (ecosystem services) or ecological integrity in a way that can maintain the ecological processes, diversity, and life support systems for the continuation and flourishing of biotic community including human beings.
2. The second principle is the satisfaction of basic human needs, which is based on the assumption that ecosystem protection and conservation is possible only through the sustenance of basic human needs while allowing the sustainable use of ecosystems.
3. The third principle is the principle of equity and social justice without which it is impossible to satisfy human needs of the present generation and maintain diversity of opportunity for future generation.
4. The fourth principle is the principle of optimum population and consumption which requires that total human consumption throughputs remain within the regenerative biocapacity of planet Earth (sustainable level) by lessening pressures on environmental resource base.
5. The fifth principle is the principle of interconnectedness and interdependence which states that *sociosphere* (human socioeconomic system) is inseparably interconnected with and dependent on the *ecosphere* (The Earth systems) and that human socioeconomic system can continue only by maintaining but not by rupturing or destroying this interconnectedness.

The health and well-being of humanity and the functional sustainability of social system are intricately dependent on and interrelated to the health and the integrity of the planetary (biospheric) ecosystem; hence, social and economic development strategies and policies must be designed to improve and promote the well-being of people in *sociosphere* and maintain the proper functioning of *ecosphere* because both are intertwined with each other determining each other's state of well-being. Policies and strategies must spring from the knowledge and the understanding of the nature of this interaction between the *sociosphere* and *ecosphere*. In other words, the policies, strategies, and development ethics must be guided by the nature of the interaction between *sociosphere* (human social system) and the *ecosphere* (planetary ecosystem). Promotion of social justice (equity and opportunity for all to satisfy their basic needs and realize their human potential) ennobles humanity to expand its moral capacity to incorporate human and nonhuman beings (biotic community) into one moral plane or one single biotic community. Without fulfilling and solving basic human needs and problems, it is impossible to expand this morality of a "single biotic community" into one moral plane. Healthy and stable biospheric ecosystem, capable of supporting the "*web of biotic life*" which is the unique character of the planet Earth, is in-itself a value and for those who cannot see this value in-itself and for it-self, perhaps, the survival dependency of humanity on the health and stability of planetary ecosystem should provide sufficient reasons for the conservation and preservation of the natural ecosystems. Hence, it follows that *biodiversity, autopoiesis, ecosystem processes*, and *organizational complexity* must be considered values for themselves, and the human actions that preserve and promote these values in Nature must not only be encouraged but also be made morally binding. Implicit in this concept is the belief that improvement and maintenance of ecosystem health (ecosystem services) and resilience can bring improvement in the health, prosperity, and the well-being of the people.

The investment on science, research, and innovation must be directed to study and understand the nature of interconnectedness and interdependence, the ecological and physical laws that are operating and maintaining the web of the interconnectedness and interdependence, and what needs to be done to sustain this web of interdependence and connectedness. In other words, scientific epistemology, approaches, and human consciousness must change their course from reductionistic to a more holistic approaches in the study and understanding of the planetary ecosystem. We can no longer afford to think that we can continue our business of abusing the planetary ecosystem as we want. Until now, all scientific methods, techniques, and approaches have been developed and diverted to achieve mastery over Nature, extract resources from Nature, and destroy its creative capacity to regenerate those resources upon which human and other life-forms depend. Exceptionally few scientific studies and methods have been directed to understand natural processes or more specifically the ecological and biophysical processes that regenerate and sustain the planetary ecosystem and that is also very recently when environmental problems began to manifest threatening the well-being of humans all over the world. This is negative feedback the planetary ecosystem has exerted to the self-destructive acts of *Homo sapiens*.

Appropriate and effective development policies and their execution to ensure conservation of biological diversity and sustainable development cannot emerge from the current dominant development paradigm. We need a new paradigm that provides a balanced and comprehensive approaches on both social and the ecological dimension of the problems and issues specifically to answer the following questions; why biodiversity and healthy ecosystems are important and valuable, what are the social and economic causes responsible for the loss of biodiversity and degradation of ecosystem services, global warming, and climate crisis, and what measures can be used to reduce or reverse the destruction of biodiversity and the degradation of ecosystem services and climate crisis. The biggest impediment to achieving sustainable development is the lack of development ethics that provides the ethical imperatives for social and ecological sustainability of the development. I submit that the direction of human socioeconomic systems in the future will inevitably be determined by two primary parameters: health and the functional stability (resilience) and the biocapacity of the planetary ecosystem and the ecologically sustainable human production and consumption patterns with minimum waste throughput. If humanity is not to be destined to meet its denouement in its evolutionary history, we need a value-based development approach that embodies these parameters, namely the resilience and the biocapacity of the Earth's systems as the determinants of sustainable development postulated above in the directive principles of sustainable development. The point is, sooner it happens, greater will be the degree of freedom for humankind to make choices to live in harmony with planet Earth, the only home of all the living entities in Nature.

Chapter 11
Buddhism, Gaia, and System Theory on Environmentalism

As we can observe there is a very dynamic, interactive, and cooperative community going on in any functioning ecosystem; Many species are dependent upon each other and contribute to other species' existence, geology, botany, soil chemistry, wind, fire, and water must all be operating in some sort of balance, which forms the basis for the interdependent web of all existence.

Arthur Thompson (1914)

11.1 Introduction

The Pali word "*Dhamma*," translated as "*Dharma*" in Sanskrit, originally meant "*the law of Nature*" or "*the truth*" and holds immense significance in Buddhism. Buddha's teachings were originally recorded in Pali, which was the *lingua franca* of the common people at that time. In the Buddhist tradition, *Dhamma* or *Dharma* refers to the law of universal truth. According to Max Muller, as reported by P. T. Raju (1939), "*Dharma*" may be correctly rendered as "*law*" in ordinary Buddhist phraseology. It can be said that Buddha's teachings are about the laws that govern our lives and all phenomena in Nature. Buddha's teachings, called *Dhamma* or *Dharma*, deal with fundamental questions about the self, human conditions, living systems, and the nature of existence, which are basic philosophical questions. Wisdom *(Pragya)*, ethical conduct or morality *(Sil)*, and mental discipline *(meditation/Samadhi)* are considered to be the three pillars of *Dhamma* or *Dharma*. Anything that manifests or expresses its own character and nature is called its *Dharma*. *Dharma* means that which imbibes and lives by—*dhāretīti dhammam*. Buddha's teachings emphasize that everything is impermanent, and therefore, one should strive for salvation through Sil (morality), Pragya (wisdom), and Samadhi/Dhyan (meditation) as explicated by this Pali Sukti "वयधम्मा सङ्खारा, अप्पमादेन सम्पादेथ".

G. Upreti, *Ecosociocentrism*, https://doi.org/10.1007/978-3-031-41754-2_11

Buddhism provides a unique perspective on environmentalism, specifically the nature of man's relationship with Nature. According to the concept of *Dharma* in Buddhist teachings, all things are interconnected in Nature. There is nothing in existence that exists as a separate and isolated entity. Things only exist in connection and relationship with other things. This implies that the boundaries between things are only useful conventions, provisionally true, but by no means absolute. This view lies at the heart of the ecological perspective known as *system theory* and the *Gaian hypothesis*, which recognize that everything in this world is connected through systems, subsystems, components, and their interactions woven into a subtle and intricate web of relationships or a network of relationships. In this chapter, we will explore how fundamental Buddhist teachings, Gaia, and system theory view the interrelatedness and interconnectedness of living and nonliving systems in Nature and how human compassion, reason, and wisdom contribute to the practice and understanding that enhance and honor the emerging ecological paradigm. Buddhism, Gaia, and system theory are enriched with an ecological worldview *(eco-dharma)* that offers great potential to guide human behavior to build a harmonious relationship with the planetary ecosystem for sustainable living.

11.2 Eco-Dharma Concept and Basic Buddhism

The term "*eco-dharma*" arises from the fusion of two concepts: "*ecology*" and the "*dharma*" concept in Buddhism. Eco-dharma concept entails the convergence of ecological principles and Buddhist teachings that emphasize human interconnectedness with the natural world and ensuing human ethical responsibilities with respect to its treatment. It encompasses a way of living and interacting with the environment rooted in mindfulness, compassion, and a recognition of the interconnectedness of all beings. Within the ecological movement, many rational and conscientious thinkers and scholars have discovered inspiration and profound parallels within the Buddhist tradition. It is noteworthy that numerous Buddhists themselves have begun to uncover the ecological implications embedded within their own Buddhist tradition and heritage. The fundamental Buddhist teachings, which center around interconnectedness, nonviolence, empathy, compassion, and conditionality, all contribute to both the practical application and intellectual understanding that amplify and honor the emerging ecological paradigm.

At its essence, *eco-dharma* acknowledges the interdependence and interconnectedness of all living beings and the natural world. It underscores the significance of living in harmony with Nature and assuming responsibility for the ecological challenges faced by humanity. Eco-dharma encourages individuals to cultivate mindfulness and compassion toward the Earth and the planetary ecosystem. Donald Swearer, a Buddhist scholar, offers enlightening insights: "*Many Buddhist practitioners have discovered, within one of the central ideas of Buddhism—the principle of interdependence—an ecological vision that integrates all aspects of the ecosphere—particular individuals and general species—in terms of the principle of mutual co-dependence. Within this cosmological model, individual entities are, by their very nature, relational, thereby challenging the notion of an autonomous self-separate from the 'other', be it human, animal, or vegetable.*" This understanding of interconnectedness finds its fullest expression in the Hua-Yen School of Chinese Buddhism. The *Avatamsaka Sutra*, a revered Mahayana Buddhist scripture central to the Hua-Yen School, presents a symbolic representation of reality through the metaphor of *Indra's Net.* This powerful metaphor describes an intricate web interconnecting all phenomena in Nature. *Eco-dharma* can be viewed as a comprehensive approach that integrates ecological awareness of this interconnectedness, ethical values, and spiritual insights to foster a sustainable and harmonious relationship between humans and the planet Earth. It holds great potential to inspire individuals, particularly the younger generation, to embrace the role of responsible stewards of the Earth and to cultivate a profound sense of interconnectedness with living systems and the planet as a whole.

11.2.1 Dependent Origination and Interconnectedness

Dependent co-origination (*Pratītyasamudpāda in Sanskrit*), also known as dependent co-arising, is a fundamental concept in Buddhism that explains the interconnected and interdependent nature of all phenomena and the cycle of existence. It describes how all things and events arise due to dependence on multiple causes and conditions. According to the doctrine of dependent co-origination, nothing exists independently or in isolation. All phenomena and events, including mental and physical experiences, and the phenomena in Nature are interconnected and arise due to a complex web of interconnected causes and conditions. These causes and conditions are not limited to immediate factors but can extend back in time and have far-reaching effects. The dependent co-origination is the Buddhist doctrine of the causality which recognizes the interconnectedness and interplay among events and phenomena, where one event or phenomenon (the cause) leads to another event (the effect) or phenomenon and so on, forming a cycle of causality. Dependent co-origination explains the sequence and interconnected nature of events and phenomena in the universe and enables us to analyze and comprehend the relationships and consequences of actions, phenomena, and processes that occur in our individual life, society, and Nature.

Indra's Net (a three-dimensional net) has been used to illustrate the phenomenon of dependent co-origination (causality) and interconnectedness. When one takes a closer look at one of the infinite jewels of the net, one can see that in each facet of the jewel, there is reflection of every other jewel in the network. As the light sparkles and glimmers in one jewel, that light change is reflected in every other jewel, and that change in each jewel is reflected again throughout the entirety of the space in the Net (Malhotra, 2014). Francis Cook (1989) presents Chinese Hua-yen Buddhism's vision of organismic interrelatedness with *Indra's Net* metaphor which depicts a cosmic web of dynamic causal interrelationship and interconnectedness.

As already described in the preceding paragraphs, according to this view, all events arise through their functional relationship to all other events and to the whole so that each thing is interconnected to everything else. It is an attempt to convey the fact that all phenomenon, all things, and all beings are intimately related to each other, intimately interconnected affecting each other. This is what the modern system theory postulates about how elements are interconnected in the network of the system.

We are profoundly connected to a *web of life* and complex social relationships stretching across the Earth, the planetary ecosystem. We should not only profoundly appreciate our own interconnectedness and the intimate relations which exist between us and the rest of things in the planetary ecosystem but also consciously and mindfully protect and preserve this wonderful web of interconnectedness that has resulted from millions year of evolutionary process. Simple question to ask is how can we *(Homo sapiens)* survive if this web of interconnectedness is destroyed irreparably? Hua-yen Buddhism envisions that everything functions as a causal condition for everything else, there is nothing which is not of value in the great harmony of Nature (Odin, 1997). Francis H. Cook (1989) describes this *metaphor* in the following words: "*Far away in the heavenly abode of the great God Indra, there hang the jewels, glittering like stars of the first magnitude, a wonderful sight to behold. If we now arbitrarily select one of these jewels for inspection and look closely at it, we will discover that in its polished surface there are reflected all the other jewels in the net, infinite in number. Not only that, but each of the jewels reflected in this one jewel is also reflecting all the other jewels, so that there is an infinite reflection process occurring.*" According to the *Avatamsaka Sutra, Indra's Net* symbolizes the interconnectedness of all phenomena in the universe. It is an elaborate web that stretches in all directions, with each jewel at every juncture reflecting all other jewels in the Net. In this metaphor, the jewels represent beings, and the reflections signify the interconnectedness and interdependence of all beings. *Indra's Net* serves as a potent metaphor in Buddhist teachings, illuminating the interconnectedness and interdependence of all beings and phenomena. It deepens our comprehension of this interconnected nature of reality and motivates us to alleviate the suffering that arises from the disruption and destruction of this web of interconnectedness and interdependence.

The Japanese Buddhist doctrine of *emptiness* postulates that there is nothing which is more real than the *interdependence* of everything in Nature. According to Japanese Buddhism, ultimate reality is to be found not in any *transcendental realm* of "*other worldliness*" as in Judeo-Christian or Hindu or Islamic religions but in the field of interrelationships in "*this-worldliness*," thus, overturning all models of transcendence and dualism. As Odin (1997) points out Buddhism provides theoretical support for environmental philosophy known as *Gaia theory* according to which the Earth has been likened as a single living organism forming a vast biotic community in which a complex grid of network of energy currents or lines of force constitutes Nature as an ecosystem of symbiotic relationships in an interconnected web of life. Gary Snyder (1990, 1995), an enthusiastic proponent of *Buddhist Ecology,* brings a very powerful and useful perspective from Hua-Yen Buddhism and also uses the metaphor of *Indra's Net* to describe the web of interconnectedness in the Nature. As is clear from his following statement: "*The universe is considered to be a vast web of many sided and highly polished jewels, each one acting as a multiple mirror. In one sense each jewel is a single entity. But when we look at a jewel, we see nothing but the reflections of other jewels, and so on in an endless system of mirroring. Thus, in each jewel is the image of the entire net.*"

Barnhill (1997) points out that mirroring for Snyder is found in the interdependencies of *nature's web*. Snyder applied Buddhist idea to ecology, or conversely, one could say he applied ecology to Hua-yen Buddhism. Odin (1997) argues that Buddhism has been invoked as a far more environmentally relevant and beneficial set of beliefs and practices than Christianity could ever be and suggests that of all the major religious traditions, Buddhism is the best equipped to form the heart of a new global *environmental ethics*. Aldo Leopold (1949), the father of the land ethics or more appropriately the father of modern environmental ethics, defines ethics in terms of his key notion of "*community.*" For him an individual is always located in the communities of interdependent parts that evolve "*mode of cooperation*" called *symbioses* by ecologists. Buddhists influenced by an *ecological consciousness* espouse Bodhisattva ideal, which teaches that the highest goal of Buddhism pertains not to personal salvation but to embody *compassionate awareness* and the *welfare* of all beings from whom we are not isolated and separated but are interconnected. Swearer (1997a) rightly pointed out Buddhadasa Bhikkhu's (Thai monk) sense of why a cooperative society is the basis for the existence of all beings: "*The entire cosmos is a cooperative. The sun, the moon, and the stars live together as a cooperative. The same is true for humans and animals, trees, and the earth. When we realize that the world is a mutual, interdependent, cooperative enterprise… then we can build a noble environment. If our lives are not based on this truth, then we shall perish.*" Socially engaged Buddhists argue that in order to be a force for social transformation, the traditional Buddhist emphasis on individual moral and spiritual transformation must be augmented with compassionate awareness to address the structures of oppression, exploitation, and environmental degradation that is occurring so rampantly on planet Earth. As Thai Buddhist monk, Sulak Sivaraksa, whose spiritual perspective center around the interconnectedness, is deeply concerned that the current neoliberal free market capitalism has effectively disrupted this interconnectedness and accelerated environmental degradation and destroyed Buddhist culture and values by insatiable consumerism in Asia. Individuals are motivated to consume and acquire more creating an endless cycle of greed and insecurity (Stanley & Loy, 2020). They recognize that *greed, hatred*, and *ignorance*, which Buddhism identifies as the root causes of suffering in the individual, need to be challenged where they are found embodied in systemic and institutionalized forms. While adhering to the Buddhist principles on the practice of mindful *awareness, compassion, nonviolence,* and a lifestyle of simplicity, Buddhist activists are applying their critique and practice to specific social and ecological issues.

Buddhism views that all beings are interconnected with one another in a web of interdependence and interconnectedness. This notion of interdependence is part of the Buddhist understanding of the law of the *cause* and *effect* which governs all events in the planet. Nothing happens apart from or contrary to the cause and effect according to Buddhism which does not allow for accidents or divine intervention into the operations of cause and effect. From Buddhist perspective, it can be clearly seen anything happening anywhere in the planetary ecosystem can have cause and effect relations on other things in the planet. Its implication can be seen in the resource extraction, utilization, and consumption in the world. For example, resource extraction and utilization anywhere in the world has repercussion throughout the entire planetary system. As Rita Gross (1997) points out to that effect: "*Often consumption of luxuries in one part of the world is directly related to the poverty and suffering in other parts of the world. Thus, the vision of universal and all-pervasive interdependence, which is so basic to Buddhism requires moderation in all activities, especially reproduction and consumption because of their impacts on the rest of planet.*" The recent development in the international law provides some optimism in the treatment of living beings. The international law has recognized the latest development in the evolutionary biology, physics, and ecology and has incorporated in it some elements of these development to justify principles and guidelines developed to promote environmental protection and sustainable development. *The law recognizes the unity of biosphere, the interdependence of humanity and Nature, the interconnectedness of all members of the larger community of life, the importance of biodiversity as well as cultural diversity.* Rockefeller (1997) informs that there has been a significant convergence of Buddhist philosophy and the contemporary physics, evolutionary biology, ecology, and environmental ethics.

A significant development in international environmental law consists of its recognition that all life-forms and species warrant respect and protection because they possess *intrinsic values* apart from whatever instrumental values they might offer to human beings. The inclusion of the statement that all species have intrinsic value in the international legal document is a significant development which cannot be overstated. Following statements by Rockefeller (1997) powerfully reflect this fact: "*It is a major breakthrough, a move beyond the traditional anthropocentric worldview that has dominated Western culture and much of the rest of the world in recent centuries. It establishes a basis for extending the community with which human identify and for which they are morally responsible to include all life forms. It means that non-human species deserve respect and care regardless of their instrumental value to humans. They are to be treated, in Kantian language, as ends-in themselves and never as a means only. In other words, nonhuman species are to be regarded as subjects with moral standing and not merely as objects to be processed and used.*" Odin (1997) observes that in the past, ethical discourse has been confined to the human community pertaining to the relation between individuals and society. Leopold (1966) extended this into the realm of the "*biotic community*" of soil, plants, and animals so as to include the symbiotic relation between humans and the land. This is clearly articulated in Leopold's concept of land ethics: "*All ethics so far evolved rest upon a single premise: that the individual is a member of a community of interdependent parts …The land ethic simply enlarges boundaries of the community to include soils, water, plants and animals or collectively: the land.*"

Donald Swearer (1997a, b) also brings a very useful perspective from East Asia Buddhist monk Buddhadasa, a prominent figure of Thai Buddhism. Swearer (1997a,

b) asserts that for *Buddhadasa*, things in their natural true state are characterized by their dynamic interdependent nature, known as the Buddhist doctrine of dependent co-origination *(Pratītyasamudpāda)*. According to this doctrine, everything is linked in a process of interdependent co-arising or as *Buddhadasa* often says: "*We are mutual friends inextricably bound together in the same process of birth, old age, suffering and death. In other words, the world is a conjoint, inter-dynamic, cooperative whole, not a collection of disparate oppositional parts.*" Consequences of today's environmental problems (global warming, climate change, and waste throughputs) have become so pervasively threatening to human existence that the humanity is beginning to realize how the destruction of Amazon rain forest or ocean dumping of toxic waste affects the entire planetary ecosystem. How we human extract and consume resources affects not only us at individual level but the health of the entire planetary ecosystem. Therefore, as Swearer argues the idea to care for *(anurak)* Nature *(thamachat)*, therefore, stems from a realization that "*I do not and cannot exist*" independently of my total environment. In Buddhadasa's terminology, I do not and cannot exist onto myself because to do so contravenes the very law of Nature. The underlying view of *Buddhist eco-dharma* is caring for Nature necessarily means not only that we care for other human beings but also that we care for ourselves. Outwardly, Nature *(thamachat)* means physical nature. But the inner truth of nature is *dhammadhatu* (the essential nature of *dhamma*,) namely the *interdependent co-arising* nature of things *(Paticcasamuppada)*. When we realize this truth, the truth of *dhammadhatu (Paticcasamuppada)*, when this law of very nature of things is firmly held in our hearts and minds, then we will overcome the *selfishness* and *greed*. By caring for this truth of interconnectedness and interrelatedness, we are then able to truly care for Nature, the living system, and ourselves.

11.2.2 Conception of the Self

Buddhism challenges the conventional notion of a fixed, permanent self or identity and offers a unique perspective on the nature of existence. It asserts that all being, and phenomena lack a permanent, independent self-entity, and clinging to the illusion of a self leads to suffering. Buddha taught that all phenomena, including the self, are impermanent, constantly changing, and devoid of inherent essence. This understanding is encapsulated in the concept of "*anatta,*" in Pali and "*anatman*" in Sanskrit translated as "*non-self*' or "*no-self*' (Malhotra, 2014). The absence of a fixed self implies that our sense of identity and individuality is a result of complex causes and conditions, rather than an inherent and separate autonomous entity. This understanding helps to alleviate attachment, clinging, and the suffering that arises from an exaggerated sense of self-importance and self-identity (Sponberg, 1997; Odin, 1997; Collins, 1982). Instead of identifying with a fixed identity, Buddhism emphasizes the recognition of the fluid and ever-changing nature of existence. Buddhists recognize the practical aspects of the uses of individuality to navigate the world. However, they emphasize that these designations should not be understood as the ultimate inherently existing autonomous self. Buddha was more concerned to characterize the nature of the *self* in terms of its end purpose rather than in terms of its functional self in everyday life. He was more concerned with the question of realizing one's potential for *enlightenment* and the elimination of *suffering*. Both these features of the Buddhist conception of the self (the dynamic and the developmental) have significant implications for the relationship of that self to the rest of existence including Nature and the environment. Perhaps, the single most distinctive and radical teachings of *Buddha* was the notion of the *non-substantiality* of the *self*, the doctrine of "*no-self*" or "*non-self*" (Collins, 1982). The doctrine of *annatta* or *anaatman* emphasizes interconnectedness, interdependence and the absence of a separate, enduring self. This doctrine of the *non-substantiality of the self* lies at the very of heart of the *Dharma*.

In the West, whether seen in religious or scientific terms, what most constitutes the nature of the self is its very specificity, usually understood as a *species-specificity* or *identity*. It may seem overly simplistic to point out such a basic fact, yet precisely because this view of the nature of the *self* is so *axiomatic,* we fail to see how much it shapes the attitudes we have toward each other, toward our fellow beings, and toward our environment. Hence, the importance of clearly identifying the striking contrast in the Buddhist conception of the *personal identity* and continuity—not just so that we understand Buddhism more accurately, but also because we may, in the process, come to a more accurate understanding of our own cultural roots as well. As Alan Sponberg (1997) observes: "*My thesis regarding Buddhist attitudes towards nature and the environment is based on the premise that our relationships with other beings, especially those of other species, are significantly shaped by the understanding of personal identity that we bring to those relationships. With a conception of personal identity that is fundamentally trans-human, Buddhists have traditionally shaped the problem of inter-species relationships in quite different terms, and as a result we should expect traditional Buddhist environmental ethics to look quite different from its counterpart in the West.*" Buddhist concept of *self* is *trans-human* and seeks its *identity* with other beings in Nature. Buddhists see themselves in other beings and other beings in themselves because of *dependent co-origination* (Buddhist doctrine of causation). This doctrine, called *Pratiityasamudpada* in Sanskrit, states that the origin of all living beings are conditioned by *dependent co-origination* and are interconnected in the web of creation and, therefore, human *compassion, intellect*, and *wisdom* must be directed to strengthen this *web of interconnectedness* but never to destroy it. Buddhist environmental ethics emerges from the fundamentals of *dependent co-origination (Pratiityasamudpada)* doctrine and transcendental identity of the self.

11.2.3 Compassion and Buddhism

Buddhadasa and Phra Prauudh, the two towering influential Thai Buddhist monks, bring valuable traditional Buddhist resources to address current environmental problems. Phra Prayudh espouses several doctrinal principles relevant to a Buddhist environmental ethics from the traditional ***Pali*** text. He is considered the most authentic scholar of the ***Pali*** literature in which Buddha's original teachings were recorded because ***Pali*** was the predominant language of the time and the region Buddha lived. Phra Prayudh attributes current environmental destruction to a Western worldview flawed by three erroneous beliefs: that *humankind is separated from Nature*, that *human beings are masters of Nature*, and that *happiness results from the acquisition of material goods* (Swearer, 1997a, b). He argues all these three are wrong views that must be changed or transformed if environmentally destructive attitude and actions are to be curbed. Phra Prayudh holds the conventional *Theravada* position that *right views lead to the right actions and until the right view prevails and until human beings are seen as part of Nature, the worldwide trends toward environmental destruction cannot be stopped.* Unlike Buddhadasa's *Dhammic biocentrism* (identification of Nature as Dhamma), Phra Prayudh emphasizes the centrality of Buddhist ethical values for an environmental philosophy. He stresses, as Swearer points out, three Buddhist moral values that promote a positive, beneficial attitude toward the environment, including plants, animals, and fellow human beings: *gratitude (katannu)*, *loving-kindness (metta)*, and *happiness (sukha)*. The following quote from the *Khuddaka Nikaya* (collection of minor dialogues) illustrates his concept of the gratitude: *"A person who sits or sleeps in the shade of a tree should not cut off a tree branch. One who injures such a friend is evil."*

Phra Prayudh links together the moral value of *gratitude* and *loving-kindness* to develop a more holistic view of Nature. He thinks the concept of *loving-kindness (metta)* originates from the recognition that humans and all other sentient beings are bound together in a universal process of *birth, old age, suffering,* and *death*. He thinks such sense of mutuality promotes cooperative and helpful feelings and actions toward everything around us rather than competitive and hostile ones. This certainly helps us in forming our sympathetic attitude toward plants and animals and constrain our reckless actions with the contemplation of the possible detrimental consequences of such actions. Unlike Buddhadasa's ontologically oriented perspective, Phra Prayudh emphasizes the *karmic* (action activities) side of mutual interdependence of all life-forms, noting that we need to carefully weigh the consequences of our actions so that we do not willingly increase the suffering of sentient and non-sentient beings. As to the third ecologically relevant moral value, Phra Prayudh espouses Buddhist teachings that *human happiness (sukha)* is dependent on our natural environment in two ways, firstly, simply living within natural setting engenders a greater sense of happiness and well-being and, secondly, Nature serves as a teacher of both mind and spirit. Forest plays a pivotal role in the development of Buddhist ideas and ideals. Monks pursued their vocation in the forest, and forest is the ideal location for training the body and mind to overcome defilements (*kilesa*) that hinder the attainment of mental freedom. Donald Swearer (1997a, b) explains that Phra Prayudh's ecological hermeneutics focuses on the life of the Buddha and Sangha as exemplifications of the Buddhist attitude toward Nature, in particular, toward forest. Chatsumaran Kabilsingh (1990), a strong advocate of "Green Buddhism" also concurs with Phra Prayudh's view: *"From the time the Buddha left his palace, Buddhism has been associated with forests. The Buddha's quest for the truth (saccadhamma) took place in the forest. It was in the forest that for six years he sought to overcome sufferings and it was under the Bodhi tree that he attained enlightenment. Throughout his life, Lord Buddha was involved with forests, from his birth in the forest garden of Lumbini under the shade of a Sal tree to his Parinirbana under the same kind of tree. Thus, Buddhism has been associated with the forest from the time of the life of its founder.*" Buddhism is profoundly enriched with resources on Nature and environment, and thus many scholars find within its rich tradition valuable perspectives that resonate with contemporary concerns regarding environment and humanity's place within Nature. Environmental ethicists, Buddhist monks, and scholars, in an attempt to address the growing environmental challenges, have proposed a new concept frequently referred to as "*Buddhist Ecology*" which apparently has a great potential to contribute to the cause of Nature protection and sustainable development. Perhaps, these two are seminally the most important agenda of this century upon which lies the fate of the Earth systems and the future of humanity in it.

11.2.4 Dimensions of Buddhism

Alan Sponberg (1997) has profoundly elaborated on two fundamental aspects of basic Buddhism, namely, *developmental dimension* and *relational dimension*. These two dimensions apparently seem quite distinct from each other but at deeper level, each complements the other to maintain the integrity of Buddhist tradition. The *developmental dimension* focuses on the transformational aspect of the Buddhist tradition as a practical means of *spiritual growth* and development. According to Sponberg (1997), Buddhism sees the spiritual life as the transformation of delusion and suffering into *enlightenment* and *liberation*. Even the Zen Buddhism (nondual form of Buddhism) acknowledges an experiential distinction between delusion and enlightenment and recognizes the existential reality of suffering. The second dimension of Buddhism, as Sponberg calls it *relational dimension*, entails the Buddhist conception of the *interrelatedness* and *interconnectedness* of all things that exist in Nature which encompasses not just all sentient beings but every aspect of the ecosystem in which they participate, ultimately, the *ecosphere/biosphere* in its totality.

Sponberg provides the graphical illustration of these dimensions by plotting them in vertical and horizontal axis, with vertical axis indicating the *developmental dimension* and the horizontal axis indicating the *relational dimension*. These two dimensions actually represent the two traditions of Buddhism, namely, *Theravada Buddhism* often called, *Hinayana* and *Mahayana*. *Theravada Buddhism (Hinayana)* places relatively more emphasis on the *developmental dimension*, while *Mahayana* including *Vajrayana* and *Zen Buddhism* places relatively more emphasis on *relational dimension* of the basic Buddhism. Similarly, the *Indo-Nepal-Tibetan* forms of Buddhism tend to be closer to the *developmental dimension*, whereas East Asian forms of Buddhism tend to be closer to the *relational dimension* of the Buddhist traditions. It is important to remember that these two traditions are not mutually exclusive, and yet they differ in their relative degree of emphasis over developmental and relational dimensions. Historically, there is no totally one-dimensional form of Buddhism. It is often pointed out that Buddha's *enlightenment* was a combination of *compassion and wisdom*. The realization of a wisdom driven by compassion by virtue of its insight into the fundamental interrelatedness of all existence, thus, inferring that Buddha's enlightenment was *interrelational*, something that could only exist in the context of *compassionate activity.*

Sponberg (1997) warns us that it is epistemologically wrong to associate Buddha's enlightenment to horizontal or relational dimension alone derived from *emptiness doctrine (sunnyavada)* of *Mahayana sutra* and ignore the importance of *developmental dimension* in the evolution of compassion and wisdom with much wider degree of relational interrelatedness. He further explicates "*We must be careful not to assume that recognition of this relational dimension of the Buddha's enlightenment was a purely Mahayana innovation. .... Is the presence of any vertical, developmental perspective antithetical to our newly gained recognition of horizontal relatedness? We miss the point that for Buddhism neither is possible without the other. The developmental and the relational are not only complementary, but they are also inseparably interrelated. Both these dimensions are inextricably linked to Buddhist ethics.*" One criticism labeled against *Green Buddhism* is that it shows a subtle tendency that significantly distorts the assimilation of the *Dharma* into the West which has the tendency to reduce Buddhism to one-dimensional teaching of simple *horizontal interrelatedness*. Some critics go even further to say that Green Buddhism has developed a tendency to disavow or even deny the crucial element of traditional Buddhism. There is no doubt that the forms of Buddhism that has apparently attracted the attention of Western scholars and environmentalists is the horizontal *relational interrelatedness* aspect of the basic Buddhism which may have initially overlooked the importance of *developmental dimension*. This is more of a cultural phenomenon. The fact is that both these dimensions are very much ingrained in the basic Buddhism, and they are, in fact, complementary to each other.

Many Buddhists reject hierarchical dominance of one human over another or humans over Nature and regard compassion as the basis for ethics which respects biodiversity and social justice. Sponberg raises an important question while discussing developmental dimension of Buddhism. He asks, "*What is the developmental dimension of Buddhism if not the teaching of the evolutionary transformation of consciousness?*" The very definition of *Buddhahood* asserts the developmental realization of a higher ethical sensibility expressed as *compassion* for all of the existence. One can argue that there is a crucial difference that distinguishes the Buddhist conception of developmental hierarchy from those forms of power hierarchy that has dominated the Western cultural history. Only then one can see the value inherent in the Buddhist notion of *developmental hierarchy* precisely for the sake of better environmental ethics, just as we strive to abandon the most Western notion of hierarchy for the very same reason. Sponberg equates Western notion of hierarchy with the "*hierarchy of oppression*" and argues the nature of this hierarchy is such that as one advances vertically, one's "*circle of interrelatedness*" becomes increasingly narrower and smaller. This is because one advances in a hierarchy of oppression by exercising one's control over and domination of all those below. Consequently, as one moves higher vertically, one necessarily becomes less and less aware of one's interrelatedness with others in the horizontal space. But from Buddhist perspective, one's actual interrelatedness is constant and absolute. The progress in the hierarchy of oppression requires that one actively deny and suppress any recognition of relatedness to those that one seeks to dominate. It is evident that in such hierarchy, one moves upward only by gaining power over those and to safeguard one's power and security one must seek ultimately to control all of existence. When one reaches the apex cone of the hierarchy, one is totally alienated not just from others but from oneself as well because one can reach apex (success) only by rejecting one's actual nature of interrelatedness.

Sponberg turns hierarchy of oppression upside down and makes it stand on its head so that the apex cone hits the bottom, and the cone broadens as it rises hierarchically. He calls this model "*hierarchy of compassion*." As one ascends in the vertical (developmental axis), something quite different happens, something that is precisely the inverse of the hierarchy of oppression as discussed previously. As one moves upward in consciousness, the circle of one's interrelatedness increases (Fig. 11.1). In fact, the only way one can move up is by actively realizing and acting on the fundamental interrelatedness of all existence. In the hierarchy of compassion, vertical progress is a matter of reaching out actively and consciously to affirm an ever-widening circle of expressed interrelatedness. Unlike the previous one, this model, quite contrarily, embodies in it an ever-increasing sense of responsibility. "*This profoundly*" as Sponberg observes: "*ethical sense of responsibility for an ever-greater circle of realized relatedness is what is expressed by Buddhist term, Karuna— compassion or 'wisdom in action.' Buddhism does offer an ethic that might be capable of transforming our current deluded environmental practice, but the developmental dimension of the tradition is crucial to that ethic, because the Buddhist virtue of compassion is something one can cultivate only by progressing up in the spiral path of hierarchy of compassion.*"

How can we expect human beings to act in an environmentally ethical manner without cultivating the capacity to act with *compassion* and *wisdom* toward interrelated and interconnected *web of life* in Nature? The hierarchy of compassion is, in fact, the hierarchy of empowerment through which human beings can cultivate higher level of compassion through higher level of consciousness which, I would like to call it, an *ecologically compassionate consciousness* (compassion-induced wisdom) that enables one to interrelate with others in the *web of interrelatedness*. As Sponberg (1997) illustrates in Fig. 11.1, human capacity for compassion increases when human *consciousness* evolves to a higher level, consequently, human beings can visualize their interconnectedness and interrelatedness with others and Nature more clearly. Once this capacity to interrelate oneself with others in the relational field is achieved through the cultivation of compassion, then one begins to see the value of interrelated web of life and, therefore, becomes actively engaged in the protection and preservation of this web of cosmic interrelatedness metaphorically expressed as *Indra's Net*. The *ecologically compassionate consciousness* can be regarded as the *emergent phenomenon* that results from the fusion

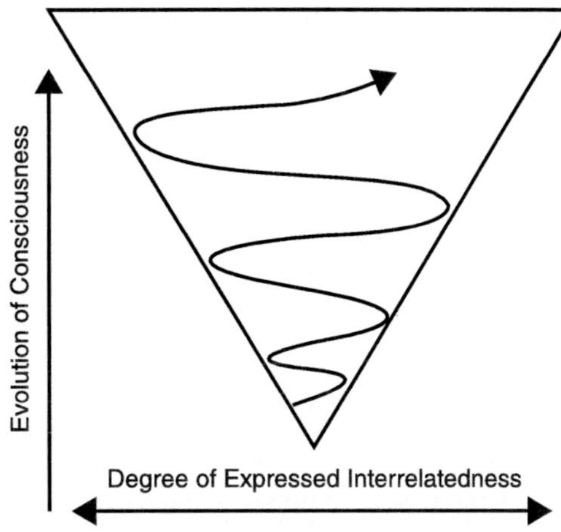

Fig. 11.1 A hierarchy of compassion. (Adapted from Alan Sponberg (1997). As human consciousness evolves to a higher level, the compassion, and the sense of interrelatedness with others increases)

of *developmental* and *relational* dimensions of basic Buddhism. We must understand the intertwined relationship between developmental and relational dimension (the degree of the interrelatedness). Higher the level of *ecologically compassionate consciousness*, higher is the degree of *cosmic interrelatedness* that one can visualize. They actually complement each other and, in essence, the dichotomy between them disappears. If we look at the *consciousness* from an evolutionary perspective, we find that human consciousness, both biologically and culturally, is gradually evolving over time. This implies that some individuals in some cultures may have advanced far along this evolutionary path pointing to a natural order or hierarchy of conscious development. Ken Wilber (1997) believes evolution of consciousness proceeds by transcendence and inclusion, where each new stage transcends and includes all those which went before. The evolution of consciousness proceeds through the identification with the whole and understanding of one's interrelation with other beings in the web of life resulting in the expansion of compassion-induced wisdom or *ecologically compassionate consciousness* as described above. Bonnett (2017) asserts that individual's being in the world is ineluctably environmental and the engagement of human consciousness with Nature presents opportunities for consciousness to go quintessentially beyond itself, to be inspired and refreshed, and to receive non-anthropogenic standards which is deeply ecological.

Spiral dynamics model views that human nature is not fixed. Humans are able to adapt to their changing environment by constructing new, more complex models of the world that can help them to resolve new problems. Beck and Cowan (2005) argue that these conceptual models are organized around "*meme*," a system of core values or collective intelligences applicable to both individuals and entire cultures. *Meme* (collective core values system) acts as an organizing principle, which expresses itself through memes (self-propagating ideas, habits, or cultural practices). The *spiral dynamics* model posits infinite stages of development dependent on the life circumstances of the person/culture and entails the potential of attaining higher stage of development. Beck and Cowan (2005) use the term "*spiral dynamics*" to describe the mechanism of how *consciousness* evolves toward higher stage with increased *wisdom* and *compassion*. Through such process, some of the great personalities in some wisdom traditions such as *Bodhisattvas* in Buddhism and *Rishis* in Vedantic tradition of Hinduism have reached a level of consciousness far in advance of the ordinary persons which can be seen as the logical outcome of the development of "*meme*"-driven consciousness within human family. *The evolution of consciousness supports the hypothesis that consciousness expands toward an ever-widening inclusiveness encompassing all the beings in the web of life, which can be described as wisdom with compassion or ecological consciousness.* This implies that every human being has the potential to eventually become a *Rishi, Saint, Bodhisattva*, an enlightened being. This *ecologically compassionate consciousness* (compassion-induced wisdom) encompassing all beings, recognized within Buddhism as the *Bodhisattva*, is increasingly evident today, and it can connect us with those who watch over the inner springs of inspiration which are needed to forge a new civilization.

11.2.5 Buddhism, Ecological Worldview, and Ethics

The most spectacular aspect of Buddhism is that it sees no human life without Nature which means life on Earth can only be conceived as an integral part of the Nature. According to Buddhism, all things on Earth exist by their interrelationship with Nature. That is why it can be said that Buddhism has an *environmental worldview* or *ecological worldview* (eco-dharma), and the concept of Buddhist reality is *ecological*. We know that change is inherent in Nature and according to Buddhism changeability is one of the perpetual principles of Nature, but Buddhism believes that such natural processes should not be disrupted or negatively affected by human actions. Human behavior and anthropogenic activities should be guided by *eco-dharma* or an ecological or environmental ethics that establishes a harmonious relationship with planetary ecosystem. Buddhists see and conceptualize Nature as a *living web* that interconnects individual beings in an interactive interdependence. According to Buddhism, humans are not separated from other living beings. *Egocentric anthropocentrism*, on the other hand, treats Nature as an object or a thing to be exploited, but the truth is that Nature is not a part of us, or it belongs to us. On the contrary, we, human beings, are a part of Nature and for our own existence and survival, we depend on Nature and its products (ecosystem services). The native American Indian proverb profoundly expresses this wisdom: "*Only when the last tree has been cut down; only when the last river has been poisoned; only when the last fish has been caught; only then you will find that money cannot be eaten.*" This American Indian proverb speaks volume on the concept of Buddhism's ecological worldview that how and why we are so interrelated and interconnected. The wisdom that emanates from this proverb has an immense implication that man-made things have no value without Nature and its services because our existence depends on the health and the proper functioning of the Nature, the planetary ecosystem, and its services. Buddhism has beautifully integrated and embraced this view which is well reflected in its *eco-dharma* (ecological view of Buddhism) and the doctrine of dependent co-origination *(Prattityasamudpada)* which focuses on causal interconnectedness of phenomena in Nature and, therefore, motivates human activities for the protection of species and ecosystems, the very basis of humanity's existence.

Our worldviews shape our basic understanding of the Nature and our attitude and behavior toward it. The current prevailing belief system, which makes us think that we are separate and superior to Nature, will only motivate us to see Nature as an object to be exploited. On the contrary, if our understanding has been shaped and informed by the knowledge of million years of biological evolutionary processes and the doctrine of dependent co-origination, as taught by the Buddhism *(Prattityasamudpada)*, we are more likely to appreciate our place in Nature and see ourselves as an elite member of this interdependent and interrelated web of the jewel of *Indra's Net*. The works of some visionary thinkers and activists (Snyder, 1990; Macy, 2017; Berry, 1990a; Thich Nhat, 2017) have shown how the Buddhist view of interdependence and the practice of insight and compassion (eco-dharma) can lead to environmental actions that not only preserve and protect the natural ecosystems but also lay the foundation of sustainable society. The environmental actions emanated from the practice of insight and compassion (eco-dharma) can heal this ruptured relationship of mankind with the Nature so that the relationship becomes reciprocal one meaning eco-dharma brings us to an authentic caring for the Nature and at the same time the sacredness of Earth and its services sustain us. If we care the mother Earth, she will reciprocate with her bountiful services on which humanity can sustain itself.

As humanity will have to increasingly confront with the adverse effects of the climate changes, biodiversity loss and many other environmental problems in the days ahead, the Buddhist teachings and practices of *empathy, nonviolence, cooperation, compassion,* and *wisdom* of interdependence and interconnectedness become not only relevant but also a stark reality for our own survival. A change in human consciousness that embodies *empathy, cooperation, compassion,* and the *wisdom* of interdependence and interrelatedness is the key for the less painful transition of humanity from current state to a more sustainable state. The cultivation of such consciousness is a natural part of the Buddhist path or eco-dharma. Buddhism taught us to understand more deeply the underlying unity and interconnectedness of life. Buddhist values such as simplicity of lifestyle, sharing with others, taking responsibility for one's actions, empathy, and compassion for all living things have always been at the heart of the *eco-dharma* (the Buddhist way of relating oneself with the other living beings). It is apparent that these values have become absolutely more important today than ever before, and the ethics of the Buddhist eco-dharma are not only identical with environmental ethics but also involve the conscious choices in the way we lead our lives and run our everyday business. From Buddhist perspective, ecological awareness is an opening of the heart to the whole web of life and deep appreciation of the natural beauty of which human beings are a part. As Thich Nhat Hanh (1997) has rightly pointed out the fate of humanity in the following words: "*We have constructed a system we can't control. It imposes itself on us, and we become its slaves and victims. We have created a society in which the rich become richer, and the poor become poorer, and in which we are so caught up in our own immediate problems that we can't afford to be aware of what is going on with the rest of the human family or our planet Earth. In my mind I see a group of chickens in a cage disputing over a few seeds of grain, unaware that in a few hours they will all be killed.*" Thick Nhat reminds us that it is not enough just to blame governments and global corporations for environmental destruction and pollutions, for the violence in our neighborhood and the mindless wars that destroyed so many lives. We need to organize ourselves and wake up to take actions in our own lives. Buddhism can greatly help us to live in peace and harmony with Nature and other living beings. Buddhism teaches us how to live with responsibility, compassion, and loving-kindness.

From the description of the developmental dimension of hierarchy as discussed above, it can be said that this dimension is about the empowerment of the self, the transformative power of the compassion. This dimension of hierarchy cultivates the conscious efforts to transform oneself and evolve one's awareness to a higher level that can benefit all life-forms including humankind. As Sponberg (1997) emphatically stressed: "*This transformative power of compassion that results from the developmental dimension of Buddhism must not be ignored because without this the assertion that all beings are interrelated becomes an empty rhetoric, devoid of any essence. It is inconceivable to act in an environmentally ethical manner without cultivating the capacity to act with compassion and wisdom.*" The "*threefold learnings" (trisiksa)*, the traditional teachings of Buddhism, provide the pathways to cultivate compassion and wisdom (compassion enriched consciousness). These are systematic cultivation of morality, meditation, and insight into the actual nature of existence. These are sequential steps meaning each step builds on the preceding one. There are two most significant implications one can draw from Buddhism. The first is an individual person can act in accordance with the basic interrelatedness of all existence once that person has cultivated a significantly higher state of awareness or consciousness. The other equally important implication of Buddhism is that both relational and developmental dimensions are inherently the essential components of Buddhism and are complementary to each other in that the relational dimension of Buddhism is necessary to secure an ecologically sound vision, but the developmental dimension is necessary for the pathways to enable one to reach that vision with elevated consciousness.

In conclusion, it can be said that in order to develop a coherent and powerful environmental ethics, Buddhism, more than any other wisdom, faiths, or religious traditions, offers the most pragmatic and useful perspective. Buddhist notion of compassion with wisdom *(ecological consciousness)* arises from the fusion of developmental and relational dimensions of Buddhism. Buddhist scholar and thinkers have eloquently postulated that the developmental and relational dimensions of Buddhism are complementary to each other and are necessary for the evolution of compassion with wisdom (an *ecological consciousness)* without which it is inconceivable for humans to relate themselves to the Nature and consider themselves as a part of Nature just like other beings in Nature. The complementary relationship between developmental and relational dimension is the foundational base for the evolution of the consciousness of compassion with wisdom to a higher enlightened level, called the *Buddha Nature*. The following Sanskrit *sutra* eloquently epitomizes the very essence of Buddha Nature. *"Buddho vabayem jagato hitaya. Buddho vabayem Prakriti hitaya."* (Be a Buddha for the wellness and the goodness of the World and the Nature).

11.3 Gaian Hypothesis and Planetary Ecosystem

James Lovelock was the principal formulator of the *Gaia hypothesis*. He "supposes the Earth to be alive" (Lovelock, 1988). This hypothesis has been hotly debated for almost two decades. For many scientists and philosophers this assertion is pure heresy, as "life" is usually considered a characteristic of organisms, and in the *Gaia hypothesis* life is attributed to a grand composite of organisms and abiotic matter and processes, the Earth itself. Wallace and Norton (1992) argue that the advocates of the Gaia hypothesis must either argue that the Earth is more than a composite and that it has an individual identity and is capable of agency or admit that the adjective *"alive"* is used in a metaphorical sense. Wallace and Norton contend that the *Gaia* hypothesis as expressed by the metaphorical interpretation provides an important and interesting departure for discussing problems in environmental policy, but that confusion of the metaphorical interpretation with claims of literal truth have diverted attention from its importance.

11.3.1 The Gaia Hypothesis

Lovelock was critical of the work of both geologists and biologists as incomplete and overly narrow. Geologists see the Earth as a rock covered with a thin layer of air, while biologists assume that because of their ability to adapt, organisms could live on an Earth of almost any physical structure (Lovelock, 1988a, pp. 11–12). Lovelock proposes to combine these two viewpoints in a new science and coined the term "*geophysiology*," which examines both the living and nonliving portions of the *biosphere*. Within this framework, the Earth and its life comprise a system "*that has the capacity to regulate the temperature and the composition of the Earth's surface and to keep it comfortable for living organisms*" (Lovelock, 1988a). It is through these *homeostatic processes* that the Earth as a whole can be envisioned as *alive*. The Gaia hypothesis is more than the sum of the hypothesis and operates on a deeper, metaphorical level in which it produces a richer "*world view*," a worldview more conducive to understanding our role in a larger scheme of thing, particularly our role in Nature. On this deeper level, Gaia is not a specific hypothesis but a meta-theory guiding hypothesis formation. Gaia can be accepted as a *useful metaphor* quite independently of whether one accepts or rejects any of the specific hypothesis that it generates. As a metaphor, the Gaia hypothesis seems mainly to imply a system approach, a consideration of human activities within a set of larger and larger, hierarchically organized systems, namely the *planetary ecosystem*.

Wallace and Norton point out that Gaian theory proposes a new scientific and management paradigm. This new paradigm (worldview) represents an alternative set of assumptions, values, and methods to unify the scientific disciplines that examine the interaction of living and nonliving systems and especially, the interaction of human activities and the larger, natural systems that constitute their context. Gaian theory, therefore, gains its power not just from the general empirical hypothesis that forms its scientific basis but also from the associated recommendation that the theory serves as an example of a new type of scientific investigation. One can adopt the new paradigm and conceptualize Nature hierarchically, without affirming the literal truth that the biosphere is alive. The justification rather can be asserted from outside any particular scientific discipline, and it can be pragmatic; the worldview provides a richer conceptualization for describing and managing planetary ecological system (Carnap, 1950; Norton, 1977).

11.3.2 Concept of a Living Earth

Scholars point out the claim that the Earth is literally alive does not follow without further argument from the scientific hypothesis that life maintains the conditions under which life flourishes. Common thermostats, for example, act homeostatically, but we do not thereby conclude that they are alive because we understand the feedback mechanism that drives the thermostat. One could demonstrate that the Earth is literally alive by showing that not only does the biosphere act homeostatically, but that it also acts intentionally. If the Earth acts intentionally, then it must be alive; many forms of life, however, lack intentionality and interpretation of the Earth as alive and acts intentionally represents a considerable strengthening of the hypothesis. This understanding of the Gaia hypothesis is sometimes suggested when the *biosphere* is personified in the personal goddess *Gaia*, and when it is implied that Gaia regulates planetary conditions with the prior purpose of maintaining conditions hospitable to life. The idea that Gaia is capable of intentional or teleological action should not be taken literally, but perhaps can be used in extended sense (Leopold, 1979). On the other hand, if Gaia hypothesis is regarded as a useful metaphor or a shift in worldview, then Gaia's role will be indirect; the hypothesis and its associated *metaphor* will locate humans within a larger context, and it will orient moral decision-makers toward particular questions as morally relevant.

Lovelock himself explicitly rejected the strongly organicist, teleological interpretation of Gaia. In response to harsh criticism of this interpretation, Lovelock and his colleague Andrew Watson developed a computer model called *"Daisy World"* to show that large systems can evolve so as to maintain conditions favorable for life without any teleological direction. While this model may answer critics' accusations that Gaia is *teleological*. It does seem to signify a retreat from the position that the Earth itself is literally a living organism, especially one personified as a Goddess. One of the most notable and significant aspects of Lovelock's hypothesis is Gaia's "*ability*" to maintain conditions for life despite enormous changes in many of the factors influencing those conditions. For example, the sun has increased in luminosity since life appeared on Earth, but the temperature of the Earth has remained relatively constant. As the sun warmed according to Lovelock, carbonate secreting plankton removed carbon dioxide (a greenhouse gas) from the atmosphere by absorption and, after the plankton died, they got deposited into the bottom of the ocean, thereby counteracting the increased heat of the sun. This process, then, works somewhat similarly to the thermostat in a house or an apartment, which uses feedback from the internal environment to regulate the ambient temperature at the set point. The Earth, however, is a dynamic system, and new equilibria are continually being reached; extending the analogy of the thermostat, in addition to responding to changes in the external environment, the thermostat also reacts to changes in the set point made by occupants. Therefore, Gaia is more properly viewed as homeorhetic having the tendency to maintain a relatively steady state about a changing equilibrium than as homeostatic (Sagan & Margulis, 1984). The self-regulatory aspect of Gaia has drawn harsh criticism. Some biologists, in particular biologist Ford Doolittle (1981), have criticized that *Gaian self-regulation* is *teleological,* that regulation of the Earth by life requires some sort of conscious action on the part of organisms to cooperate and to know how to evolve so as to best benefit life (Joseph, 1990, p. 113). Supporters of Gaia recognize the importance of this criticism *"if only the results and not the feedback processes were stated, it would look as if the organisms had conspired to ensure their own survival"* (Sagan & Margulis, 1984, p. 66) and have responded with the Daisy World model.

In addition to this model, work in system theory suggests that order can indeed arise out of *chaos*. Contemporary physical theorists have shown that some random and chaotic physical processes can (when a particular threshold is passed) evolve into highly ordered system (Prigogine & Stengers, 1984). In a chemical clock, for example, a system with only two types of molecules in random motion approaches a limit cycle with stable, periodic behavior. Instead of seeing a mixture of the two types of molecules, an observer looking at one part of the vessel containing molecules would see all of one type of molecule, then all of the other in a regular cycle. They surely do not possess any intelligence, but "*to change .... all at once, molecules must have a way to communicate*" (Prigogine & Stengers, 1984, pp. 146–48). The evidence of chemical clocks and other self-organizing systems then strengthens Lovelock's argument that even systems lacking an inherent intelligence or teleological purpose can display *self-regulating behavior.* Lovelock and Watson have demonstrated that holistic action to protect conditions favorable for the maintenance of life need not involve teleology or activity according to a prior intentional plan. Therefore, it is possible to move beyond the unproductive controversy that centers around a literal interpretation of the Gaia hypothesis. Gaian theorists have implicitly agreed that Gaia should be most basically viewed as a productive *metaphor* rather than as a literal truth. If Gaia is viewed as a *metaphor* or *analogy* to explain the functioning of various processes that are necessary for the existence of life, then perhaps the hypothesis can provide a basis for understanding and suggest to scientists which of the Earth's many processes and systems are most vital for the maintenance of the life-forms, including our own species, to which we have become accustomed.

Living organisms and their surrounding environment form an intricately intertwined system. This interconnected system can be likened to a superorganism, and as it evolves, a remarkable attribute can emerge—the capacity to self-regulate. This notion provides the foundation of the Gaia hypothesis, which perceives the Earth and its living inhabitants as a unified system capable of regulating the temperature and composition of the planet's surface to maintain a hospitable environment for life. In the words of Margulis, who proposes the Gaia hypothesis, one can view the Earth as a living entity. However, it is important to note that Gaia is a hypothesis concerning the planet Earth, encompassing its surface sediments and atmosphere. Recognizing that the Earth's atmosphere cannot be fully comprehended through the lens of physics and chemistry alone, biology plays a crucial role in elucidating its exceptional characteristics, such as chemical disequilibrium (Margulis, 1988). While the notion of Earth as a self-regulating superorganism remains a topic of heated debate, the remarkable influence of minuscule organisms on the Earth's atmosphere despite their relatively small size is undeniably intriguing.

11.3.3 Gaian Holism and System

The Gaia hypothesis has not as yet had a significant impact on policy discussions both because of questions concerning its scientific verifiability and because its exact implications are unclear. What would be the policy implications of Gaian theory interpreted metaphorically? It is helpful to begin with Lewis Thomas' statement in the "Foreword" of Lovelock's book, "The Ages of Gaia": "*If Lovelock turns out to be right ... we will be viewing the Earth as a coherent system of life, self-regulating, self-changing, a short of immense organism.*" This statement presents Gaian holism on a planetary scale, with the Earth itself as the largest holon of interest to geophysiologists, but not, of course, the largest holon in an absolute sense (i.e., there is the solar system, the Milky Way), and Gaia depends, for example, on the sun for energy and the moon and the stars for its rhythms (Weston, 1985, p. 230). It is important to notice that Thomas's description represents the change in question as a change in worldviews and yet he capitalizes "Earth," implying personification or deification. This ambivalence is important; the implications derived from the Gaia hypothesis depend almost entirely on the version of holism that is adopted. In what ways, or to what extent, is Gaian theory holistic? Gaia is unquestionably holistic in the sense that it focuses attention away from individuals and even individual species. Humans are given no special importance in Gaia; rather, they are seen as elements in the larger system of the Earth. "Gaia denies the sanctity of human attributes" (Sagan & Margulis, 1984, p. 70). In this respect, Gaia is concerned primarily with the well-being of the living systems of the Earth as a whole and is holistic in this sense.

From system perspective, Gaia is holistic in that it treats lower-level systems as subsystems of larger holons; the analysis cannot, therefore, be reductionistic, but rather must treat low-level holons as parts of a larger system. Lovelock emphasizes, for example, the workings of large systems in relationship to the relative constancy of global temperature and atmospheric composition over millions of years. During this period, millions of species evolved and became extinct, but the temperature and the atmospheric content of the Earth remained relatively constant, providing the setting for the continuance of life. The relative constancy of the Earth's temperature and atmospheric composition also contributes to another holistic aspect of Gaia; higher holons exercise controls over lower holons. Because life, as we know it, can exist only within specific physical limits, Gaia provides the constrains for keeping the global climate within these limits. This allows species the flexibility to evolve and adapt to the slow-changing macro-environment of the Earth. Without the homeostatic regulation that occurs on the higher levels, life may very well have already ceased to exist on Earth. *The idea that Gaia is holistic can be derived from understanding of the symbiosis, connectedness, and interdependence that exists between various life-forms for the purpose of maintaining the global environment.* The development of more complex species could never have occurred if phytoplankton had not removed excess carbon dioxide from the atmosphere three billion years ago. Lovelock points out that individual cells contain organelles, many of which (e.g., mitochondria) once lived as independent organisms (Lovelock, 1988a, p. 18). This provides a vivid illustration of how each individual is both part and whole at the same time. Taking this idea of symbiosis to an extreme, Sagan and Margulis suggest that "*the brain may be a special case of symbiosis among modified bacteria*" (Margulis & Sagan, 1997, p. 70).

As we understand the Gaia hypothesis, however metaphorically, this holism need only imply a contextual, ecological, and systems-oriented conceptualization of the Earth as the context of human life; it emphasizes that all living things are interdependent (Weston, 1985, p. 225). It seems important to avoid personification of Gaia because in its extreme form, literal holism leads directly to "moral holism"—the belief that the good of this superorganism overrides, completely or, at least in important situations, the good of its parts. This moral holism in Gaia has been described as meaning that individuals are irrelevant, the homeostasis of the Earth is of primary importance (Weston, 1985, p. 221). Viewed in this way, not only are humans merely a part of the whole, but their welfare is of no concern unless it significantly affects the larger system. Because *microorganisms seem to play the major role in the self-regulating mechanisms of life*, it has even been suggested that perhaps animals, which are covered by and full of microbes, exist simply to provide living space for the microorganisms that are so essential for the health of the Earth (Sagan & Margulis, 1997 p. 68). Such assertions, however, which seem to discount the sufferings of individual humans and animals, rest on the discredited assumption that evolution proceeds with the prior intention of maintaining life on Earth. If we reject this literal organicism and at this time not even Lovelock argues for the literal conception of Gaia, we can see the Gaia hypothesis more as an orientation than as a specific locus or set of substantive values. Viewed in this manner, Gaia can guide human thinking about values without dictating specific normative rules.

As can be inferred from the previous discussion, the scientific basis of the Gaia hypothesis is rooted in the descriptions of the activity of large systems. It is within systems that Gaian processes, such as *self-regulation*, occur. In fact, it may be easier to think of Gaia as one large, Earth system (planetary ecosystem) that encompasses numerous, less extensive processes, such as the one that regulates global temperature. Open systems receive inputs from the external world, while closed systems do not. The Earth, it should be obvious, is not an isolated system because, if nothing else, the sun provides the Earth and its life with a continuous supply of solar energy. If Gaia were an isolated system, then the Earth would soon become a frozen ball of ice and rock, Lovelock explicitly acknowledges the open nature of the Gaian system when he recognizes that Gaia consists of more than just the biota: "*Gaia, as a total planetary being, has properties that are not necessarily discernible by just knowing individual species or populations of organisms living together*" (Lovelock, 1988a, p. 19). More importantly, Lovelock also recognizes that Gaia is affected by forces external to the Earth; for example, he has written that "*solar energy sustains comfortable conditions for life*" (Lovelock, 1988a, p. 19).

11.3.4 Environmental Problems and Gaia

Having examined the holistic, self-regulatory, and systemic nature of Gaia, and having reasoned that the hypothesis should prove its value as a guide to conceptualizing problems in environmental management, it is reasonable to wonder next how the Gaia hypothesis can be put to actual use in addressing contemporary environmental problems? After all, if the Gaia hypothesis is nothing more than a beautiful and wondrous story, then like its ancient namesake, it is probably best to be catalogued as mythology. Given the opportunity, however, the current incarnation of Gaia has the potential to offer scientists, philosopher, educators, and policy-makers a unique perspective from which to analyze environmental problems.

Before starting to examine environmental problems from a Gaian perspective, it is important to briefly outline the approach that would be used by a hypothetical Gaian policy analyst. First, this analyst would be concerned with the *effect of environmental problems on self-regulating systems; do the problems represent a threat to the system's ability to return to some equilibrium after being disturbed? If not, then how, if at all, will the new equilibrium differ from the previous one?* The answer to these questions will in part require addressing a second question of concern to the Gaian analyst; on what scale (spatial and temporal) should environmental problems be addressed? Are the problems encountered on a regional, global, or local scale? And, how long into the future ought we project solutions? Note that reasoning of this sort can be useful even to scientists who do not believe in the literal existence of Gaia.

In attempting to extract useful information from a scientific theory or model, it is necessary to apply that theory or model at the proper level of analysis. Biologists, for example, do not use a telescope to examine communities of bacteria; they use a microscope. Similarly, Gaia cannot be expected to provide meaningful insights at every level of environmental policy-making. Global environmental problems, I believe, appear to be appropriate for Gaian analysis, but local, and perhaps regional environmental problems appear to be less appropriate for such analysis. The main reasons for this view are related primarily to the large-scale, holistic nature of the Gaia hypothesis. First, Lovelock has emphasized repeatedly that Gaia is a control system that operates on a global level. The slow-changing macrosystems of the Earth provides the context for organisms on lower levels of the hierarchy to change and adapt. If dramatic changes were to occur on the global level, for example if global temperature were to rise suddenly, then species would face rates of change too rapid for them to proceed with their slow evolution, and more complex and highly adapted life-forms might be wiped out. Conversely, the holistic nature of Gaia also implies that the Earth, the Gaian super system, is relatively impervious to deleterious changes in small, isolated portions of the planet. For example, in the history of life on Earth, millions of species have become extinct, but life continues to flourish. Because it emphasizes whole-system characteristics, it provides too coarse-grained an analysis to allow determination of policies on a local level. The Gaia hypothesis has little to say, for example, about policy on the levels at which Superfund sites are chosen or dioxin emission standards are set. Small systems may, like Gaian systems, also display homeostatic activity, and such systems may be altered at a magnitude and rapidity far exceeding their capacity to reach a favorable equilibrium. A lake, for example, can assimilate only so much pollution at a time; if the rate of pollutant inflow exceeds the lake's pollutant assimilation rate, then the lake may jump to a new equilibrium point that is not hospitable to life. Because such changes occur on such a small scale and on a short frame of time, Gaia provides little guidance in analyzing such changes. Having made the case against using the Gaia hypothesis as a model for understanding local environmental problems, however, I would like to qualify this argument by recognizing that Gaian hypothesis does emphasize the importance of systems analysis and, therefore, it can also serve to remind scientists, policy analysts, and others concerned with local environmental problems that their local systems are embedded inside larger systems that provide the local context. Local decision-makers may make good use of Gaian theory in attempting to understand how their decisions may affect the larger surrounding systems and how the larger systems may constrain the effectiveness of proposed measures to resolve the problems at hand. In such situations, systems analysis on all levels may be a natural complement to Gaian reasoning, but Gaia alone is inadequate.

Based on these arguments, then it seems apparent that the Gaia hypothesis will not dictate solutions to local environmental problems. Gaia may be capable of ensuring that life will continue to exist on Earth, but humans shoulder a large portion of the responsibility for deciding which life-forms will continue to exist. Unfortunately, however, humans all too often fail to recognize that in their actions they are implicitly making such choices. Gaia is indifferent to individual species, but Gaia may be sensitive to losses of functions as they might be represented in groups of species (such as blue-green algae) or by keystone species.

11.3.5 Policy Implications of the Gaian Perspective

In the remainder of this chapter, I will illustrate some of the effects of accepting a moderate version of Gaian theory, minimal Gaia, one might call it with particular emphasis on the effects on public policy and its analysis. One advantage of the Gaian image is its validation of medical analogies for understanding management science. In this view, environmental science, conservation biology, and restoration biology should be treated as normative sciences like human and animal medicine which are guided by an ideal of maintaining a healthy patient. Conceptualizing natural systems as self-regulating and self-maintaining, but capable of diminution of these qualities ("illness") therefore, has important consequences for environmental science. *Good environmental science must fulfill not only the normal epistemological criteria of the scientific methods, but it must also be useful in protecting the self-organizing capacity of the systems it studies.* This conclusion has important implications for science policy. Gaian theory and its attendant medical analogies strengthen the case for mission-oriented science over pure science and weakens the case for the current hegemony of traditional scientific disciplines. The medical analogy also suggests a multitiered system for policy analysis. A physician often treats the organ but must worry also about the impacts of treatment on the well-being of the person (an integrated whole of the organs). Similarly, elements of natural and seminatural systems can be "*treated*" for economic reasons, but only if these activities are of a sufficiently limited scale so that the organizational capacity of the larger ecosystem is unimpaired. In other words, such activities should be permitted only up to the extent that they do not degrade the organizational and functional capacity of the planetary ecosystem (the Gaian system).

Even though scientists are debating whether Gaia theory can be accepted as a valid scientific theory that can explain the earthly phenomenon, particularly the natural environmental and atmospheric phenomena, the question of whether *Homo sapiens* possesses the capacity and the will to maintain the Earth's atmospheric system in a relative homeostatic balance is intriguing. The Montreal Protocol on substances that deplete the ozone layer, which has resulted in the phasing out of chlorofluorocarbons (CFCs), a group of industrial compounds that react with and disassociate Ozone molecules, is a collective adaptive response by humans to a perceived and predicted threat to life from stratospheric Ozone depletion. The Environmental Protection Acts ratified by the United Kingdom and Australia and the Kyoto Protocol to the United Nations Framework Convention on Climate Change are some examples of attempts to combat deleterious environmental change associated with the release of additional carbon dioxide into the air. If humans are to maintain the Earth's systems (hydrosphere, lithosphere, atmosphere, or biosphere), it is through the combined proactive actions (consciousness) of political, social, scientific, and educational institutions that will make the decision-makers recognize climate and environmental threats and convince them to make wise decisions which will pave the way to the stewardship of the planet.

Like any other perspective from which policy can be formulated, the Gaia hypothesis provides no single, correct choice in response to a particular problem. Lovelock himself admits to having misjudged the significance of the release of CFCs into the atmosphere for years before coming to believe that these chemicals pose a serious threat to many of Earth's life-forms (Joseph, 1990). If, however, humans decide that they want to maintain the type of world to which they have become accustomed, one with abundant stretches of forest and moderate temperatures, for example, then Gaia appears to provide a useful framework for analyzing policy options. We think that it is safe to predict that over the coming years and decades, questions of scale and perspective within hierarchically ordered systems will be at the center of the environmental policy debate. From this viewpoint, Gaia is a useful metaphor, but not a literal truth. We also expect that the Gaia hypothesis will help us to ask productive questions about global environmental problems and to explore systems-oriented approaches for resolving these problems. The concept of the Earth as an animate entity with life-giving properties was there in every great ancient civilization. Plato envisioned the world as "*a living being endowed with soul and intelligence. .... A single entity containing all other living entities, which by their nature are all related.*" Chief Seattle saw it the Great Web of Life "*Man did not wave the web of life. He is merely a strand in it. Whatever he does to the web, he does to himself.*"

James Lovelock's hypothesis that Earth is a self-regulating system and has the capacity for homeostasis—implying that it carries its internal adjustment through self-regulation (positive and negative feedback mechanism) in response to the changes to the outer world. Gaia hypothesis is very close to "*System Theory*," and apparently many scholars do not find difference of any significance between the two. As Cashford et al. (2017) observes: "*Gaia, the Goddess, has become a symbol of new mode of consciousness, something called 'Gaia Consciousness' that expresses a reverence for the planet as a living being who is home to all other living beings, all of whom share in and give form to her own original and dynamically changing life. It follows that Earth can no longer be seen as dead matter, collection of objects, and 'it' merely a resource for human being to plunder at will.*" Campbell (1986) also beautifully expressed similar views: "*A profound sympathy for Gaia as our beautiful living Earth, and a compassion for all of her creatures, might then inspire us towards the next evolutionary phase, one in which Earth, and all who live on Earth, become sacred as before. It is up to us to participate imaginatively with this symbol and this story, and to assist the process of realizing the 'one harmonious being', which for so long we, as a species, have disrupted and, ever increasing, harmed. In this way, perhaps, humankind live again 'the symbol of life' so the souls of the world and World Soul might be reunited at a new conscious level through the imagination.*" Gaia does not provide a clear value system for determining what should be the goals and objectives of environmental policy. Examining the world from a Gaian perspective does not, for example, mandate that current levels of pollution be lowered or that emission of greenhouse gases be dramatically reduced. What Gaian theory can do, however, is aid in predicting how global systems and processes are likely to respond to human-induced perturbations by introducing a large systems analysis of the biosphere and by helping the public to understand the meaning, value, and importance of systems analysis. Humans, then, must choose the kind of world in which they desire to live. Gaia appears capable of maintaining life on Earth whether humans want a world characterized by abundant forests and mammal populations, including ourselves, or whether we prefer a world in which colonies of microorganisms are the most visible living communities. As the ancient Greeks realized, "*Gaia would reward mankind with her bounty when treated well but equally she would revenge abuse*" (*Lovelock, 1988*).The choice is ours as to how we want to treat Gaia?

11.4 System Theory and Autopoiesis

System theory and autopoiesis are two concepts that have gained significant attention recently in various fields of study, including ecology, biology, and social sciences. Both concepts focus on the organization and behavior of complex systems, and they share some fundamental similarities. We will explore the relationship between system theory and autopoiesis and how they contribute to our understanding of the organization and dynamics of living systems.

11.4.1 System Theory

System theory is a general theory that describes the behavior of complex systems in terms of their components, interactions, and feedback. It is a holistic approach that emphasizes the interconnectedness of different parts of a system and their influence on each other. Systems can be classified into different categories, depending on their nature and scope. One of the fundamental concepts of system theory is feedback, which refers to the process of information exchange between the different components of a system. Feedback loops can be positive or negative, depending on whether they reinforce or counteract the effects of a disturbance. In ecology, feedback is a crucial mechanism that regulates the dynamics of ecosystems and helps maintain their stability. For example, when a predator population increases, it can lead to a decrease in the prey population, which in turn reduces the pressure on the resources that the prey depends on. This can allow the prey population to recover, which can then support the predator population again. This feedback loop is an example of negative feedback, which stabilizes the system. Another key concept of system theory is emergence, which refers to the phenomenon of new properties or behaviors that arise from the interactions of the components of a system. Emergence is a common feature of complex systems, and it can lead to surprising and unpredictable outcomes. In ecology, emergence is particularly relevant because it can help explain the complexity and diversity of ecosystems. For example, the interactions between different species can lead to the emergence of new ecological niches, which can support the growth of new populations and increase the biodiversity of the ecosystem.

One of the most significant contributions of system theory to ecology is the development of network theory, which focuses on the structure and dynamics of ecological networks. Ecological networks are complex systems that describe the interactions between different species in an ecosystem. Network theory provides a framework for analyzing the structure and properties of these networks, such as their connectivity, modularity, and resilience. It has also led to the identification of keystone species and ecosystem services, which are critical for maintaining the stability and function of ecosystems. System theory is a powerful tool for understanding the complexity and dynamics of ecological systems. It provides a holistic approach that emphasizes the interconnectedness of different components of a system and their interactions. By applying system theory to ecology, scientists have gained insights into the feedback mechanisms, emergence, boundaries, and hierarchies that underlie the functioning of ecosystems. Network theory, in particular, has revolutionized our understanding of ecological interactions and has led to new insights into the importance of biodiversity and ecosystem services. As such, system theory remains a crucial framework for ecological research and a valuable tool for addressing the challenges of environmental sustainability.

Gregory Bateson (1972), a twentieth-century system theory thinker, argues that human beings have acted in ways that are destructive to fragile ecological systems because humans do not see the interdependencies between natural systems and their lives. Bateson postulated that bio-ecological and social systems are composed of interdependent network of components and sub-opponents and are maintained in more or less stable condition (homeostasis) by negative and positive feedbacks. The system is resilient and maintains its stability through some reversible adjustment within it unless it is impacted by forceful perturbations from external source. As it is clear from Bateson's following statements: "*All biological and evolving systems (i.e., individual organisms, animal and human societies, ecosystems, and the like) consist of complex cybernetic networks, and all such systems share certain formal characteristics. Such systems are 'conservative' in the sense that they tend to conserve the truth of propositions about the values of their component variables—especially they conserve the values of those variables which otherwise would show exponential change. Such systems are homeostatic, i.e., the effects of small changes of input will be negated and the steady state maintained by reversible adjustment.*" He warns that civilization is on its road to extinction unless there is a change in the current thinking of linear material and wealth accumulation. It seems paradoxical that on the one hand, we want to preserve our natural environment and on the other hand the current *egocentric consumerism* has irreparably disrupted the natural environment and our relationship with it. This is a contradiction in itself. It is important to ponder into the implication of Einstein's famous quotation "*No problem can be solved from the same level of consciousness that created it.*" We must raise our consciousness and replace current pathological *modus operandi* with a new enabling and benevolent *modus operandi* based on higher level of ecological consciousness.

11.4.2 Autopoiesis

Autopoiesis is a concept that was first introduced by Maturana and Varela in 1970s to describe the self-organizing and self-maintaining nature of living systems. Autopoiesis focuses on the organization and structure of living systems and how they maintain their identity and integrity over time. *Autopoiesis* means "*self-creating*" or "*self-organizing*" systems. Humberto Maturana's (1980, 1981) work on "*Autopoiesis*" and "*Patterns of Life*" to be found inside of all living systems is also very much in line with Bateson's works. Maturana's theory of *autopoiesis* (self-production) shows how organic cells organize themselves in producing inner organic self-components needed for maintaining homeostasis, natural barriers, and remaining cognitive at the molecular level (Maturana, 1981; Podgorski, 2010). *Autopoiesis* is a creative process behind how systems create, sustain, and generate life while maintaining their overall structure and organization. Autopoiesis explores the internal occurrences that happen within a system and the parts that make up the system; the relationships between those parts; the boundaries that surround and contain the parts; how information emerges from the system via cognition; and how external information triggers the structure of the overall system. According to Maturana (1981), the "organization" of living system represents its identity, while the "structure" represents the components that make up the system. A system may change its structure without loss of identity, as long as the organization remains the same.

The central characteristic of an *autopoietic system* as described by many system thinkers (Capra, 1996, 1999; Podgorski, 2010; Bateson, 1972; etc.) is that it undergoes continual structural changes while preserving its web-like pattern of organization. The components of the network continually produce and transform one another, and they do so in two distinct ways. One way is through the process of "*self-renewal.*" Every living organism continually renews itself. The second type of structural changes in a living system is changes in which new structures are created, thus creating new connections in the *autopoietic network*. "*Over the past thirty years,*" Capra (2014) says, "*a new systemic conception of life has emerged at the forefront of science. New emphasis has been given to complexity, networks, and patterns of organization, leading to a novel kind of 'systemic' thinking.*" The system thinking has caused the reappearance of key concepts such as *autopoiesis, dissipative structures, social networks*, and a *systemic understanding* of evolution. The systems view of life brings a new and powerful perspective that can have far-reaching implications for healthcare, management, and the global environmental and economic crises.

System theory and autopoiesis both focus on the organization and behavior of complex systems, and they emphasize the interconnectedness of different components of a system. They also share a holistic approach that considers the system as a whole, rather than just the sum of its parts. One of the key similarities between system theory and autopoiesis is the emphasis on feedback mechanisms. Feedback is a critical mechanism that regulates the dynamics of systems and helps maintain their stability. In system theory, feedback refers to the process of information exchange between the different components of a system. In autopoiesis, feedback mechanisms are essential for maintaining the organization and structure of living systems. For example, the feedback between a cell and its environment allows the cell to adjust its behavior and maintain its internal organization. Another similarity between system theory and autopoiesis is the concept of emergence. Emergence refers to the phenomenon of new properties or behaviors that arise from the interactions of the components of a system. In system theory, emergence is a common feature of complex systems, and it can lead to surprising and unpredictable outcomes. In autopoiesis, emergence is critical for the self-organizing and self-maintaining nature of living systems. New properties and behaviors can emerge from the interactions between the different components of a living system, leading to the formation of new structures and functions.

From scientific and system theory perspective, a process view of the self has emerged that tends to establish that humanity is inseparable from the web of the relationship that sustains it. Science tells us that a separate and distinct self (soul) from the world it observes and acts upon is just an illusion. The biggest problem of humanity is its alienation from Nature. Current egocentric consumerism has alienated human self from Nature and has erased from its memory that humanity cannot exist in isolation from the biosphere or ecosphere. The biggest challenge is how to dismantle this *egocentric self* and reestablish its relationship with Nature. Einstein demonstrated that a person's or self's perception is determined by its position in relation to other phenomena. Heisenberg, according to his uncertainty principle, demonstrated that self's perceptions are changed by the very act of observation. System theory has debunked the old assumption about a separate and continuous self. I could not help borrowing Joana Macy's (2017) statements to illustrate this false notion of separability of the individual self: "*There is no logical or scientific basis for construing one part of the experienced world as 'me' and the rest as 'other'. That is so because as open self-organizing systems, our very breathing, acting, and thinking arise in interaction with our shared world through the currents of matter, energy, and information that move through us and sustain us. In the web of relationships that sustain these activities, there is no line of demarcation.*" The Buddhist eco-dharma is based on the core principle of inseparability of human self from this web of interdependence and interrelationship. It is important to realize that the present crisis, that threatens planetary ecosystem, derives from a dysfunctional and pathological notion of the self that stemmed from the misperception of our place in the order of things as Joana Macy eloquently stated above.

As Norton and Ulanowicz (1992) argue the health of large ecological systems is best measured over many generations, it follows that sustainability criteria are stated on a different system scale than are economic criteria. These criteria must be characterized in the paradigm of ecology, which has been described as "*the science of self-organization in nature*" (Cook, 2008; Faber et al., 1992), not in the paradigm of economics. Multitiered systems of analysis require that different concerns and criteria be applied in different situations. Accepting Gaian theory and system theory (both have hierarchical structure) require a pluralistic approach to decision-making and, therefore, opens the entire question of how to determine which rules to follow on particular occasion. An integrated approach to environmental policy will have multiple rules to follow and consequently will require meta-rules means to determine which rule to emphasize in various situations. Because the Gaian system and system theory are hierarchical, it encourages us to think of holistic causation, what is sometimes misleadingly described as *top-down causation*. Larger, contextual systems constrain smaller systems, and these constraints set the context for policy decisions in the sense that any adaptation, cultural or genetic, gains meaning within a natural context. But stability is an illusion. All levels of the system are dynamic and environmental management must therefore embody dynamic models rather than equilibrium models. Relative stability or resilience is nevertheless important. Since larger systems change on a slower scale, they provide relative stability or resilience for elements of the system.

This reasoning suggests that there exist independent scales on which human activities may be understood. According to hierarchy theory, quite different dynamics may drive the different levels of the system affecting the human species (Norton & Ulanowicz, 1992). For example, it is useful to distinguish a scale of individual action that unfolds in the temporal perspective of economics, and another that unfolds across multiple generations. We are thinking on this scale, for example, when we show concern to protect biological diversity over multiple generations (Norton & Ulanowicz, 1992). It may also be useful to posit a third level, a global perspective, that we can adopt when we are concerned about such issues as global climate change. Much work will be necessary to refine and define these different scales so that they can yield precise policy implications, but one need not await those developments to recognize a conceptual breakthrough. Because different dynamics are posited to drive the different levels of the systems, different tools of analysis and different criteria for success will likely apply. Because we cannot measure the individual preferences of future people, the focus shifts to maintaining the complex system, the context on which individual choices will unfold. Maintaining the relative stability (health) of these larger, complex systems, therefore, provides a general guide to intergenerational policies.

One consequence of the shift from a unitary to a pluralistic system of analysis (which is already implicit in the epistemology of Gaian and system theory paradigm) is the rejection of the ideal of value-free descriptive science. Both epistemology and physical theory have been driven, despite enormous intellectual energy expended to avoid it, toward the conclusion that there exists no single, uniquely correct description of the physical world (Norton, 1991; Capra, 1999, 2014). The problem, however, is not that no consistent and accurate descriptions of the world exist; rather, there are many. The world of experience is unavoidably complex, and there are many valid perspectives and scales upon which to describe and evaluate Nature. There is no unitary picture of reality against which a paradigm can be compared. *To choose a paradigm is to choose one way of describing the world, and this choice cannot be determined descriptively.* A determination of scale and perspective first requires a determination of a goal, a value-laden decision. Choosing a "perspective" for management, a scale on which to conceive a given problem and a paradigm to describe it, is therefore a value-driven decision. Until the public develops and articulates both socioeconomic and ecological goals, science alone is inadequate for choosing a perspective from which to study the system. Science guided by a deeply felt obligation not to destroy the options of future generations, can, however, delineate a horizon of concern and a time perspective in which to address a given problem.

It is often pointed out that if hierarchy theorists (both Gaian and system theorists) can establish three hypotheses: (a) that smaller systems change at a different rate than do the larger systems that constitute their context, and (b) that the boundaries between subsystems, systems, and supersystems represent quantum differences in scale, and (c) that changes on these different scales are driven by different dynamics, then it should be possible to describe a characteristics dynamics associated with various levels of the system and to associate these dynamics with social values of high priority. Therefore, Gaian and system theory (understood hierarchically), entail a whole new approach to the analysis of environmental policy, an approach that is both value-driven and fact-driven that attempts to integrate "ethics" from both *anthropocentrism* and *ecocentrism.* We cannot help but view environmental problems from a human perspective, but there is not just one human perspective in the context of concern about intergenerational equity, human values are expressed on the ecosystem level also. Here, we go beyond the position of Allen and Starr (1982), for example, who interprets the choice of descriptive scales only as epistemologically useful. One can argue that policy should guide us to scales and descriptive systems that will be useful in protecting these very complexities. Guided by such a vision, the goal of policy is to isolate levels of hierarchical systems that are associated with important human values and to develop policies that will have positive impacts on various levels. For example, incentives to reduce waste in the use of fossil fuels would have positive economic impacts by reducing trade deficits; they would also reduce acid deposition on ecological systems, and they would also help to stabilize atmospheric carbon dioxide levels.

11.5 Convergence of Buddhism, Gaia, and System Theory

Love and empathy transcend the paradigm of separation and interconnect with others. This is what *Buddhism* teaches us. Love is an experience of our true nature, an experience of our interconnectedness. Primal love is the love we feel toward a limited number of people (our own family, family members, siblings, relatives, etc.), but it can be extended to a higher universal love that connects everything and everyone (universal love). When people transcend regularly, they enliven and strengthen this unity in their awareness experiencing this connection to others and to Nature. The practice of transcending positivity and love to others will transform ourselves, and we begin to feel a paradigm shift in our consciousness. When we lose our ability to transcend positivity and love to others, we lose our ability to train our minds to develop deeper layer of understanding and begin to live from the most superficial layer of our minds. Human mind is comparable to the ocean, which has wave on the top surface and the deep, calm, and vast ocean in the subsurface. The top or the conscious part of the human brain (conscious mind) is analogous to the wave, and the subconscious mind is the ocean part that is so deep, calm, and vast. The ocean part of the mind, the *subconscious mind* where we are all connected, is now hidden. Carl Jung called this "*collective unconscious*," the part of our unconscious mind that we share collectively. The paradigm of separation (reductionism) relegated the field of interconnectedness to Nature and to each other from the conscious mind down to the subconscious mind. Consequently, the conscious mind no longer remembers this interconnectedness to Nature and to each other and no longer experience it. It is there and it must be retrieved, invoked, and brought to the conscious layer of the mind. The field of interconnectedness of which we are all a part must be revived and brought to the conscious layer of our mind from the subconscious layer.

Buddha was the first human being who discovered and understood this field of interconnectedness to Nature and to each other and developed the teachings and techniques to dive into the deeper understanding of this interconnectedness. His teachings and techniques were directed to understand and elevate the deeper understanding of this field of interconnectedness to the conscious mind, which can also be called an *enlightened mind*. Actually, Buddha was not satisfied with the *God paradigm* that postulated that there was a higher omnipresent being separate from rest of the beings that controlled everything in the universe. The consequence of God paradigm can be seen in the development of the religious practices, rituals, and the technologies to please a higher being (e.g., praying, confessions, religious practices, and cult rituals). Buddha moved away from God paradigm to the paradigm of interconnectedness in which he saw everything, and everyone interconnected to each other through the field of interconnectedness and to think one's existence separate from others is simply an *illusion*. This is well articulated in his principle of *dependent co-origination* (*Pratītyasamudpāda*).

For Buddha, there was no higher being necessary to explain the worldly phenomenon around us such as why the Sun shines from the east, why weather patterns change, and why we get old and sick. Buddha conceived that there is a law in the universe that interconnects everything with everything and what we observe in Nature are the phenomenon determined by the field of interconnectedness. *He concluded that nothing exists separately, and nothing can escape from the field of interconnectedness in the Nature and, therefore, human happiness and the happiness of other beings lies in the proper understanding of the working of the laws of this field of interconnectedness in the Nature.* His teachings, lifestyles, and the techniques of meditation all revolve around developing a deeper understanding of this field of relational interconnectedness that ultimately results in the wisdom with compassion, the *enlightened human consciousness*. Edward Wilson (1998) thinks that the greatest challenge of science is the accurate and complete description of the *complex system* (planetary ecosystem). While explaining the highly misunderstood chaos theory, he states "*extremely complicated, outwardly indecipherable patterns can be determined by small, measurable changes within the system.*" Wilson points out the challenge to understand more thoroughly the operation and the products of human brain and how it relates to culture and its evolution. Wilson thinks that the combination of complexity theory and evolutionary approach is the best way to study neuroscience. As he states "*Brain scientists have vindicated the evolutionary view of mind …. that passion is inseparably linked to reason. Emotion is not perturbation of reason but a vital part of it.*" Wilson has elaborated how "*gene-culture coevolution*" are particular value to the humanity. He hypothesized that "*the human being has evolved genetically by natural selection in behavior, just as it has in the anatomy and physiology of the brain…. natural selection has added the parallel track of cultural evolution, and the two forms of evolution are somehow linked.*"

To solve present environmental crisis including the climate change, there is no single bullet approach. The convergence of different approaches from multiple perspectives can help us see more clearly the potential solution to resolve present crisis. There is no doubt political economy and ecology provides the main arena of the interplay of these approaches, and human spirituality and ethical perspectives are also equally potent that can help resolve this crisis. Such approaches may consist of political economy, science and technology, religious, spiritual, and ethical-based approaches. The eclecticism of combining diverse approaches and perspectives, perhaps, hold a far greater potential to address and resolve present crisis humanity is facing today. *Ethical environmental pragmatism*, in my view, should be more concerned about bringing the practically useful perspectives from diverse sources (Buddhism, Gaian and system theory, political economy and ecology, etc.) that can help political and power elites change their mindset (consciousness) and realize that humanity cannot survive by destroying its very *niche* environment in which it evolved and flourished.

Buddhism, Gaia, and system theory have one fundamental unity; that is, they are all founded on the concept of *interdependence* and *interconnectedness*. Many scholars apparently do not find much difference between Gaia hypothesis and system theory. Gaia and system theory are both holistic as they treat lower-level systems as subsystems of larger holons and the parts of a larger system. Gaian hypothesis and system theory both emphasize systems analysis and remind scientists, policy analysts, and others concerned with local environmental problems that their local systems are embedded inside larger systems that provide the local context. The scientific and system theory perspective entail a process view of the life that asserts humanity is inseparable from the web of the relationship with other beings in Nature that sustains it. The Buddhist *eco-dharma* has been founded on the core principle of inseparability of human self from web of *interdependence* and *interrelationship* with other entities in Nature because of the condition of *dependent co-origination (Pratītyasamudpāda)*. It is important to realize that the present crisis has resulted from a dysfunctional and pathological notion of the egocentric human self that stemmed from the misperception of our place in the order of things.

It can be said that for the development of a coherent environmental ethics, Buddhism, more than any other faiths or religious traditions, offers the most pragmatic and useful perspective. The physicist James Lovelock postulated a hypothesis that Earth was a self-regulating system and has the capacity for homeostasis—implying that it carries its internal adjustment through self-regulation (positive and negative feedback mechanism) in response to the changes to the outer world. Gaia hypothesis is very close to "system theory." System theory (theory of living system) provides the most logical formulation of the ecological paradigm. Living systems entail a wide range of phenomena encompassing individual organisms, ecosystems, and human social systems. System theory provides a common framework for ecology, biology, neuroscience, social ecology, economics, medicine, and other sciences, including organizational management and information network. Buddhism, Gaia, and system theory all emphasize the interconnectedness of different components of a system, the importance of feedback mechanisms, emergence, boundaries, and hierarchies. Gaia and system theory framework with Buddhist philosophical view of interconnectedness and dependent origination and the ethical conducts of nonviolence and reverence for life has the potential of liberating humanity from present predicament.

Chapter 12
Power of Collective Human Consciousness

We have the power to decide the destiny of our planet. If we are awakened to our true situation, there will be a change in our collective Consciousness. We have to do something to wake people up. We have to help the Buddha to wake up the people who are living in a dream.

Thich Nhat Hanh (1997)

12.1 Introduction

The collective Consciousness of the people has great potential to change the direction and behavior of the political institutions and power centers. With collective ecological awakening, it is possible to bring desirable political outcomes and orient the political economy of the nation-states toward maintaining functional integrity and resilience of the planetary ecosystems based on ecological laws and wisdom. Unless a new breed of informed and environment-friendly politicians and managers take over the powerful decision-making institutions with a vision of creating a sustainable society in which human behavior is shaped and guided by the conscious efforts of meeting the essential needs of all rather than satisfying the greed and self-aggrandizement of a few, the planetary ecosystem cannot be protected and is impossible to maintain the regenerative and resilient capacity of the Earth systems.

Philosophers and scientists from ancient times have pondered into the age-old question of what consciousness is. They attempted to define and understand consciousness from various perspectives but had no clue as to how consciousness originated in the human brain. With the advancement of science and research in neuroscience in recent years, neuroscientists began to unravel some aspects of consciousness. Evolutionary history informs us that it took billions of years for primitive life to evolve into life with simple brains, hundreds of millions for life with simple brains to evolve into primates, a few million more for primates to evolve into humans, then only hundreds of thousands for language to appear and tens of thousands more for written language, a few thousand years to get into enlightenment

G. Upreti, *Ecosociocentrism*, https://doi.org/10.1007/978-3-031-41754-2_12

period. This chapter brings perspectives on the complex subject of what human consciousness is, how it evolved, and what potential it has to resolve the planetary environmental crisis humanity is facing today.

12.2 Human Consciousness

At one time in evolutionary history, genes were more dominant than now as they are in other animals and plants. But as times passed and human cultural evolution consolidated, things changed faster, and culture appeared to play a more dominant role in human history. Through religion, political ideology, and other mechanisms, cultures dominated human minds and dictated what happened for over 10,000 years. Culture is unique to humans in that humans happen to evolve the language instinct, which may rarely be found in other animals. Richard Dawkins (1976) invented the term "*memes*" to describe the symbolic unit comprising any culture. With the advent of scientific discovery and inventions in the seventeenth and eighteenth centuries, a major shift occurred that changed the cultural evolutionary landscape. The "*enlightenment period*" of scientific discovery and technological progress accelerated, governments and societies changed, and "*memes*" (unit of culture) of progress logic and reason were selected favorably in a new equilibrium called *modernity*. Cultural selection dominated the process, but it paved the way for *conscious selection*. Thus, consciousness relying on logic and reason began to defy cultures and has taken an increasingly more dominant role in human minds.

Genes and *memes* are the units of biological and cultural evolution, respectively, and, therefore, biological and cultural evolution are interrelated and interactive, and one cannot be studied effectively in isolation from the other. What implications can be drawn from interactive biological and cultural evolutionary perspectives on the direction of the evolution of *human consciousness*? Advances in neuroscience inform us that the human brain functions as the hardware infrastructure, and the human mind or consciousness is the superstructure that emerged from it. Consciousness evolved from human brain, and gradually, consciousness started influencing human brain. This interaction between brain and consciousness (infrastructure and superstructure) is *dialectical* in nature, constantly influencing each other and creating conditions for their further evolution. *Dialectics* is a pervasive phenomenon in *the ecosphere* and *sociosphere*, including human brain–mind–consciousness nexus. Since it is the *human consciousness* that characterizes and motivates human actions and behavior, the solutions to the problems created largely by humans must be sought within the purview of *human consciousness*.

Consciousness is the most remarkable and complex evolutionary development in *Homo sapiens,* and this is the only attribute that uniquely distinguishes humans from other beings in the animal kingdom. It is a complex phenomenon that consists of many apparently opposing elements, such as human intellect, rationality, cooperation, and positive and negative emotions (love, empathy, compassion, selfishness, hatred, arrogance, and greed). Human emotion is a very powerful component, and perhaps this is what makes us human and makes us distinctly unique from robots or artificially intelligent machines. Human intellect and rationality can overpower or direct human emotions, or conversely, human emotions can overpower human intellect and rationality. The human consciousness that consists of the balanced mix of human intellect and emotions (human rationality and reason, cooperation, love, compassion, and limited greed and selfishness) has the potential to extricate humanity from the present predicament and can direct humanity toward the path of *eco-civilization*. The current social consciousness has been hijacked by greed, arrogance, unhealthy competition, and egocentric consumerism propelled by mercantile capitalism. This egocentric paradigm needs to be replaced by a new paradigm of consciousness that entails cooperation, healthy competition, compassion, and wisdom (intellect) to extricate humanity from the current unfortunate predicament.

The perspective from integrated information theory (IIT) is very helpful in understanding how human *consciousness* emerges, expands, and rises to a higher level. It is crucial to know how human consciousness helps us to understand our place in the universe and what we do with our lives. According to Tononi's (2016) integrated information theory, the more information that is shared and processed between many different components of the brain, the higher the level of consciousness. Although the concept of group consciousness may seem like a stretch, Tononi's theory certainly helps us to understand how large number of people living in different parts of the world sometimes begin to think, feel, remember, decide, and react similarly as one entity. This *collective Consciousness* needs to be organized, consolidated, and channelized toward the stewardship of planet Earth so that Earth can still be a hospitable home for humans and other living entities.

Evolutionary history informs us that it took billions of years for primitive life to evolve into life with simple brains, hundreds of millions for life with simple brains to evolve into primates, a few million more for primates to evolve into humans, then only hundreds of thousands for language to appear and tens of thousands more for written language, a few thousand years to get into enlightenment period. The story continues to develop, and consciousness increasingly determines the fate of the world through scientific and technological progress. As Andres Campero (2018) succinctly articulates: *Humans, along with the first RNA molecules, the first life forms, the first brains, the first conscious animals, the first societies, and the first artificial agents, constitute an amazing and crucial development in a path of increasingly complex computational intelligence. And yet, we occupy a minuscule time period in the history of Earth, a history that has been written by genes, by cultures, and by consciousness. If we abandon our anthropomorphic bias, it becomes obvious that Humans are not so special after all. We are an important but short and transitory step among many others in a bigger story. The story of our computational minds, which is ours but not only ours.*

Neuroscientists argue that consciousness is an *emergent property* of the human brain. However, the dynamics of the neural events associated with information processing, precise timing, and location where consciousness arises are not clearly and unequivocally determined. Consciousness is a complex mental phenomenon consisting of everything we experience. Philosophers call these experiences "*qualia*." It is often stated that the origin and Nature of these experiences have been a mystery from antiquity to the present. Some philosophers and scientists believe that consciousness is an illusion and deny that such *qualia* exist. They argue that consciousness can never be meaningfully studied by science.

12.2.1 Evolution, Consciousness, and Rationality

The evolution of human consciousness has profoundly shaped our species and the world around us. It is an intricately woven tapestry of cognitive development, cultural evolution, and biological adaptation. In this context, understanding the evolution of *human consciousness* and its impact on behavior can provide crucial insights into who we are, how we interact with each other and the world, and how we might direct our future, especially regarding environmental stewardship and sustainable living. One might begin by understanding that consciousness, as a phenomenon, is intimately intertwined with the evolutionary process. Over hundreds of thousands of years, the human brain has undergone significant changes, particularly in the neocortex, the part of the brain responsible for higher-order functions such as sensory perception, cognition, generation of motor commands, spatial reasoning, and language (Rakic, 2009). This evolution has not only given us the capacity for abstract thinking and problem-solving but also a level of self-awareness or "consciousness" that appears to be unmatched in the animal kingdom.

The transition from early *hominids* to anatomically modern humans witnessed crucial shifts in cognitive abilities and consciousness. Early evidence of these shifts can be found in the archaeological record, from the creation of complex tools to the development of art and symbolic thinking, all of which suggest a growing cognitive sophistication and conscious awareness (Mithen, 1996). For example, cave paintings dating back around 40,000 years found in Indonesia and Spain represent not just an artistic expression but an embodiment of human consciousness and its evolutionary progression (Aubert et al., 2018; Hoffmann et al., 2018). This evolution of consciousness had significant implications for human behavior, leading to the development of language, social structure, moral and ethical codes, and culture. Language, one of the key milestones in this journey, is thought to have emerged between 50,000 and 100,000 years ago (Pinker & Jackendoff, 2005). This development facilitated complex communication and the sharing of abstract concepts, contributing to the formation of social bonds, shared norms, and collective identities. Furthermore, it allowed for the codification and transmission of knowledge across generations, an essential driver of cultural and technological evolution.

As human consciousness evolved, so too did our capacity for moral reasoning and ethical consideration. This moral consciousness, which arguably separates us from other species, can be traced back to our early human ancestors who lived in cooperative groups (Tomasello & Vaish, 2013). The need to maintain harmony within these groups likely required an understanding of *fairness, empathy*, and *cooperation*, which subsequently formed the basis of human moral behavior. In more recent times, the impact of evolved human consciousness on behavior is evidenced in the multifaceted societies we have built. It underpins our legal and political systems, our scientific and artistic endeavors, and our economic structures. Our capacity for self-reflection, introspection, and empathy, enabled by our consciousness, fuels our pursuit of justice, equality, and well-being.

The evolution of human consciousness also impacts our relationship with the natural environment. Our awareness of self and others extends to an awareness of our place in the natural world and our influence upon it. This environmental consciousness has grown particularly strong in recent decades, sparked by the realization of our collective impact on the planet and the necessity of sustainability (Gifford & Nilsson, 2014). Understanding this consciousness–environment interface opens up the possibility of harnessing our evolved consciousness to foster more sustainable behavior and environmental stewardship. Our conscious awareness, our capacity for foresight and moral reasoning, and our ability to learn and adapt can all be leveraged to respond to environmental challenges and reshape our behavior toward Nature and its creation. The evolution of human consciousness, a remarkable journey through time, has fundamentally shaped human behavior and our interaction with the natural world. As we navigate the environmental challenges of the twenty-first century, an understanding of this evolution and its impact on our behavior will be crucial in fostering the necessary shift toward sustainable living.

The classical evolutionary theory teaches us that the evolution of flora and fauna, as well as the evolution of the physical and chemical environment, occurred separately on planet Earth, but the study of the global evolutionary changes gives us a different picture that the evolution of all these components occurred in close interconnected relationship, and they together constitute a unified evolutionary process (Trubetskova 2010). From this perspective, it seems logical to think that the biosphere is the unified functional system that evolved from the co-evolution of the Earth's component systems through an interconnected, interactive process. This becomes clear from the major events (Precambrian and Phanerozoic stages) of ancient life, such as eukaryotic cells, photosynthesis, multicellular organism, and the complex system that are responsible for global biosphere transformations. If we thoughtfully analyze the web of the biotic community, it becomes obvious how we are interconnected with other life forms and how we are interrelated and interdependent on each other.

This amazing web of interrelatedness and interconnectedness of biotic communities in the planetary ecosystem is the result of billions of years of evolutionary process. This evolutionary creation is not only amazingly beautiful but also wonderfully complex. This web of interdependence and interconnectedness is the very basis upon which all members of the biotic community, including human beings, exist and continue to exist. *Homo sapiens* has become the biggest threat to its own existence by its own very destructive acts of rupturing this web of interrelatedness and interdependence. The most disappointing fact is that it is becoming increasingly difficult for our species to realize that we have evolved from Nature and are dependent on it and ultimately will return to it. Only the ecological wisdom consciousness can help us to recognize and realize this web of interconnectedness and interdependence and also empower us to live in harmony with Nature.

It may be helpful to bring some perspectives from evolutionary biology to understand how the development and selection of certain attributes (traits) in human primates by natural selection resulted in what *Homo sapiens* has become today. Biologists, including Edward Wilson (1999), have postulated that three major traits favored by natural selection define the character attributes of *Homo sapiens* (the human species). These are self-concern, cooperation, and altruism. Self-concern (self-centeredness) was favored within a group by natural selection, but natural selection between groups favored the traits of cooperation and altruism, which serve the human spirit and our remarkable social intelligence. Human social behavior is motivated and very much shaped by the instinctive interplays of these traits, and within individuals and within society, struggles between these instinctive behaviors are inevitable. Recognizing that the dominant motivating force of human behavior is self-concern or self-interest, or self-preservation, all human societies have developed rules of socialization that inhibit or limit the scope of self-centeredness (greed and aggression). The culture of modern consumer society created by the free-market economy glorified the egocentric individual self-interest or self-centeredness and also generated a mental consciousness of endless competition undermining cooperation, symbiosis, empathy, and altruism. This has promoted hyper-individualism, undermining the value of cooperative and symbiotic relationships.

In view of the erratic climate-related events with their enormous impacts all over the world and the conclusion of the scientific community about global warming, environmental and climate crises, it is appalling that the political power centers, politicians, and corporate establishments continue their business as usual. The global business corporate world has become the most powerful institution for shaping the development policies of powerful governments, including the United States. As John Stanley and David Loy (2016) have rightly pointed out: "*The dominant institution of our age is no longer religion, government, or academia. It is a global business corporation. Well-documented examples of corporate behavior correspond to traits of psychopathy, a condition of zero empathy with others. Of course, not all corporations act as if they were functional psychopaths. Yet the record shows that leading fossil fuel companies possess a truly dangerous combination of wealth, power, and destructive intent. They spend huge sums of money to undermine climate science, subvert political institutions, and corrupt governments. Where can we find the power to dismantle such a systemic problem?*". Evolution has established that the traits comprising empathy, cooperation, and altruism (compassion) evolved and were selected by natural selection in the course of collective human adaptation against adverse environments.

These are the very special resources that have survival values. Otherwise, they would not have been selected in men by natural selection. If these values are expanded and used judiciously to address the current ecological crisis, including climate change and the destruction of the web of life (species loss), humanity could have some hope and optimism. This is highly imperative in view of an unprecedented crisis of biological extinction that includes the very real possibility of the disappearance of *Homo sapiens* itself. If these values are overshadowed by greed, egocentric consumerism, and individual self-aggrandizement, then we are doomed to be extinct, and perhaps, we deserve that nemesis. The epistemology of the evolutionary process indicates that it progressively gives rise to a greater level of organizational complexity, from atomistic microcosm to the Earth's biosphere system with the myriad of life forms forming a web of interdependence and interconnectedness with diverse ecosystems, millions of species, and innumerable life-forms. Among those species, *Homo sapiens* is uniquely positioned in this web of interdependence because of the cultural evolution it was crowned with. The cultural evolution in the sociosphere accelerated the development of creativity, intelligence, cooperation, and wisdom, on the one hand, and greed, egocentrism, aggression, and domination, on the other hand. The struggle and tension (contradictions) between these two opposing tendencies are growing, but I am optimistic that the the ecological wisdom consciousness and values will emerge from the dynamics of the cultural evolutionary process itself. Evolutionary history has informed us that evolution has moved forward from inorganic matter to organic life and from single-celled organisms to multicellular plants and animals and eventually complex planetary ecosystems and humanity in it. Though the transition from one stage to another has always been difficult and costly, however, it has resulted in new forms of creativity.

Humanity right now is passing through a transition called the *Anthropocene epoch*. The urgency and the necessity of ecological wisdom consciousness in the Anthropocene epoch cannot be undermined, given the magnitude and dimension of environmental and climate crises created by anthropogenic activities. In fact, human behavior shaped by ecological wisdom consciousness is the only means that will form the foundation of a new civilization that can be called eco-civilization because this consciousness will inevitably replace the hyper-anthropocentrism, individualism, and ecologically hostile consumerism with interdependences, interconnectedness, the kinship of biospheric life systems, and fundamental needs. If we analyze how deeply we are embedded in the complex biospheric ecosystem and how deeply dependent we are on other life forms and the life-sustaining services of the biospheric ecosystem, it is not difficult to see how we are destroying the very biophysical base that ensures our own continuity as a species. It becomes clear that we are destroying our own distinctive *niche* in the evolutionary process. Politicians, corporate executives, and development experts have completely shut their eyes and minds on this and have become blindfolded by the greed and their short-term interest undermining the future of humanity and the biosphere.

Currently, we are in the midst of the sixth-extinction period due to the current massive loss of species. Paleontologists (Burton, 2016) inform us that those earlier periods of extinction were caused by a variety of factors, including meteoric collisions and climate change, but the present extinction is different in that it is being caused exclusively by humans. This raises an important question why did *Homo sapiens*, the most intelligent and conscious of all the species, become blind to its own behavior that is causing such an extinction? The time is ticking, and every moment is so precious for humanity to change its direction from the trajectory of its present course. Human cultural wealth, including religion, ethics, spirituality, and scientific understanding, must be called to contribute to the resolution of this crisis. As Mary Tucker and Brian Swimme (2016) argue: "*The challenge for religion and ethics is both to re-vision our role as citizens of the universe and to reinvent our niche as members of the earth community. This requires reexamining such cosmological questions as where we have come from and where we are going. In other words, it necessitates rethinking our role as humans within the larger context of natural processes of life on Earth. What is humankind in relation to 13.7 billion years of universe history? What is our place in the framework of 4.6 billion years of Earth's history? How can we foster the stability and integrity of life processes?*" These are critical questions underlying the new consciousness of the universe story. This is not simply a dynamic narrative of evolution; it is a transformative cosmological story that engages human energy for a future that is sustaining and sustainable.

The present dominant paradigm founded on excessive consumerism has not only caused alienation and fragmentation of humanity but also endangered the survival of the humanity itself. "*Our present economic system,*" as Jay Carter (2017a) emphatically pointed out: "*is destroying the Earth's natural environments and, at the same time, destroying the lives of millions of people. We must restructure our economic system to the point where it is compatible with the Earth's limited capacities. Every person who identifies him or herself as an economist or has plans to run for a major political office must first be educated to the point of understanding basic ecology and Earth science. ..... It is time to make "Earth First" much more than a bumper sticker. It must become a way of life for all living things, including humankind.*" Eventually, it comes down to conscious human rationality (an intellect informed by knowledge, understanding, and wisdom). Generally, humans do what they consider to be rational based on knowledge, reason, and understanding of natural and social phenomena. What they consider to be rational depends upon what they believe to be their own good, including their future good, and what actions, policies, attitudes, and behavior lead to that good based on the epistemology and the understanding of life phenomena. Human rationality leads to accepting certain moral principles and values to guide their behavior that result in their own greater good and the good of other entities that sustain the system. Human rationality, informed by knowledge, understanding, and wisdom acquired over the past thousands of years, can change and shape the faulty worldview or replace it with the one that can help resolve the crisis, the environmental and climate crisis of the day.

The philosophical challenge we are facing today is to carefully examine the nature of morality and the scope of our responsibility toward our fellow beings and the planetary ecosystem in which humans are the most dominant actors. The current hyper-anthropocentric moral value has been criticized as having too narrowly restricted only to fellow human beings. Environmental ethics requires us to see our moral responsibility in a much broader scope and dimension, which must include future generations, nonhuman life, or Nature as a whole rather than just fellow human beings. Unless the dimension of the moral scope and responsibility is extended from our immediate fellow humans to the nonhuman plane of life on the planet that includes the planetary Earth ecosystem and ecosystem processes, humanity faces its own existential threat. I hope human intelligence and rationality informed by science and wisdom traditions will lay the foundation of new ecological ethics by liberating humanity from the current dominant hyper-anthropocentrism that feeds on human greed and unhealthy competition rather than fundamental needs, cooperation, and symbiosis.

Though *Homo sapiens* evolved with a higher level of consciousness or self-awareness and is capable of assessing the consequences of its own actions and changing the trajectory of its actions if the consequences are detrimental to its own good; however, we have seen very little realization among those who are in the control of political, economic, and technological superstructures. Unless world leaders, particularly the leaders of powerful countries like the United States, Europe, China, India, and other large economies of the world, come up with a clear vision and plans to deal with the challenges of current environmental problems (global warming, climate change, destruction of ecosystems and ecosystem services, depletion of natural capitals, and biodiversity loss), the tragedy of the Earth's commons will continue and become irreversible rapidly accelerating humanity's nemesis. There is no limit to human greed and self-aggrandizement, which is the chief cause of this anomaly. As Mahatma Gandhi rightly said, "*The world has enough resources to satisfy human needs, but not enough to satisfy human greed.*" Unless the collective ecological consciousness of the people across the globe can free the greedy minds, which are controlling power structures and institutions, one cannot expect any significant change in the affairs and the manner of the current resource extraction and consumption, the main culprits of today's environmental destruction and crisis.

The development of science and technology should be directed toward the sustainable uses of Earth's systems resources while maintaining the regenerative biocapacity of the Earth's systems. This is possible only when humanity can be guided by a transcendental ecological consciousness that integrates ecosphere and sociosphere that, ultimately, gives rise to an ecological civilization that is the basis for an ecologically sustainable and equitable global society. If we dream of such a world alone, it will be just a dream, but if we dream it together, it will be a reality. That is the power of collective ecological consciousness.

12.2.2 Theories of Consciousness

Human consciousness has intrigued philosophers, psychologists, and neuroscientists alike, leading to diverse theories on its nature and origins. What is the material basis of consciousness? What particular piece of brain matter gives rise to consciousness? These were the questions that neuroscientists wrestled with for a long time. They were trying to come up with a theory that could answer and explain some complex questions pertaining to the origin of consciousness. While the nature and origins of human consciousness remain a complex mystery, recent advances in neuroscience and cognitive science offer promising directions for future research. It is crucial to continue integrating insights from philosophy, psychology, and neuroscience to progress our understanding of this fascinating phenomenon. As Neuroscientists (Koch, 2018; Tononi, 2016; Campero, 2018) point out: "*Ultimately what we need is a satisfying scientific theory of consciousness that predicts under which conditions any particular physical system—whether it is a complex circuit of neurons or silicon transistors—has experienced.*"

Two theories, namely, global neuronal workspace (GNW) and integrated information theory (IIT), have been advanced by neuroscientists to explain the complex Nature of consciousness (Koch, 2018). GNW theory begins with the observation that when we are conscious of something, many different parts of our brain have access to that information. According to GNW, consciousness arises from a particular type of information processing, and it emerges when incoming sensory information is inscribed and broadcast globally to multiple cognitive systems which process these data, store, or call up a memory or execute an action.

Giulio Tononi et al. (2016) and Christof Koch (2018), renowned neuroscientists, have advanced a theory called the integrated information theory (IIT) of consciousness, which postulates that the extent of integration within a computational system (e.g., the brain) is a determinant of the system's degree of consciousness. The starting premise of IIT is the experiential state itself, which has inherent fundamental properties. The cerebral cortex, a sophisticatedly folded and interconnected layer of neural tissue, is intimately related to conscious awareness. A distinctive set of neuronal activities, termed neuronal correlates of consciousness (NCC), corresponds to each experiential state within the cerebral cortex. The minimal neuronal mechanisms that together suffice for a given conscious experience are defined by neuroscientists as the NCC. IIT purports that consciousness is the intrinsic causal power that is associated with the complex mechanisms inherent in the human brain. In the words of Andrew Campero (2020): "*As our comprehension of the mind improves, consciousness is becoming less dependent on genes and culture, which evolve through differing computational processes and at slower time scales. The narrative of minds may now transition to the narrative of consciousness.*" The human brain, a product of millions of years of evolutionary processes, is a vast computational hardware system.

Throughout the twentieth century, the question of consciousness was deemed beyond the scope of scientific investigation due to its perceived complexity, subjectivity, and ambiguity. However, the advent of novel experimental methodologies and brain imaging techniques in recent years has transformed consciousness studies into a typical experimental science. *The history of the world has been driven by the history of the development of the mind, and the study of the mind has become the study of consciousness.* Everything we *experience is consciousness.* The entirety of our experience is embodied by consciousness, and thus, we are, by definition, consciousness. Genes and cultures are relevant only as far as they influence consciousness—a means to an end. The "*qualia*" or subjective experiences of consciousness render life meaningful. Yet, the paramount question of our time is the nature of the "*qualia*" that imbues our life with worthiness.

The teachings and the ideas inherent in Buddhism (Theravada and Mahayana) are becoming more relevant and significant in dealing with current problems humanity is facing in *the ecosphere* (environment) and *sociosphere* (socioeconomic sphere). The resolution and reconciliation of current development predicament, that resulted from faulty human–Nature relationships and reckless profiteering, is simply not possible within the framework of the current development paradigm dictated by market capitalism. The magnitude of environmental and associated social problems are not only hugely detrimental to the very existence of humanity but are also complex and multidimensional in nature. These problems can be analyzed and put into the right perspective only in the light of a new *social consciousness* that embodies *cooperation, compassion*, and *ecological wisdom*. The consciousness propelled by cooperation, *compassion* and *ecological wisdom* can be collectively termed as *ecological wisdom consciousness*. The cultivation of such consciousness in peoples' mind is possible and is highly imperative to move forward on our evolutionary path.

The emergent paradigm resulting from the integration of cooperation, compassion and ecological wisdom will enable us to see how we are interrelated and interconnected with others in the *web of life* in the planetary ecosystem. If we look into our own evolutionary and cultural history, we find that over time, we have evolved mental faculties with a greater capacity for learning and understanding. If we can collect, organize, and direct this consciousness, the strength of the collective Consciousness (intelligence) would be very powerful and revolutionary that can open a new vista for the quantum leap to *eco-cultural enlightenment*, a new milestone in the evolutionary history of *Homo sapiens*. Albert Einstein's (Popova, 2016) words impeccably evoke this vision: "*A human being is part of the whole called by us "the universe," a part limited in time and space. We experience ourselves, our thoughts, and our feelings as something separate from the rest—a kind of optical illusion of our consciousness. This delusion is a kind of prison for us, restricting us to our personal desires and affection for a few persons nearest to us. Our task must be to free ourselves from this prison by widening our circle of understanding and compassion to embrace all living creatures and the whole of Nature in its beauty.*" No doubt, the biggest challenge of humanity today is how to free itself from this prison of *optical illusion* to preserve the web of interconnectedness and live in peaceful coexistence with other entities in Nature.

Unquestionably, humanity's survival is under siege from a formidable foe—the culture of industrial corporate consumerism and the matrix of institutions that promulgate a hyper-anthropocentric worldview. An overarching ethos places man at the center of all ecological equations, neglecting the delicate symbiosis we share with Nature. The wheels of this machine are tirelessly turned by the iron fist of financial, corporate, and political elites, who utilize their unyielding power to influence society's dominant communication channels—from cable television and talk radio to the seemingly ubiquitous social media. What is even more disturbing is the reciprocal relationship between these two elements; an unending cycle that continuously fortifies the *status quo*, systematically thwarting any sensible efforts to combat imminent existential threats coming from climate chaos, ecological collapse, and rampant social and environmental injustices.

The problem, however, is not insurmountable. The key lies in overcoming the monolithic mindset of these powerful global corporate and political elites that drive the global corporate economy in a bid to amplify their authority, often disregarding the plight of common people. As they persist in exploiting Earth's resources to increase their power, people worldwide bear the brunt, precipitating an impending environmental catastrophic collapse. Now, more than ever, we need a seismic shift in our *collective consciousness*. We must rise, unified in purpose, to usher for the environmental sustainability of development and environmental justice that respects the natural rights to exist for every living being on planet Earth. This compelling vision is neither fanciful nor idealistic; it is an imperative that requires us to subvert the current norm and demand change. It is time to realize that our survival is intricately linked to the health of planet Earth, and by extension, if we embrace "*The Earth First Paradigm*," we can rescue humanity from this unfortunate predicament. We must, therefore, strive to dismantle the reigning culture of *hostile* and *excessive consumerism* and influence the attitudes of our leaders to guide humanity away from the brink of disaster and toward a future of *environmental sustainability* and *equity*. Only then can we save ourselves from the looming catastrophic collapse with "The Earth First Paradigm" becoming our way of life.

12.3 Consciousness and Spirituality

The environmental crisis confronting us today is not simply a question of technological or economic solutions, though those are certainly critical components. Instead, the crisis also calls for an understanding and transformation of the fundamental aspects of the human experience—*consciousness, spirituality*, and *moral imperatives*. Our relationship with the environment is fundamentally determined by our consciousness—our perception of ourselves and the world around us. Our consciousness shapes our actions and behaviors, influencing the way we treat our environment and the choices we make about resource use, consumption, and disposal. Understanding the role of consciousness in shaping human behavior toward the environment is, therefore, a crucial task for environmental stewardship (Gifford & Nilsson, 2014).

In the modern world, however, consciousness alone may not be enough. We have all the information we need about the environmental crisis, yet behavior and attitudes seem slow to change. This is where *spirituality* and *moral imperatives* come into play. Spirituality can foster a sense of interconnectedness with the natural world, promoting a greater sense of responsibility for its well-being (Kinsley, 1995). Moral imperatives, meanwhile, define the ethical dimensions of our actions, establishing what is "right" and "wrong" in our treatment of the environment (Sandler, 2012). Hence, an attempt has been made to shed light on the complex interplay between these three elements—consciousness, spirituality, and moral imperatives—and exploring how they can collectively contribute to environmental stewardship and sustainable living. The goal is to demonstrate that these elements are not only crucial for understanding our current environmental predicament but also for informing solutions.

This exploration is not an abstract academic exercise. It has critical practical implications. By understanding how these elements interact, we can better design policies, educational programs, and public messages that effectively promote environmental stewardship. For example, understanding how moral imperatives can shape environmental behavior can inform the design of environmental regulations and policies (Schultz & Zelezny, 1999). Likewise, integrating spirituality and a sense of connectedness with Nature into educational curricula can cultivate a sense of environmental responsibility from a young age (Chawla, 2006). Moreover, this book aims to push the boundaries of our understanding and challenge the status quo. It invites us to question the dominant narrative of "Man versus Nature," to consider alternative paradigms of human–environment relationships, and to envision new ways of living sustainably on this planet. It presents an opportunity to reflect on the way we live, the choices we make, and the world we want to leave for future generations. Importantly, this book is also a call to action. It urges us to not only understand but also to act upon our understanding. As we grapple with the complexities of the environmental crisis, it is imperative that we draw on all available resources, including our consciousness, spirituality, and moral imperatives, to guide our actions. Only then can we hope to transition toward more sustainable ways of living.

It is difficult to find a universal definition of *spirituality*. Historically spirituality had its roots in religion and religious practices intended to establish communication with the "*spirits*" of the dead, but I am not talking about this notion of spiritualism. For me, spirituality is very close to the Buddhist practice of cultivating compassion and wisdom necessary to alleviate the suffering of the living world caused by greed, arrogance, and selfishness. Every human being is capable of being spiritual, and the kind of spiritualism we need at this difficult time is a *spiritual ecology* with compassion and wisdom in our treatment of other living creatures and Nature.

Let us begin by understanding consciousness, a concept that seems abstract and elusive but is integral to our interaction with the world around us. Consciousness, as defined in psychology, is our awareness of our surroundings, our own internal states, and the relationship between the two (Block, 1995). From an environmental perspective, our consciousness plays a pivotal role in shaping our attitudes toward the environment and our subsequent behaviors. This perspective, known as environmental consciousness, is the understanding and acknowledgment of the interdependencies between humans and the natural environment (Nisbet & Zelenski, 2013). It is this understanding that often acts as the precursor for environmentally friendly behavior, a vital component of environmental stewardship.

Consciousness alone, however, is not enough to spark the profound behavioral changes needed to address the environmental crisis. The information about climate change and environmental degradation is widely available and accessible, yet many continue with unsustainable practices. This is where the roles of spirituality and moral imperatives become crucial. Spirituality, as it pertains to the environment, often involves a sense of interconnectedness with the natural world. This connection fosters a deep respect and reverence for the natural environment, prompting individuals to protect and preserve it (Emmons & Paloutzian, 2003). Spiritual practices and teachings, ranging from Buddhism's principles of interconnectedness to Native American beliefs of Earth as a living entity, can provide moral and ethical frameworks that promote environmental stewardship. By viewing Nature as sacred, these spiritual traditions encourage behaviors that are less likely to harm the environment.

The words *spiritual* and *spirituality* are difficult to define in a way that can have universally acceptable meanings. They may have variable meanings to variable individuals belonging to variable cultural, intellectual, and wisdom traditions. One may define the word spiritual as connecting with a source of ultimate meaning. Spiritual feelings and meaning may arise from the recognition of the *intrinsic connectedness* between us and the natural world. The dynamical system theory and the Buddhist doctrine of dependent co-origination (*Prattityasamudpada*) recognize this connectivity within and between all living entities in Nature. Systems biologists and system thinkers have identified the principles of self-organization (autopoiesis), which enriches not only the scientific understanding of life but also the spiritual wisdom of mankind with respect to Nature. The Buddhist and Confucian traditions of the East have been well recognized for their rich reference to the principle of interconnectedness, self-organization, the unity of life, and rich spiritual wisdom.

The Buddhist concept of dependent co-origination (*Prattityasamudpada*) has a profound implication in the understanding of interconnectedness and interdependence. One of the prominent Buddhist scholar and system thinkers, Joanna Macy (2017), has eloquently expressed this concept as "*the radical interdependence of all phenomena... everything arises through mutual conditioning in reciprocal interaction.*" For Buddhists, compassion, wisdom, and loving-kindness *(metta)* constitute the central component of their *spiritual path*, the origin of which is rooted in the *intrinsic connectedness* with all other sentient beings. Buddhists believe that the fixed sense of self is illusory. Buddhists see the self as ever-flowing dynamic creation of human consciousness. Buddhist *spirituality* is fundamentally concerned with eliminating suffering through the understanding of the causes of suffering and its reality through the cultivation of compassion and wisdom. Wisdom constitutes the core element of *Buddhist spirituality*. Buddhists claim that one cannot eliminate suffering and achieve spirituality only with compassion and love alone. With compassion and love, a person can become kindhearted but remains an ignorant individual. When wisdom is blended with compassion and love, only then can one achieve enlightenment. Compassion and love are natural reactions to a situation. However, when compassion and love are blended with wisdom, this, in turn, leads to eliminating suffering through enlightenment. In Buddhism, wisdom is a spiritual component that seeks to diminish, if not completely eliminate, human beings' need to be materialistic and self-centered.

Spiritual ecology is reawakening our awareness or being consciously aware of how we are inseparably interconnected with other being in Nature, being aware of what is good and great in the act of the creation of Nature and knowing that only by working in harmony with Nature's law, can we redeem what we have destroyed through our greed, arrogance, and selfishness. It is possible only through the cultivation of *compassion* and *wisdom* to understand this interconnectedness and interdependence of life and the natural and ecological processes operating to sustain the planetary ecosystem on which our own existence and the existence of the living system depend. *The longer humanity lives in the wasteland of the separation from Nature, consumed by its ego-driven arrogance devoid of compassion and wisdom, the faster it falls into the ditches of darkness and ignorance.* It is foolish to think that science and technology can bail us from this predicament and that we can continue to irreparably damage planetary ecosystems. Scientific and technological innovations can certainly help us in some limited ways in increasing the efficiency of Earth's resource uses but cannot increase the size of the Earth, expand its biocapacity beyond certain limit, or create new material resources that can substitute Earth's systems resources.

Einstein eloquently stated that our task must be to free humanity from the prison of optical delusion, without which it is not possible to adopt a new vision that can guide human thoughts and behavior toward transformation. The need for transformation is perhaps more urgent than ever before. The pace of history has accelerated the transformation of the world on a dramatic scale. *Homo sapiens* has reached the height of its powers to a dangerous level, and its might has outpaced human intellect and wisdom. The life on planet Earth, the creation of four billion years of the evolutionary process, may be on the verge of disappearance. The vast majority of scientists, ecologists, and philosophers converge on the idea of the primacy of *spirit* and the indivisibility of wholes (Kauffman 1991; Orr 1989). We can be optimistic that the *ecological wisdom consciousness* will eventually emerge from the convergence of the thought process that humanity cannot prolong its own survival with the destruction of Earth's systems, the very basis for the existence of living system. There is an interesting parallel between ecology and spirituality. Interconnectedness, interdependence, and interrelatedness are the fundamental concepts in ecology, and similarly these very concepts are the very essence of spiritual experience. In a way nature's ecology serves as the source of human spirituality.

The consciousness paradigm will bring a revolution in global mind change for transformation to a new vision of ecological civilization which entails cooperation, compassion, and wisdom as its directive principles. This will certainly be a great turning point, a leap toward humanity's understanding of itself in relation to the planetary ecosystem and, ultimately, the cosmos. Spirituality is the positive emotion that originates from human consciousness enriched with wisdom and compassion. Perhaps, this can also be called the *enlightened state of mind*. It may also be called *being spiritual*. For one to be spiritual requires a consciousness enriched with wisdom and compassion. The important thing to remember is that no human being can achieve spirituality or develop spiritual feelings without being awakened (conscious) with compassion and wisdom. *Enlightened human Consciousness is not only the basis of human spirituality but also the foundation that can ultimately unite and connect us to the unified whole, the Cosmos.*

Cultural evolution has led humankind to accumulate excessive material objects of desire but very little to free the human *spirit* from the desire for excessive material possession. The breakdown of Earth's systems (planetary ecosystem), ecological catastrophe, and the suffering of all beings, including humans, are entangled with the human desire to possess excessive material objects, desire for excessive consumerism, and hedonism. In the final analysis, consciousness enriched with compassion and wisdom (Buddhist spirituality) is the only way that can save humanity and the planet Earth and transform the prevailing egocentric paradigm of excessive material possession and consumerism into a new vision of ecological civilization, a great turning point in human history.

12.4 Environmental Stewardship

In many spiritual traditions, the idea of stewardship, or caretaking, is a common theme. Many indigenous cultures consider themselves stewards of the land, tasked with its care and protection. This belief, rooted in spirituality, inherently respects the interconnectedness of all life and acknowledges the human responsibility to the environment (Carmody & Carmody, 1993). Moral imperatives provide the ethical dimension to our relationship with the environment and set the standards for our treatment of the environment. Often, moral imperatives are influenced by social, cultural, religious, and philosophical beliefs, leading to diverse interpretations of what constitutes ethical environmental behavior (Sandler, 2012). Environmental ethics, a branch of philosophy that studies the moral relationship between humans and the natural environment, posits that our ethical responsibilities extend beyond our fellow human beings to encompass all living beings and the natural world (Brennan & Lo, 2020). This broadened moral responsibility leads to a strong emphasis on stewardship and care for the environment.

Taken together, these three elements—*consciousness, spirituality*, and *moral imperatives*—interact in complex ways to shape our attitudes and behaviors toward the environment. A high level of environmental consciousness, for instance, can lead to greater moral obligations toward the environment, which, in turn, can be reinforced by spiritual beliefs that emphasize respect for Nature (Klöckner, 2013). This interaction informs environmental stewardship, a concept that involves responsible use and protection of the natural environment through sustainable practices, protection and conservation. As stewards, we are charged with the task of safeguarding the Earth for future generations, ensuring its sustainability and resilience. This task is not merely an ethical duty but also a spiritual one, connected deeply with our consciousness and understanding of the world.

The nexus between consciousness, spirituality, moral imperatives, and environmental stewardship forms a robust framework for understanding and addressing the environmental crisis. By unraveling this complex interplay, we can identify effective strategies for fostering sustainable behaviors, promoting environmental education, and shaping policies that prioritize the environmental protection, preservation and conservation.

12.5 Noosphere and Collective Consciousness

The word "*Noosphere*," which literally means "*sphere of mind*" or *Earth's mental sheathe*, was jointly coined by a French paleontologist, Pierre Teilhard de Chardin, and a Russian geochemist, Vladimir Vernadsky in 1926 (Levchenko et al. 2017; Vidal 2010). The concept of *Noosphere* consists of dual perception: that life on Earth is a unity constituting a whole system known as the *biosphere* and that the *mind* or *consciousness* of life—the Earth's thinking layer—constitutes a unity that is discontinuous but coextensive with the entire system of life on Earth, inclusive of its inorganic support systems. In simple terms, *the Noosphere* is the *planetary sphere of mind* or *thinking layer of planet Earth* as envisioned by Pierre Teilhard de Chardin (1959) and Vladimir Vernadsky (1926). To understand the idea of *the Noosphere*, it is necessary to elevate our consciousness and open ourselves to the most general, elemental, and cosmic principles of life on Earth. According to the Foundation for the Law of Time, as described by Trubetskova and Arguelles (2010) states: "*As the mental sheathe of the planet, the Noosphere characterizes mind and consciousness as a unitary phenomenon. This means that the quality and nature of our individual and collective thoughts directly affect the Noosphere and create the quality of our environment—the biosphere. As the Earth's 'mental sheathe', the Noosphere represents the breakthrough to a new consciousness, a new time, and a new reality arising from the biospheric crisis. This is known as the biosphere-noosphere transition. Just as the biosphere is the unity of all life and its support system, the Noosphere is the unity of all minds and its thinking layer.*"

The most defining characteristic of the *Anthropocene epoch* is the creation of huge techno-metabolism or the Technosphere, which consumes natural resources faster than they can be replaced or regenerated and create more waste than the Earth can assimilate. Thus, the Technosphere operates at the expense of the biosphere or ecosphere. Vernadsky visualized Noosphere as the emergent universal phenomenon of the human mind poised for the reconstruction of the biosphere as Trubetskova (2010) articulates: "*The historical process is changing dramatically before our eyes ... Mankind taken as a whole is becoming a powerful geological force. Humanity's mind and work face the problem of reconstructing the biosphere in the interest of freely thinking of mankind as a single entity. This new state of the world we are approaching without noticing it is the 'Noosphere'.*"

The work of Jose Arguelles (1996, 2010) forms an essential foundation for the theory of *Noosphere*. The fundamental premise of Arguelles' theory of the Noosphere is that it can only come into manifestation through a *conscious effort* on the part of an informed and enlightened human minority as well as comprehending a type of *dialectical process* evolving toward a *unified field of global consciousness*. This perspective of the Noosphere as an evolutionary *dialectical process* has been anticipated by Pierre Teilhard de Chardin, the noted French paleontologist, as well as his colleague, Russian geochemist Vladimir Vernadsky. Masani (2005) argues that the ecological movement has ignored the Earth's ionospheric layer (thinking layer) and how the life-destroying interactions stem from this layer. "*The major noospheric pollutants,*" as Masani points out, "*are the marketing sector of capitalism, miseducation, and the promotion of idolatry by the judicial system.*" The ecological consciousness movement, therefore, needs to be directed to act on economic, educational, communications, aesthetic, and political fronts to reverse.

Humanity today is standing at a crossroad choosing either the path that can lead it to the sphere of cooperation, healthy competition, symbiotic relationships, and empathy, minimizing hyper-individualism (self-centeredness) and egocentric consumerism or sticking to the current path that will inevitably lead to it its own misery and annihilation. Given the current state of ecological destruction and crises, the necessity of shaping human behavior by cooperation, social and natural symbiosis, empathy, and altruism (helping others to actualize their potential) is highly imperative for the survival of the human species and the living system on planet Earth. It cannot be done without unified collective global ecological consciousness movement. The power of collective ecological wisdom consciousness is tremendous. When large number of people begin to think, feel, decide, and react as one entity against or for any issue concerning their lives, seemingly impossible changes can happen, and such changes can create a new civilization. We can see the sign of ecological awareness (consciousness) gradually increasing in the mindset of common people, especially the younger generation. This awareness must manifest in the election of political offices. Can ecological consciousness significantly affect and impact the election of political offices in the United States? This is a great question in that if a substantial number of congressmen, senators, and governors in the US Congress and delegates and senators in the States are elected, the United States can become the leader and a major player in environmental protection and nature conservation. The United States' role in the protection of the planetary ecosystem is far more important and pivotal than any other country in the world simply because United States is the biggest contributor to the pollution and destruction of the planetary ecosystem and at the same time the most powerful country in the world to effect a positive change toward environmental sustainability. It is sadly disappointing to see President Trump and his administration becoming arrogantly anti-environment and pushing the agenda of special interest groups that are bent on destroying the environment for reckless profiteering. President Trump's decision to pull out from Paris Climate Protocol, Biodiversity Convention, and many other environmental protection and nature conservation international covenants and agreements is not only deplorable but also equally myopic and foolish.

12.6 The Path Forward

Human collective ecological consciousness can be a powerful instrument to effect social, political, and cultural changes and direct such changes to the greater benefit of humanity and the biotic community in Nature. Human collective Consciousness is the only means that can reconnect and reestablish humanity's ruptured relations with Nature, the planetary ecosystem. When this consciousness expands from a few visionary people to the masses of the people and assumes formidable magnitude, one can see the cracks and cleavage in the walls of ecologically hostile consumerism propelled by neoliberal corporate market capitalism. When voices of reason would be loud enough to be heard everywhere, then happens a revolution by consciousness, a revolution that can provide the moral impetus to direct humanity's trajectory toward a just and sustainable global society in which humanity not only reconnects itself with Nature but also enters into healthy stewardship with the planet Earth, our only home, the home of the *Homo sapiens* and the rest of the biotic community.

The power of collective consciousness is a highly significant force that can be harnessed to address our time's pressing environmental and climate crisis. This phenomenon can be seen as an effective means that allows us to tap into our full potential as individuals and as a species and to wield our collective power for the greater good. At the heart of the environmental crisis lies a fundamental disconnect between humans and the natural world. Our modern way of life, emphasizing consumerism and materialistic values, has led us to exploit the planet's resources in grossly unsustainable ways. However, by tapping into humanity's collective consciousness power, we can shift our collective mindset to bring about meaningful changes. This involves a shift away from the ego-driven, individualistic mindset that has dominated our society and toward a more holistic, interconnected worldview that recognizes our inherent interdependence with the natural world.

One way to tap into the power of this consciousness is through mindfulness practices such as meditation and yoga. These practices can help us cultivate a deeper sense of connection to the natural world and to develop a greater awareness of the impact of our actions on the environment. By becoming more mindful, we can make more conscious choices about how we live our lives and take actions that are in harmony with the planet. Another way to harness the power of this consciousness is through collective action and activism. By coming together as a community to advocate for policies and practices that promote sustainability and protect the environment, we can use our collective power to effect real change. This might involve participating in protests, engaging in political advocacy, or supporting environmental organizations working to make a difference. Ultimately, the power of collective consciousness offers us a path toward a more sustainable and harmonious relationship with the natural world. It is up to us to embrace this power and use it to create a better future for ourselves and the planet.

We must rise, unified in purpose, to usher for environmental sustainability of development and environmental justice that respects the natural rights of living being to exist on planet Earth. This compelling vision is neither fanciful nor idealistic; it is an imperative that requires us to subvert the current norm and demand change. It is time to realize that our survival is intricately linked to the health of our planet, and by extension, our actions, and choices. We must, therefore, strive to dismantle the reigning cultureand influence the attitudes of our leaders to guide humanity away from the brink of disaster and toward a future of environmental sustainability and equity. The only way out for our survival is to learn to live within the regenerative biocapacity of the planet Earth with ecological consciousness. If we dream alone, it will be just a dream but if we dream collectively, it will be a reality. Let us dream with "The Earth First Paradigm" for sustainable living.

Chapter 13
Ecosociocentrism: The Earth First Paradigm for Sustainable Living

The economy of nature and ecology of man are inseparable and attempts to separate them are more than misleading; they are dangerous. Man's destiny is tied to Nature's destiny, and the arrogance of the engineering mind does not change this. A man may be a very peculiar animal, but still a part of the system of Nature.

Marston Bates (1960)

13.1 Introduction

We have arrived at the final and most important chapter of this book. This chapter is the synthesis of the previous chapters that covered a wide range of interrelated topical themes, including the current neoliberal development model, its underlying principles, assumptions, and values primarily responsible for today's environmental destruction and climate crisis. This chapter is devoted to the value-based development worldview or paradigm that argues for the necessity of cultural change in the patterns of production and consumption of material goods and services, recognition of some intrinsic values along with instrumental and the uses of science and technology to rebuild and restore the resilience and the health of planet Earth's systems for the sustainable living of humanity including other living beings on planet Earth. The chapter brings perspectives on the need for a new paradigm that embodies values and concepts such as diversity and ecosystem health, interconnectedness, interdependence, autopoiesis and organizational complexity, and integrity that can be applied to keeping Earth's systems and socio-economic sub-system in a sustainable state. Building upon the previous chapters, this chapter proposes a new paradigm called "*Ecosociocentrism: The Earth First Paradigm*" to reconcile instrumental, relational, and intrinsic values in Nature and provide the foundational basis for environmentally sustainable development. While the term "*Ecosociocentrism*" does not currently exist in English dictionaries, it is a potent term coined by the fusion of two

G. Upreti, *Ecosociocentrism*, https://doi.org/10.1007/978-3-031-41754-2_13

seemingly opposite concepts: "*Ecocentrism*" and "*Sociocentrism*." This term aims to reconcile with the apparent gap between the *ecosphere* and *sociosphere*, two domains often considered at odds with each other. "*Ecosociocentrism*" symbolizes a harmonious synthesis of these two opposite tendencies. It embodies the notion that these seemingly opposite tendencies can coexist in a state of complementarity, rather than stark negation of each other.

I have used two terminologies: *ecosphere* to denote the biosphere or planetary ecosystem or Earth's system and *sociosphere* to denote the human social-economic subsystem. Any development that is not informed and guided by ecological wisdom consciousness and does not operate under certain moral and ethical frameworks will ultimately lead to its own destruction because of the inherent contradiction embedded in it. The essential task of humanity in the *Anthropocene* is the resolution of this contradiction through integrating political economy in the *sociosphere* with the ecology of Nature in the *ecosphere*. The proposed paradigm envisages the harmonious integration of political economy in sociosphere with the ecology of Nature (sociosphere). Such integration and synthesis essentially requires a cultural change in *hyper anthropocentrism* and the recognition of the *intrinsic value* of the web of interconnectedness and interdependence that is so pervasive in the *ecosphere* and *sociosphere*. "*Ecosociocentrism: The Earth First Paradigm*" proposed in this chapter, attempts to integrate the *ecosphere* and *sociosphere* through value-based scientific epistemology and ecological consciousness. In essence, "*Ecosociocentrism*" encourages us to perceive the health and sustainability of both natural ecosystems (ecosphere) and human societies (sociosphere) as intimately interconnected and interdependent. It considers that the vitality of *sociosphere* (human social system) hinges on the functional health (well-being) of *ecosphere* (natural ecosystems) and vice versa. With this perspective, humanity can redefine its relationship with Nature fostering a more holistic, sustainable, and mutually beneficial coexistence of humanity, and the living system in Nature. Only from the conceptual framework of such paradigm can emerge policies and strategies for the protection of Earth's systems, conservation and protection of living entities, and the maintenance of the planetary ecosystem in a manner that ensures proper functioning of both *ecosphere* (biosphere) and *sociosphere* (social-economic system).

Stephen Boyden (1987) argues that the moral values and cognitive beliefs inherent in culture play an essential role concerning how human societies adapt to the natural environment and what kind of economic and political systems they create and maintain. This can be seen from the historical fact that many native and traditional societies' cultural heritage and religious traditions possess ecologically sound knowledge that originated from their interaction with the natural world. This environmentally sound knowledge, which is a part of the cultural traditions of many native societies, was the basis for the development of sustainable use systems of natural resources. On the contrary, as Osvaldo Sunkel (1988) points out, modern dominant cultural values have destroyed sustainable land and natural resource use systems and patterns. Therefore, the task of ethics is to understand and evaluate the moral codes woven into cultures, what they are, and how they enhance or distort human relationships with one another and the planetary ecosystem.

Only the political, social, and economic determinants of human behavior received the attention of scientists, policymakers, and professionals. This is primarily due to the assumption of value neutrality and objectivity in science and policy formation. There has been a shift in rejecting these assumptions in the post-modern era of development discourse following the paradigm shift in classical physics, biology, and ecological sciences. Values' role in human activity must be added to the modern development discourse. It is argued that there is no independent observer of reality, and the values cannot be separated from the act of observing. As Fritjof Capra (1990) states: "*The patterns scientists observe in nature are intimately connected with the patterns of their minds, with their concepts, thoughts, and values.*" IUCN and WWF (1980), the two seminal organizations involved in nature conservation with the historic pronouncement, stated that ethics is an explicit concern to the international conservation movement: "*Ultimately the behavior of entire societies towards the biosphere must be transformed if the achievement of conservation objectives is to be assured. New ethics, embracing plants, animals, and people, are required for human society to live in harmony with* the *natural world on which they depend for survival and well-being. The long-term task of environmental education is to foster or reinforce attitudes and behavior compatible with the new ethics.*"

Stephen Toulmin and Goodfield (1982), a philosopher of science, argues that scientists can no longer treat themselves as merely an observer and remain disinterested in the moral significance of their actions while doing science. "*Instead of viewing the world of nature as onlookers from outside,*" Toulmin (1982) says: "*we now have to understand how our own human life and activities operate as elements within the world of nature. So, we must develop* a *more coordinated view of the world, embracing both the world of nature and the world of humanity, a view capable of integrating, not merely aggregating, our scientific understanding and capable of doing so with practice in view. Nowadays, scientists always consider themselves agents, not merely observers, and ask about the moral significance of the actions that comprise even the very doing of science.*" Ronald Engel (1990) maintains that classical disjunction between subject and object, value, and fact, is invalid; the knower is implicated in the known, and there can only be relative objectivity. How facts are investigated, selected, and interpreted depends on one's values, which are affected by one's worldview (how one sees the world).

13.2 Science, Values, and Ethics

Capra (1993, 2014) argues that the understanding of a *self-organizing system* is essential because it provides the ideal framework for ecologically oriented ethics, which is urgently needed. After all, what scientists are doing today is not life-furthering and life-preserving but life-destroying. Since the scientific revolution in the seventeenth century, facts have been separated from values. The scientific community tended to believe that scientific facts are independent of what they do and, therefore, independent of their values. In reality, scientific facts emerge from a *paradigm*—the constellations of human perceptions, values, and actions from which they cannot be separated. Scientists are intellectually and morally responsible for their research. One critically important insight rendered by the *system theory of life* is that life and cognition are inseparable. The epistemological process is also the process of self-organization, the process of life. The conventional model of knowledge is an image of independently existing facts which is the model derived from classical physics. The system theory views knowledge as part of the process of life, a dialogue between object and subject. Knowledge and life, and therefore, facts and value, are inseparable from each other (Capra, 2014; Bahg, 1990; Bertalanffy, 1972; Merchant, 1990). No system of ethics expresses the same ecological awareness as the system theory of life. This is a tall task for scientists and philosophers to establish an ethical system based on *ecological wisdom*. Marston Bates (1960) thinks that any attempt to separate *nature's economy* and *human ecology* is misleading and dangerous since human destiny is entangled with the destiny of Nature. Human arrogance cannot change it.

A critical task of ethics is resolving the apparent conflict between resource conservationists and eco-centrists, reflected in a heated argument, often resulting in harsh criticism. Resource conservation reflects an anthropocentric and technocentric worldview (current dominant paradigm) for the adherents of ecocentrism. Conservationists stress the efficient long-term utilization of natural resources and recognize only the *instrumental values*, whereas ecocentrists emphasize preserving *intrinsic values* inherent in Nature. On the other hand, conservationists consider the ecocentric view as romantic and misanthropic. The international environmental movement attempted to resolve this conflict by affirming both the values as reflected in the statement of the World Charter for Nature, which states that *every form of life warrants respect regardless of its worth to man and that ecosystems and organisms shall be managed to achieve and maintain optimum sustainable productivity*. Engel (1988) argues that ethically informed leaders in environment and development communities recognized the underlying moral concern of conservationists, which is the *stewardship of natural resources* on behalf of distributive justice within and between generations. Moreover, thus, the fundamental issue is not between *instrumental* and *intrinsic* values but between two kinds of *intrinsic values—social justice* and *ecological integrity*. Both values require efficient, long-term instruments of support and each of which can contribute to the enhancement of the other.

Do we need a new *paradigm*, and if so, can such a paradigm embrace ethical/moral consideration, is the central concern of this book? We need a new paradigm that integrates *instrumental* and *intrinsic* values in Nature, universally promotes ecocivilization, and lays the foundation of *sustainable living*. Ethics is essential in critiquing and reforming the *dominant social paradigm*. Understanding how our social and ecological values are determined and shaped by our worldviews within the framework of which we perceive and interpret the worldly phenomena around us is fundamentally important. A brief revisit to current dominant and evolving alternative paradigms (worldviews) is necessary to understand the origin of the present environmental crisis and conceptualize the complexity of sustainable living and intergenerational equity.

13.3 Dominant Worldview

The current dominant worldview is a Western Worldview, as discussed in the preceding chapter. The seventeenth- and eighteenth-century scientific revolution was instrumental in developing the Western worldview. A secular and mechanistic worldview gave rise to the growth of mercantile capitalism that largely determined the mainstream thought and actions of individuals, societies, and institutions to the present day (Sterling, 1990). The critical characteristics of this worldview are separation and dissociation. Descartes' logic laid the foundation for the scientific paradigm by differentiating mind and body, subject and object, value and fact, spirit, and matter. These distinctions helped flower scientific inquiries, but the deep schism between these dualisms (opposites) is now at the heart of the current crisis. Thus, there is an inherent bias in favor of facts over values, object over subject, quantitative over qualitative factors, analysis over synthesis, instrumental over intrinsic values, competition over cooperation, individual greed, selfishness over community needs and interests, etc. As Stephen R. Sterling (1990) points out: "*Most importantly, Cartesian duality* sets *human beings apart from and over Nature, thus opening* the *way for* a *primarily exploitative and manipulative relationship. We have faithfully enacted Descartes' belief that humans should be "the masters and possessors of Nature.*"

Our worldview (paradigm) has conditioned our perception and understanding of the role of ethics in the ideas and issues of development, environment, and conservation. The positivist influence of Descartes has effectively suppressed the growth of ethical thought in Western societies. Positivists argue that value judgments are subjective and unreliable and, thus, they do not constitute "proper knowledge." The Ottawa Conference on Conservation and Development (1986) exhumed the inherent flaws and contradictions in the current dominant paradigm and recognized the need for structural transformation. As it becomes clear from the conclusion of that conference: "*We need an alternative society, another type of development linked with structural transformation. While new solutions will incorporate mixtures of the old ... concrete solutions to environmental problems will largely depend on a new organizational capacity of society as a whole, based on the cultural values of different communities, their creativity, and their potential for innovation.*" The Ottawa Conference was the first conglomerate of international environmental and development professionals to critically examine and challenge the market-centric dominant paradigm whose central premise is unlimited economic growth, profit, and a machine view that technology can fix all human and environmental problems.

13.4 Alternative Worldview

Philosophers and religious leaders for centuries have provided their sermons and percepts on how one should live on this planet. This question has become more paramount at this juncture of human history because how we live on Earth today directly impacts what kind of future we want and what kind of Earth we want to live in. Whether humankind has any long-term future on planet Earth is questionable, given the current state of environmental, ecological, and climate crises. Adopting a new ecological worldview is necessary to create a sustainable living culture to have a long-term sustainable future. This culture can recognize that ecological processes and the integrity of planetary ecosystems must not be tampered with by the greedy pursuit of unlimited economic growth and consumerism, undermining the perpetuation of the human species and other species in Nature. *Treating Earth with respect and reverence implies maintaining the ecological processes and integrity of the planetary ecosystem in a functionally healthy state. This systemic or holistic ecological worldview is the only genuine hope for a sustainable future for humankind and other living beings on planet Earth.*

The fundamental assumption of the ecological worldview is that we are interconnected to the rest of Nature, materially and spiritually. This understanding helps us recognize Nature's *instrumental* and *intrinsic* values and conceptualize the justification for *conservation* and *preservation*. The scientific, mechanistic worldview does not recognize such a connection with Nature. The ecological worldview profoundly acknowledges that we must act judiciously to restore our ruptured relationships with the planet Earth and reinvigorate her biospheric ecosystem processes. Evolution embodies diversification and integration as complementary processes (Axelrod, 1984; Gould, 1982; Kauffman, 1990). However, the mechanistic scientific worldview is fundamentally at odds with the evolutionary process. This has caused severe disruption in the ecological processes, reduced the resilience of the Earth's systems and their ability to self-regulate, and has severely increased the vulnerability of the *biosphere* and the planetary ecosystem.

If facts are understood in the proper context, what we call facts can become values. For example, if we consider it to be true that human existence and the existence of other beings depend on the ecological processes and integrity of the planetary ecosystem as facts based on ecological and scientific knowledge we have acquired so far. Then, we must value the ecological processes and integrity of the planetary ecosystem. In this case, ecological processes and the integrity of planetary ecosystems are both descriptions of facts and values, inseparable from each other. Capra (1990) argues that the classical dualism between subject and object, value, and fact, is invalid, and how facts are investigated, selected, and interpreted depends on one's values which are again determined by one's worldview through which one sees the world. As Heisenberg (1958), one of the founders of quantum mechanics stated: "*By its intervention, science alters and refashions the object of investigation; in other words, method and object can no longer be separated. The scientific worldview has ceased to be scientific in the true sense of the word.*"

Classical reductionistic models attempt to understand the whole by observing the constituent parts and parts of the parts. System thinking maintains that the concept of "part" is an illusion that precludes us from understanding the dynamics of the relationship involved in the system. This is especially true in a bio-physical world where the survival of living organisms cannot be separated from their respective environment. The unit of survival is not the organism *perse* but the organism and its environment (the larger whole). Suppose we assume the biosphere is evolving toward a complex system of interacting networks of components. In that case, this necessitates a certain degree of cooperation, interaction, and mutual support among the components, which implies fewer degree of freedom for components to operate beyond the boundary of the system. This concept is most strikingly expressed in Lovelock's Gaia theory (1986) which views the planet as a *complex living organism* that optimizes conditions for its survival: "*When an organism benefits the environment as well as the organism itself, then its spread will be assisted. Eventually, the organism, and the environmental change associated with it, will become extent. The reverse is also true: any species that adversely affect the environment is doomed, but life continues.*" It is sadly disappointing to remind that humanity has adversely affected its *niche* environment (Earth) and is marching towards its doom day. The ecological worldview has given rise to the concept of *ecodevelopment*, which recognizes the complementarity between preservation/conservation and development and attempts to integrate them into the practice of sustainable development. It is based on the premise that humans and Nature can exist in a harmonious symbiotic relationship. Holmes Rolston III (1990), a strong proponent of environmental ethics, advocates that animals need to be valued *intrinsically* for what they are and *instrumentally* for the roles they play in ecosystems. As Rolston (1990) states: "*A metaphysically based, culturally derived values that run contrary to science-based value will not be intellectually sustainable over time. Hence, it will not be socially functional either. To survive, value must be made complementary to the facts of science. If from biology, we learn that the various species are what they are primarily due to biological determinants. Those with oriental or native philosophies will have to decide what the operational value of their metaphysics is.*" Conservation biologists and environmental ethicists are concerned about losing vital genetic information (information created over millions of years of the evolutionary process) when species become extinct and worry about terminating the speciation process. They have shared their concerns and frustrations about the ecosystem's loss of *stability, diversity,* and *resilience*. It can be argued from the utilitarian point of view that there is a human interest to protect nature for the sake of humanity itself. Recognizing the *intrinsic values* of diversity, the richness of life on Earth, and their protection is compatible with utilitarianism.

Arne Naess (1990), an ardent proponent of the deep ecology movement, criticized the conservation organizations like World Conservation Strategy (WCS) for not amply recognizing the *intrinsic value* of nonhuman life. Deep ecology thinkers and activists think nonhuman life is valued independently of human life. Naess (1990) argues: "*Every living having the right to live and flourish," the foundational value of deep ecology. It requires a change in the perception and understanding of the societies for the transition to a new way of doing things. Societies need to re-inhabit instead of occupying their region, the bioregion, or* the *ecoregion where they live. Ecosystem people evolve social behavior that will enrich the life of that place, restore its life-supporting systems, and establish an ecologically and socially sustainable pattern of existence within it.*" Planet Earth's ecology cannot be separated from human social justice. The *per capita* degradation of the planet Earth is directly dependent upon the social-cultural lifestyle of the individuals. Social lifestyle depends upon class, social stratification, social services, and the protection received. The reduction in the per capita degradation of the conditions of planet Earth requires and demands changes in the lifestyle habits of individuals and their moral and ethical behavior. If there is no change in lifestyle and ethical behavior, the wealthy class power elites in developing and developed countries will be judged ecological and ethical misfits because they are mainly responsible for the degradation and destruction of Earth's ecosystem and ecosystem processes.

One of the valid and critical issues Guha (1989) raised in his criticism of deep ecology is that the dichotomy between *biocentrism* and *anthropocentrism* provides little understanding of the dynamics of worldwide environmental degradation. Guha maintains that overconsumption by the West and urban elites in the developing world and militarization are the two fundamental ecological problems, and none of these problems has any tangible connection to anthropocentric-biocentric distinction. I would like to add that though it is seemingly anthropocentric, primarily it is *egocentric ideology*, deeply grounded in neoliberal capitalism, responsible for creating an ego-satisfying (self-aggrandizement) *ecologically hostile consumer society*, which is the primary cause of environmental destruction. It is corporate egocentric anthropocentrism that is destroying the environment today. Consciously or unconsciously, the proponents of deep ecology have evaded this corporate egocentrism and broadly tainted it with anthropocentrism, a powerful umbrella term that encompasses all human beings and makes all human beings equally responsible for creating today's environmental problems without any reference to nature, context, and the point of their origins. Deep ecology or ecocentrism has ignored the social dimension of natural resource development issues, inequities in resource distribution, and, hence, the social roots of the problems of environmental degradation. The biggest problem with ecocentrism is its mysterious silence on human welfare, social justice, and equity in resource distribution and their uses. The essence of ecocentrism is that humans are part of "*biotic life*," not at the top of creation but equal to many other aspects of creation. This image inspires respect for natural biodiversity and evolution but completely ignores human social needs and problems, particularly the "*social web*" of human life and institutions. Ecocentrism and deep ecology cannot provide a sound and practical ethical basis without addressing intricately interrelated economic, social, and political issues of modern consumer society. It is one thing to be romantic about an idea, but it is quite another to derive practical norms for human conduct from those ideas. What practical relevance does deep ecology or ecocentrism have for the behavioral conduct of the people in general and the behavioral conduct of the people of the developing world in particular? Deep ecology or any environmental ethics must incorporate appropriate social ethics into its conceptual frameworks to remain viable and offer some practical guidelines for human conduct. We need some *pragmatic* environmental ethics that can solve human problems and save and maintain the health and integrity of the biospheric ecosystem. Henryk Skolimowski (1990) argues that the resolution of environmental problems lies in the matrix of our values, the values we hold, and how those values control our behavior. Unless new and sound sustainable values are established, and human behavior is motivated by those values, all expertise and technological fixes will signify nothing. Sokolowski talks of foundational values as the bedrock of ethical systems. For him, the foundation value of ecological ethics (eco-ethics) is rooted in the *sanctity of life*. He argues that such foundation value for Gandhi was *ahimsa* (nonviolence). For Schweitzer, it was *reverence for life*, and for Aldo Leopold, it was the *sacredness of the land. Thus, for him, ecological ethics is based on the sanctity of life.* Finally, these ethical principles underlie and justify our rational strategies and practical choices.

The existence of individuals in the community is the essential phenomenological feature of life on Earth. This is true both in society and Nature, and this has been the way things are. As many scholars have rightly pointed out that Earth can best be described as a mosaic *of coevolving, self-governing communities* consisting of diverse forms of life with intricately interconnected and interdependent parts and processes (Axelrod, 1984; Kerr, 1988; Boucher et al., 1982). *The emergent ecological* worldview *considers this as the description of the facts and values.* All human cultures are guided by certain codes of conduct and moral duties to individual human beings. Ronald J. Engel (1990) thinks the ecological worldview expands the concept of moral obligation to include all organisms and the interactions between individual organisms in the ecosystem and the ecosystem's order. Holmes Rolston III (1988) elaborates on this concept by arguing that to deny moral value to ecosystems because they are not integrated in the same way as a centralized individual organism is to make a category mistake. In thinking about the value of social and biotic communities, we should be concerned with a *"matrix" of interconnections between centers, not a single center*. He further enunciates (1988): "*Ecosystems are in some respects more to be admired than any of their component organism because they have generated, continue to support, and integrate tens of thousands of member organisms… We want to love "the land." as Leopold terms it, "the natural processes by which the land and living things upon which it has achieved their characteristic forms (evolution) and by which they maintain their existence (ecology)." The appropriate unit of moral concern is the fundamental unit of development and survival. Loving lions and hating jungles is a misplaced affection. An ecologically informed society must love lions in jungles, organisms-in-ecosystems, or fail in vision and courage.*"

Is it possible to talk about *intrinsic values* without falling into the trap of *moral relativism* or *moral absolutism*? Perhaps, it is possible if we proceed from one foundational value, that is, *life has value for itself* and in itself because of its *self-organizing* or *autopoietic* attribute that drives it toward realizing its potential. Accepting *intrinsic value* provides a more substantial basis for conservation strategies. Environmental ethics (eco-ethics) must recognize the *intrinsic value* of life. This also demands that intellectual and moral insights complement each other and be integrated into the ecological worldview. "*Our intellectual consciousness,*" as Henryk Skolimowski (1986) explains, "*can declare something of intrinsic value only after it is informed by the axiological level, by the values which we cherish in* the *depth of our hearts and souls. The fact that it is our value or axiological consciousness that informs and guides our cognitive consciousness regarding values is of great importance, for it leads to a new clarification of intrinsic value. There are no intrinsic values beyond our consciousness as a species and independent of it. It is our consciousness that makes things valuable.*" We need a moral insight that recognizes the *intrinsic values* of other life in Nature and that we are a part of Nature, and all living beings are our fellow creatures in creation. Aldo Leopold (1990) long ago recognized the *intrinsic value* of the biotic community that formed the basis of his famous land ethics, which states: "*A land ethic, then, reflects the existence of an ecological conscience, and this, in turn, reflects a conviction of individual responsibility for the health of the land. Health is the capacity for self-renewal. Conservation is our effort to understand and preserve this capacity.*" It can be argued that environmental/ecological ethics can be considered the extension and the continuation of Leopold's land ethics and Buddha's *reverence for life (nonviolence and ahimsa).* One can argue that the ultimate end of all development is life because development serves and positively contributes to life; however, it is not merely the life but the quality of living for all. Ultimately, the end goal of development is life with meaning, dignity, fulfillment, and self-actualization.

13.5 Autopoiesis and System View

From the principle of *interconnectedness* and *autopoiesis*, it is not difficult to recognize the fact that the health and the well-being of humanity and the living world are inextricably linked to the integrity and the health of Earth's systems (planetary ecosystem). The dynamical system view recognizes the embedded nature of humankind's existence within Nature. Each living organism can be considered a system within a *holarchy*—a metasystem of irreducible wholes that are part of the larger whole, ultimately comprising all life forms on Earth from a single cell to the entire planetary biosphere (ecosphere). This implies that humans can only thrive sustainably within a healthy or properly functioning planetary ecosystem or biosphere. Organic complex and dynamic interrelationships, a web of cyclical interconnections across time and geographical space, characterize the system approach (in contrast to the mechanistic model). System theory postulates that the whole is greater than the sum of its parts; thus, systemic understanding becomes a new model of reality. This *holistic* understanding of the systemic processes and interconnectedness gives rise to an ethical framework. Since the ecological worldview is based upon systemic processes and relationships, human values and actions should be consistent with the systemic reality or the system view of life. If human activities are inconsistent with the systemic processes, the consequences of human actions will disrupt the systemic processes, and everything will suffer. We must infer prescription with description, fact, and value cannot exist as separate categories but are in close interconnected relationships.

System theory (theory of living system) provides the most logical formulation of the ecological paradigm. Living systems entail various phenomena encompassing individual organisms, ecosystems, and human social systems. System theory provides a common framework for ecology, biology, neuroscience, social ecology, economics, medicine, and other sciences, including organizational management and information network. The system framework is the most appropriate framework for explicitly manifesting ecological phenomena and perspectives (Bertalanffy, 1972; Bahg, 1990). Fritjof Capra (2014), a prominent system theorist, has identified five criteria of systems thinking that can be applied to both natural and social sciences. These are the *shift from part to the whole, from structure to process, from objective to "epistemic" science, from "building" to "network" as* a *metaphor for knowledge, and from truth to approximate description.*

The old mechanistic paradigm employs a reductionistic approach in studying the complex dynamic system and infers the properties of the whole from the properties of the parts. The system paradigm reverses the relationship between parts and whole and asserts that the properties of the parts can be understood only from the dynamics of the whole (Capra, 1991; Maturana, 1980; Lovelock, 1991). This shift from the parts to the whole was the central aspect of the conceptual revolution of quantum physics in the 1920s (Heisenberg, 1971). More recently, it has been increasingly realized that the view of physical reality is a web of relationships demonstrated by S-matrix theory which gave rise to *string theory* and *holographic principles* (Stapp, 1971). Likewise, the old paradigm asserts that there are fundamental structures, forces, and mechanisms through which they interact and give rise to the processes. The system paradigm asserts that every structure is the manifestation of an underlying process, and the entire web of relationships is intrinsically dynamic. In his classic book From Being to Becoming, Ilya Prigogine (1980) illustrated this shift from structure to process. The conventional scientific paradigm considered scientific descriptions to be objective independent of the human observer and the knowledge process. The system paradigm integrates epistemology (the process of knowledge) in explaining natural phenomena. Heisenberg (1971) introduced this concept into physics which views physical reality as a *web of relationships*. Whenever we isolate a pattern from this interrelated network and define it as a part or object by cutting through some of its connections to the rest of the network, the fragmented part or object cannot give us an accurate picture of reality. As Heisenberg (1971) eloquently puts "*What we observe is not nature itself, but nature exposed to our method of questioning.*" This method of questioning, in other words, the epistemology inseparably becomes part of the theory.

In Western science and philosophy, the metaphor of knowledge as a building has been used for thousands of years. The phrases such as fundamental laws, principles, and basic building blocks have been used as the foundation of knowledge on which the Tower of Science had been built. The old paradigm still clings to this metaphor of knowledge as a *building*. The system paradigm rejects this metaphor and replaces it with that of the *network*. For the system paradigm (system approach), the *network* becomes the metaphor of knowledge, and reality is perceived as a *network* of relationships. The descriptions of the perceived reality form an interconnected *network* of concepts and models in which there is no foundation of building blocks. It is explicitly expressed in Geoffrey Chew's bootstrap theory of particles in physics. Chew (1999) explains that Nature cannot be reduced to any fundamental entity but must be understood entirely through self-consistency. Physical reality is conceived as a dynamic web of interrelated events. "*Things exist by virtue of*" Chew (1999) states: "*their mutually consistent relationships requiring that its components be consistent with one another and with themselves. The bootstrap theory is a theory of networks of subatomic particles in which the property of each particle is derived from its relationships to the others. This is systemic thinking par excellence.*" These system thinking criteria, as discussed above, are all interdependent. Nature is conceived as an interconnected dynamic web of relationships in which the identification and study of specific patterns as "objects" depends on the human observer and process of knowledge. System thinkers (Capra, 1999, 2014; Chew, 1999; Stapp, 1971) have pointed out this web of relationships, described in interconnected networks and models, none of which is more fundamental than the others. System theorists postulate that there is no such thing as "*absolute truth*" and that there is only an *approximate description of reality or truth*. This insight is crucial to all modern sciences because it recognizes that all scientific concepts and theories are limited and approximate. Science can never provide any complete and final definite understanding. Scientists do not deal with the truth. They deal with the limited and approximate description of reality. As Heisenberg (1971) stated, "*The often-discussed lesson that has been learned from modern physics is that every word or concept, clear as it may seem to be, has only a limited range of applicability.*"

13.5.1 Autopoiesis and System Complexity

The living system's self-organizing attribute is the system theory's central concept. A living system is defined as *a self-organizing system* whose order is not imposed by the environment but is established by the system itself, indicating that self-organizing systems exhibit a certain degree of *autonomy*. This is not to suggest that living systems are isolated from their environment; on the contrary, they continuously interact with their environments, but such interactions do not determine their organization. For self-organizing systems, the pattern of organization (the totality of relationships that define the system as an integral whole) is characterized by a mutual dependency on the system's parts, which is necessary to understand the parts (Jacobs et al., 2015; Capra, 2014; Kauffman, 1990). Humberto Maturana and Francisco Varela (1980a, b) studied and precisely described the pattern of self-organization. They called it *autopoiesis*, which means *self-creating* and *organizing processes*.

Ilya Prigogine (1980) extensively studied the structure of *self-organizing systems* and called them *dissipative structures*. He described the two main characteristics of the dissipative structure: First, a dissipative structure is an open system that maintains its pattern of organization through a continuous exchange of energy and matter with its environment. Second, the dissipative structure operates far from thermodynamic equilibrium and thus cannot be described in classical thermodynamics. Prigogine has been credited for his contribution to creating new thermodynamics to describe living systems. However, it is equally important to ponder into the comments by mainstream scientists and scholars (Ehrlich & Ehrlich 1993): "*Occasionally, it is suggested erroneously that the process of biological evolution represents a violation of the second law of thermodynamics. The catch is that Earth is not an isolated system; the process of evolution has been powered by the sun, and the decrease in entropy on Earth represented by the growing structure of the biosphere is more than counterbalanced by the increase in the entropy of the sun.*"

Humans are integral to Nature and cannot be separated from the natural world. They can separate themselves from the rest of Nature only when they can substitute the life-supporting and sustaining natural system services, processes, and the products, such as air, water, land, food, and energy, with the processes and products of their own creation. Humans have not developed the technological capability that can separate them from their dependency on Nature. What humans do can have implications and consequences that inextricably pull us back. Humans are constantly changing and recreating themselves, irrespective of whether they know it or not. Thompson and Varela (2001) describe the emergence of a higher level of organization through reciprocal causality: *Emergence through self-organization has two directions. The upward direction is the local-to-global causation, through which novel dynamics emerge. The downward direction is a global-to-local determination, whereby a global order parameter "enslaves" the constituents and effectively governs local interactions. There is no supervisor or agent that causes order; the system is self-organized. The spooky thing here, of course, is that while the parts do cause the behavior of the whole, the behavior of the whole also constrains the behavior of its parts according to a majority rule; it is a case of circular causation. Crucially, the cause is not one or the other but is embedded in the configuration of relations.* Thomson et al. (2001) describe that emergent processes consist of the collective behaviors of large ensembles, in which positive and negative feedback interactions give rise to nonlinear consequences. First, there is local-to-global determination or upward causation, as a result of which novel processes emerge that have their own features. Second, there is global-to-local determination, often called downward causation, whereby global characteristics of a system govern or constrain local interactions. Although usually called *circular causality*, this reciprocal relationship between local and global levels is better described as *reciprocal causality*. It is generally perceived that the behavior of a system arises from a combination of positive feedback, response thresholds, and negative feedback. Sumpter (2006) studied collective animal behavior to determine the algorithms that produce the collective behavior and to understand the principles that underlie these algorithms. Natural selection is the reason for the evolutionary origin of animal behavior and, thus, becomes a powerful principle in obtaining such understanding. Like natural selection, rationality is only one of the principles governing collective behavior in humans (Schelling, 1978; Milgram, 1992). Sumpter (2006) advanced the hypothesis that many collective human behaviors are similar to their animal counterparts and that the same mathematical models can describe collective patterns of behavior in humans and animals.

13.5.2 Autopoiesis and Intrinsic Values

Dynamical systems theory views that the natural world can only be understood appropriately by recognizing and identifying the organizing principles of the nonlinear systems by which everything interconnects as opposed to focusing only on the things themselves. This holds true for our climate, ecosystems, organisms and cells, and our own body systems: our immune system, cardiac and respiratory systems, the nervous and neuronal cognitive system, including *human consciousness*, which can be seen as an *emergent phenomenon arising from the complex interactions of neuronal networks in the human brain.* Johnson (1992), arguing for the intrinsic value of species and ecosystems, states that ecosystems are not just the aggregation of the species of fauna and flora occurring in particular places. They are living systems with their own organic unity and self-identity and character. Johnson considers ecosystems as living systems that maintain their identity, unity, or homeostasis and argues that the interaction of genes or genotypes with the environment and the manifestation of this interaction at higher levels makes it possible not only for individual organisms but also species and ecosystems properly to be considered living entities in their own right. Though species and ecosystems have evolved in such a way that their life processes tend to maintain the viability of the whole. Their interests suffer to the extent that their life processes cannot do so. Because of these interests, the species and ecosystems can meaningfully be considered for moral standing.

Recognizing our fundamental interconnectedness with other life forms, the *autopoietic* (self-organizing) nature of the life process, and the interdependent nature of our existence, we must maintain this shared and embedded existence's stability, integrity, and beauty. Does it not qualify to be a value in itself, an *intrinsic value*? I argue it does provide a sufficient basis to be called *intrinsic value*. It is imperative to realize that values and ethics do not originate from any external source but emerge naturally from experience and understanding of our inseparable interconnectedness with life forms and living systems that manifest as the increased complexity of self-organization which, we can safely call an *intrinsic value* without digging deeper into any metaphysical abyss. *Human actions and behavior that tend to preserve such values (intrinsic values) in Nature should be considered ethical and necessary.* The planetary ecosystem or Earth's biosphere arose from this self-organizational complexity of interconnectedness driven by the energy input from the sun, ultimately comprising all life forms on Earth from a single cell to the entire planetary ecosystem (ecosphere). Recognizing that life is an emergent form of self-organized complexity, leading physicists Schrodinger (1944) and Prigogine (1980, 1984) have described life as negative entropy or *negentropy*. This has important implications for the intrinsic value of life; life as negentropy is the ultimate source of meaning. Evolutionary history informs us that over four billion years, life has been evolving on Earth, reaching ever greater levels of complexity, from individual cells to complex organisms (human beings) to communities and biospheric or planetary ecosystems. An amazing saga of the self-organizational complexity of the life process on planet Earth.

The ecological services and the products generated by ecological processes necessary for the existence of human and nonhuman life forms fundamentally depend on the proper functioning of the Earth's systems or planetary ecosystem. The proper functioning of the ecosystem is analogous to the concept of a healthy ecosystem. The quality of ecological services and products depend on the ecosystems' health (Haskell et al., 1992; Upreti, 1994, 1996; Bark et al., 2016). Proper functioning of the Earth's systems or the health of the planetary ecosystem has never become more critical before than now, given the perilous state of the degradation and disruption in its functioning as manifested by global warming, adverse climate changes, the rapid rise in temperature, carbon dioxide, and sea level, melting of Himalayan and arctic glaciers, the prevalence of drought and appearance of more virulent and resistant crop, animal and human diseases and pests. Human prosperity and social system depend on the health and the proper functioning of the planetary ecosystem, its self-organizing and regenerative biocapacity. This is the description of both the fact and the value, which cannot be separated from each other. I believe this provides a sufficient basis for considering the *intrinsic value* of the self-organizational complexity of the biosphere that must be protected. Even if we consider the "*intrinsic value*" of the planetary ecosystem and its *autonomous self-organizing* property a *naturalistic fallacy*, a change in human consciousness and behavior is necessary for the planetary ecosystem to restore its resilience for its proper functioning. The *intrinsic value* of the resilient planetary ecosystem encompasses and embodies the interests and values of the *Homo sapiens* and other beings in nature.

13.6 Paradigm Shift

In recent days, Cartesian anthropocentrism has come under severe criticism for all the excesses it has committed and the environmental degradation and destruction it caused. A rational thinker, John Dietrich (1933), criticized the Christians and humanists who spearheaded this separation: "*The mistake that all these critics make is in separating man from Nature and looking upon Nature as antagonistic to man's purposes and ideals. They do not seem to realize that we are an inseparable part of the universe. We are not alien children in a strange and foreign land. We are a product, the natural development of its forces and condition. Every human function – physical, mental, and moral- has developed due to constant and successful adaptation to natural conditions. .... We are a part of the universe and cannot be separated from it.*" We are inseparably embedded in the natural environment, and our dependence on it is much deeper than we have discovered; therefore, humans must prevent any endangered species from going extinct. The web of interdependence and interconnectedness has value in itself; humans must try to protect and prevent this web from breaking down. Aldo Leopold (1949), the most ardent advocate of environmental ethics, argued for a holistic, biocentric morality called "*the land ethic,*" which affirms that the life forms that share the planet with people should be allowed to live as a matter of biotic right regardless of the presence or absence of advantage to humans.

Leopold's ethical system recognizes this web of interdependence and interconnectedness. It includes the whole of Nature (the integrity of the land, plants, animal, water, the air, and everything that exists) and human obligations to respect and maintain this integrity. David Ehrenfeld (1978) argues that humanism's narrow focus on using reason to serve human interests leads to a dichotomy between man and Nature in which ecological factors are subordinated to the satisfaction of human wants. He believes that long-standing existence in Nature carries with it the unimpeachable right to the continued existence of even those species that have no significance to humans. He calls it an "*existence value*" because it was there in the ecological system. He recognizes the intrinsic rights of all species and assumes a moral duty to protect and preserve them (Naess, 1989). The phrase "*web of life*" was first coined by J. Arthur Thompson (1914), a Scottish biologist, to describe this phenomenon of interdependence and interconnectedness so prevalent in the ecosystem. Thomson (1914) eloquently elaborates to that effect: "*As we can observe, there is a very dynamic, interactive, and cooperative community going on in any functioning ecosystem; Many species are dependent upon each other and contribute to other species' existence, geology, botany, soil chemistry, wind, fire, and water must all be operating in some sort of balance, which forms the basis for the interdependent web of all existence.*"

Edward Wilson (1990), a prominent biologist (entomologist), advanced the concept that there is an inborn affinity that human beings have for other forms of life, which he calls *biophilia*. This inborn affinity in humans forms a subconscious kinship with other life forms and the life processes embedded in the billions of bits of genetic information, even in the tiniest living creature. This forms the foundation of human's reverence for life and ethics for conservation. As Wilson (1999) eloquently postulates: "*The most important implication of innate biophilia is the foundation it lays for an enduring conservation ethic. If concern for the rest of life is part of human Nature, if part of our culture flows from wild Nature, then on that basis alone, it is fundamentally wrong to extinguish other life forms; Nature is part of us, as we are part of Nature.*" There is an interesting parallel between Wilson's notion of *biophilia,* Buddha's concept of *compassion* and *nonviolence,* and Albert Schweitzer's (1933) "*reverence for life*" which forms the basis of Schweitzer's ethics in which he says, "*A man is ethical only when life, as such, is sacred to him, that of plants and animals as well as that of his fellow men*". Similarly, many thinkers have embraced the notion of reverence for life and eloquently expressed that reverence which can be summed up in the following paragraph: It is a sense of the whole, a capacity for wonder, a respect for the intricate universe of individual life. It is the supreme awareness of awareness itself. It is pride in being. This reverence for life begins culminating in Rachel Carson's (1962) moral philosophy: "*Life is a miracle beyond our comprehension, and we should reverence it even when we struggle against it.*" Sarah Oelberg (2002), a prominent humanist thinker, has succinctly summarized the anthropocentric humanist oversight into Nature: "*If humanism remains stuck in a purely naturalistic 'man is the measure of all things' mode, if it cannot embrace the larger concept of interrelationship and interdependence with Nature and grant it worth; if its adherents cannot experience the awe and wonder at the inherent marvels of Nature and see themselves as only one tiny part of the grand universe; then it deserves all the criticism leveled at it.*"

Organicism rejects mechanistic reductionism. Organicism was born from the quest to eliminate the Cartesian picture of reality, a view that has been the most destructive paradigm from science to politics. Several biologists in the early to mid-twentieth century embraced organicism. The larger organization of an organic system has features that must be considered to explain its behavior. Alfred North Whitehead (1929), a prominent philosopher, championed the idea of interdependence and interconnectedness in the framework of his *process philosophy* which posits that everything in the universe is characterized by ongoing, dynamic processes of becoming. Whitehead argued that reality is not composed of isolated entities, but rather a network of interrelated processes and events. This interconnectedness extends from the microcosmic level of subatomic particles to the macrocosmic level of galaxies and encompasses all levels of existence in between. This interconnectedness implies that no entity or event can be fully understood without considering its relationships and interactions with other entities. Whitehead's concept of "*organicism*" abandons objectivity and leads humankind to recognition of the intrinsic worth of every component of the environment. In recent years, one of the most promising applications of Whitehead's thought has been in the sphere of ecological civilization, sustainability, and environmental ethics. Whitehead's holistic metaphysics of value lends itself readily to an ecological point of view as a promising alternative to the traditional reductionistic and mechanistic worldview, providing a detailed metaphysical picture of a world constituted by a web of interdependent relations (Capra, 1996).

Natural history informs us that many of today's desert landscapes were once part of productive natural ecosystems. Inappropriate anthropocentric activities have predominantly led to the conversion of these once productive ecosystems into less productive, desert-like environments. This transformation is evidenced by the ancient remnants found on Easter Island and the surviving oak trees from the old climax vegetation in Lebanon and Syria, which serve as testaments to this process (Cliff, 1990). This anthropogenic alteration of once productive natural ecosystems into degraded desert environments is rapidly and extensively occurring worldwide, particularly in Africa, Asia, and South America. The literature is replete with accounts documenting signs of rapid desertification across numerous regions within these continents (Erickholm, 1975; Upreti, 1987, 1994). Suppose even if it is difficult to argue for the intrinsic values of diverse ecosystems, ecosystem processes, and entities, it is in the interest of humankind to prevent these rich and diverse ecosystems and their services from being degraded and destroyed, consequently leading to the emergence of desert ecosystems.

Instrumental values include all those direct or indirect use values of resources. The opportunity cost of a resource is the natural measure of its value in a particular use, and the more valuable or productive the resource is in that use, the greater its value. Usually, the price of a resource is the approximation of its value. However, as

Perring et al. (1992) argue, this is not the case with the ecosystem processes and the organisms because markets for their services, particularly the ecological services, do not exist. Lack of appropriate valuation theory and techniques to assess life support and environmentally vital services rendered by biodiversity and natural ecosystems, on the one hand, and lack of practically pragmatic environmental ethics, on the other hand, have contributed to unprecedented loss of biodiversity, degradation, and destruction of ecosystems.

Economists take a quantitative approach to valuing ecological services rendered by ecosystems and biodiversity. They ask questions such as how much a particular organism or species has contributed to total ecological services. How an organism produces ecological services? Quantifying and partitioning the ecological services generated by an individual organism or species proves to be an arduous task due to the inherent complexity of these services. They do not arise solely from the efforts of an individual organism or species, but rather result from intricate interactions among diverse populations of organisms and species, as well as their respective environments, over extended periods of time. These interactions encompass a wide range of dynamics, including competition, mutualism, communalism, coevolution, synergism, and feedback mechanisms. An individual's contribution to ecological services cannot be quantified in a linear additive model because the reductionistic linear model cannot study the complex nature of biological and ecological interaction. Given this intricate web of interactions, it is imperative to view biological diversity as a cohesive system comprising a diverse array of species, organisms, and populations engaged in multifaceted interactions with both one another and their environment. Merely cataloging individual organisms and species falls short of capturing the true essence of biodiversity. This comprehensive understanding, as advocated by Gee (1992), presents a more objective foundation for the conservation of natural landscapes and habitats.

The significance and magnitude of the role held by environmental planners and development professionals have reached a critical juncture. Environmental planning, inherent in its nature, necessitates a deep comprehension of ecosystem dynamics, and the physical and biological environments, as well as an understanding of the socioeconomic factors and the intrinsic value of natural systems and their various applications. The urgency to safeguard and enhance the environment demands a fundamental shift in the worldview, conceptual frameworks, perceptions, and mindsets of political decision-makers, planners, and professionals. Only through this transformative process can appropriate development and environmental policies come to fruition. Decision-makers must envision and wholeheartedly embrace the principles that promote a flourishing planet and assume moral responsibility for the well-being of all members of society, both presently and for generations to come. Environmental destruction is deemed morally reprehensible due to its adverse impact on the security and viability of all living organisms, encompassing the human species. Moral and ethical principles play a pivotal role in shaping human conduct, yet their efficacy hinges upon their ability to harmonize with the fulfillment of fundamental human needs. If we genuinely want to prevent the destruction of planetary ecosystems, we must alter our faulty conceptual and perceptual understanding of the biosphere (Earth's systems). This change in worldview presupposes a change in our assumptions, value systems, and ethical and moral philosophy. In other words, it requires a shift from the prevailing egocentric machine paradigm to a new paradigm that recognizes and embraces the values of interconnectedness, organizational complexity, cooperation, and interdependence.

The prevailing economic growth under the current paradigm has not only compromised the ecological sustainability of our resource base, production methods, control mechanisms, and consumption patterns, but it has also severely eroded the social sustainability of human institutions. This erosion has been fueled by the promotion of inequitable development models, the marginalization of certain groups, and the concentration of resources and wealth among a select few, all justified under the guise of competition and a free market economy. On one hand, this paradigm has pushed a significant portion of humanity to the brink of social catastrophe and dehumanization. On the other hand, it has fostered an *ecologically hostile consumer society* that idolizes and prioritizes consumption for its own sake. Consequently, this has fostered a culture of greed, unhealthy competition, and a desire to exert dominion over nature, extracting from it without consideration for protection, sustainability, cooperation, equity, or social justice. In essence, this paradigm exhibits both negative anthropocentrism and ecosystemic breakdown. It is negatively anthropocentric because its consequences ultimately pose a threat to the survival of humanity rather than its preservation and longevity. Simultaneously, it is negatively ecosystemic as it fails to acknowledge that the existence of all life forms, including humankind, hinges upon the health, integrity, and resilience of the Earth's systems, encompassing ecosystems and their associated services. This paradigm clearly cannot solve today's human and environmental problems. It must be replaced by a new paradigm that inculcates and nurtures the notion of interconnectedness, diversity, and ecosystem health in the ecosphere and social integrity and justice in the sociosphere in which each individual is a member of a highly interactive community of interdependent parts.

A paradigm can be defined as a dominant belief system within the framework of which individuals perceive and interpret worldly phenomena around them. Paradigms become the fundamental parts of social architecture and the social construction of reality (Upreti, 1994). Kuhn (1962) first developed this concept to describe the structure and process of the scientific revolution. Since then, this concept has been increasingly extended to describe the structure of dominant belief systems (values, beliefs, and shared wisdom), particularly in social science. Pirages (1977) coined the term *dominant social paradigm* to describe a society's prevailing belief system (worldview), which embodies commonly held values, beliefs, and wisdom about natural and social environments. It is imperative to understand the dominant social paradigm of a society to understand why certain social practices, behaviors, and social institutions exist in a particular society because a social paradigm legitimizes and justifies such social practices, behavior, and the existence of social institutions. In essence, a social paradigm or worldview becomes an *ideology* that governs peoples' behavior in a society. The most interesting aspect of our sociosphere and ecosphere is that they constantly interact, thereby impacting and changing each other and evolving into a new relational state. *Homo sapiens* is both an actor and part of this *dialectical* social and natural dynamism. Humans must constantly recreate, redefine, and reestablish themselves in changing ecosphere. In this *dialectical process*, there comes a time when the dominant paradigm becomes no longer functionally useful because the perception and interpretation of social and natural phenomena that follow from such paradigms do not conform to the newly evolved state of reality in the sociosphere and ecosphere. In other words, they become no longer functionally relevant in adapting themselves to changing social and natural world. They must be replaced by a new, more relevant, and adaptive paradigm that offers a better framework for human perception and interpreting worldly phenomena in the sociosphere and ecosphere. When it happens, we call it a paradigm shift or scientific or social revolution.

13.6.1 Ecological Wisdom Consciousness

The current development paradigm's drive for infinite growth and reckless profit has accelerated the destruction of planetary ecosystems and environmental problems beyond the resilience of Earth's systems. Mark Hertsgaard (1999) succinctly points out: "*The profit motive is what makes capitalism go, but it is so basic to the working of the system that it tends to override other social goals……Capitalism needs and promotes constant expansion, yet the evidence that human activity is already overwhelming the Earth's ecosystems is all around us.*" Vaclav Havel (1998) asks an important question, and he passionately makes the statement of great significance in attempting to answer that question when he says: "*What could change the direction of today's civilization? …. It is my deep conviction that the only option is a change in the sphere of the spirit, in the sphere of human conscience. It is not enough to invent new machines, regulations, and institutions. We must develop a new understanding of the true purpose of our existence on this Earth. Only by making such a fundamental shift will we be able to create new models of behavior and a new set of values for the planet.*" Similarly, Aldo Leopold, the father of the land ethic, also saw a fundamental antagonism between the philosophy of dominant development paradigm and the philosophy of the conservationist and believed that nothing could be done about nature conservation "*without creating a new kind of people*" with a new kind of development ethics.

Noted scientists Paul Ehrlich and Donald Kennedy (2005) also strongly reinforce the view of Aldo Leopold as they say: "*It is the collective actions of individuals that lie in the heart of the environmental dilemma and that "analysis of individual motives and values should be critical to the solution.*" In their call for a Millennium Assessment of Human Behavior, they emphatically pointed out the importance of cultural changes in the following words "*to conduct an ongoing examination and public airing of what is about how human culture (especially their ethics) evolve, and about what kinds of changes might permit transition to an ecologically sustainable, peaceful, and equitable global society… We are asking for a cultural change; we know that cultures evolve, and we hope that the very process of debate will speed that process and encourage change in a positive direction.*"

Paul Raskin and his Global Scenario Group (2002a, b) are critical of the global economic and environmental scenarios that lack fundamental changes in consciousness and values. They strongly advocate the New Sustainability worldview where society turns to "*non-material dimensions of fulfillment .... the quality of life, the quality of human solidarity and the quality of the Earth.... Sustainability is imperative that pushes the new agenda.*" Raskin and his colleagues envision a revolution in values and consciousness. Humankind must revisit and re-evaluate the current developmental paradigm and ask some critical questions; what is the purpose of our existence on Earth? How long can we go on consuming planetary resources at current rates? Does humankind have any responsibility towards other communities of living beings? What are the ecological and thermodynamic consequences of extreme anthropocentric activities? Has humankind reached a turning point in its evolutionary history that requires a new social, cultural, and ecological consciousness? Can we transition to a Just and Sustainable Society with a new ecological consciousness and cultural and ethical values? Erich Fromm (1977) calls for a radical change of the human heart as he eloquently articulates: "*The need for profound human change emerges not only as an ethical or religious demand, not only as a psychological demand arising from the pathogenic Nature of our present social character but also as a condition for the sheer survival of the human race....Only a fundamental change in human character from a preponderance of the having mode to predominantly being a mode of existence can save us.*"

There is no doubt that humans evolved on this planet as an integral part of the biotic community and, indeed, must not consider that they are the sources of all rights and values. It is essential to realize that humans cannot survive without the biotic community's healthy planetary ecosystem and ecosystem services. Charles Reich, professor at Yale University (1970), in his popular bestselling book *"The Greening of America,"* talks about the need for a revolution by consciousness in the United States and possibly in the Western world where political revolution is not possible. He succinctly elaborates, "*Revolution by consciousness requires two basic conditions. First, a change of consciousness must be underway in the population – a process that promises to continue until it reaches most people. Second, the existing order must depend on its power on an earlier consciousness and therefore be unable to survive a change of consciousness. Both of these conditions now exist in the United States.*" Public environmental awareness in the United States has increased substantially over the decade. However, it has not yet reached the level that can significantly impact the dominant development discourse.

Development professionals and scientists from wide-ranging areas have warned that today's environmental challenges require new thinking and a new social and ecological consciousness because the prevailing dominant worldview is entrenched in *anthropocentric egocentrism* and *reductionism* and cannot resolve the current environmental and ecological crisis (Kellert & Speth, 2009). It needs to be replaced by a new worldview with new outlooks and values that can protect and preserve the biotic community, their ecosystems, the ecosystem services, and the natural capitals upon which humanity depends. The new worldview must embody the values that direct human activities and behavior to live harmoniously with Nature and the biotic community. The following statements by Paul Raskin (2002), a visionary of our time, are worth pondering: "*The emergence of a new suite of values is the foundation of the entire edifice of our planetary society. Consumerism, individualism, and domination of Nature – the dominant values of yesteryear – have given way to a new triad: quality of life, human solidarity, and ecological sensibility……Love of Nature is complemented by a deep sense of humanity's place in the web of life and dependence on its bounty. Sustainability is a core part of the contemporary worldview, which would deem any compromise of the integrity of our planetary home both laughably idiotic and morally wrong.*" David Korten (2007), in his book "The Great Turning," also sees humanity reaching a turning point when he says: "*The Great Turning begins with a cultural and spiritual awakening – a turning in cultural values from money and material excess to life and spiritual fulfillment, from a belief in our limitations to a belief in our possibilities and from fearing our differences to rejoicing in our diversity. It leads us from policies that raise those at the top to those at the bottom, from hoarding to sharing, from concentrated to distributed ownership, and from the rights of the ownership to the responsibilities of stewardship.*"

We must recognize and challenge the current mode of production and consumption and strive to restructure our social-economic relations in the sociosphere to produce ecologically conscious societies that discourage the accumulation of man-made capital at the cost of irrecoverable depletion and extinction of Nature's capital. Justin McBrien (2016) points out that the logic of accumulation cannot outrun extinction because accumulation and extinction are the same processes and cannot be decoupled. However, human beings can be decoupled from capital. Decoupling from capital accumulation requires a critical appraisal of our current model or paradigm and a dramatic reimagining of humanity's mode of existence. This means we must move away from this flawed paradigm of capital accumulation to a new paradigm informed by ecological wisdom consciousness that is capable of reconstructing a society wherein new knowledge and ways of being can be formed independently from the logic of capital, a new way of experiencing the essence of human existence. Mary Evelyn Tucker and John Grim (2007), leading authorities on religions and ecology, argue that ethical values, religion, and spirituality are essential in transforming human consciousness and behavior. As is clear from what they say: "*We are called to a new intergenerational consciousness and conscience and that values and ethics, religion and spirituality are important factors in transforming human consciousness and behavior for a sustainable future.*" Concerns for planetary ecological crisis emanate from the essential human capacity to transform its history globally. Humans are thinking, historical and esthetic beings with language, memory, empathy, compassion, and a massive capacity for imagination. These are the unique capacities of human beings from which the ultimate salvation of planetary crisis must emerge. Human consciousness can restructure the political, economic, technological, and cultural superstructures that created this crisis. The natural and ecological processes that sustain the proper functioning of the entire biosphere of the Earth have already been disrupted, the consequences of which have been manifested in various forms, including that climate change, mass species extinction, unprecedented destruction of forest ecosystem (recent wildfire in Australia), cyclones, storms, and tornadoes, etc. The amount of energy and material resources extracted, consumed, and throughput produced by humankind today has far exceeded the regenerative and absorptive capacity of the planetary ecosystem. The pollution in soils, water bodies, oceans, and the atmosphere has reached a catastrophic level. Tropical and Amazonia deforestation, loss of terrestrial vegetation, and the release of carbon dioxide, methane from the forested land, beef industry, and terrestrial biomass have reached a horrifying level. This indicates that humanity has already entered a geological conflict with the planetary ecosystem of the Earth, signifying a direct existential threat to human civilization. The current dominant growth paradigm has rapidly accelerated the environmental and climate crisis. Therefore, unless the current paradigm brings structural changes in its assumptions and *modus operandi*, the crisis will be further exacerbated to the point of no return. Perhaps, this is the time for environmentalists, conservationists, scientists, ecologists, religious and spiritual leaders, development thinkers, and philosophers to converge their collective consciousness and intellectual endeavor on changing the mindsets of the ruling elites of the powerful countries (USA, EU, China, UK, Japan, India, Brazil, etc.), corporate worlds and power centers so that an ecologically sustainable living paradigm can be constructed.

Thomas Berry (1999), a visionary cultural historian, talks about forging a new consciousness and rightly points out that "*The deepest cause of the present devastation is found in a mode of consciousness that has established a radical discontinuity between the human and other modes of being and the bestowal of all rights on the humans…. We have difficulty accepting humans as an integral part of the Earth's community. We see ourselves as a transcendental mode of being. We don't really belong here. But if we are here by some strange destiny then we are the source of all rights and all values. All other beings are instruments to be used or resources to be exploited for human benefit.*" Berry further elaborates that what is required of us is "*a reversal in our perspective on ourselves and the universe about us…. What is demanded of us now is to change attitudes that are so deeply bound into the basic cultural patterns that they seem to us as an imperative of the very nature of our being.*" A transition from an ecologically hostile production and consumption culture to an ecologically enabling production and consumption culture that remains within the safe operating boundary of planet Earth is necessary to save Earth and humanity. This may be the beginning of a new civilization called "*ecocivilization.*"

What social values can complement and enhance natural ecosystemic values and vice versa? Is there mutual reciprocity between such values coming from entirely different spheres? It is generally assumed that human freedom, equality, and social justice are necessary for realizing every human individual's potential and should be the development goal. Ronald Engel (1990) argues that reconceptualization of moral values is necessary for sustainable development to become a new paradigm to do justice to the full complexity of interactions within and between ecological and social communities. The biggest challenge of development ethics is envisioning the development that can maintain ecological and social integrity. It requires scientists, development professionals, thinkers, and ecological and social activists to collaborate closely to make sustainable development practically feasible. International Council of Voluntary Agencies (ICVA) and Rome Development Forum (1988) strongly emphasized the need for holistic moral understanding that can lay the foundation of true "sustainable development." As is clear from their statement: "*We affirm both the integrity, stability, and beauty of the ecosystem and the imperative of social justice. We recognize that poverty, environmental degradation, and population growth are inextricably related and that none of these fundamental problems can be successfully addressed in isolation ... We will succeed or fail together.*"

To address the current planetary ecological/environmental crisis, we need a new scientific epistemological approach and science agenda that can integrate the ecosphere (natural world) and sociosphere (social world) to chart the pathways for sustainable living and the stewardship of planet Earth. The scientific inquiries and agenda should focus on reconstructing the current unsustainable socioeconomic system operating in the sociosphere. Such reconstruction must be based on socially and ecologically sustainable and value-based development practices and institutions that discourage capital accumulation in the hands of few and promote equitable development for all in the sociosphere. The reconstruction of the socioeconomic production and consumption system based on understanding social and ecological sustainability requires drastic changes or modifications in the current capitalist mode of production and consumption patterns. Such changes demand a new culture based on ecological knowledge, values, and wisdom. It is challenging, but cultural change is necessary because the human socioeconomic system is a subsystem of the planetary ecosystem, and it must remain within the carrying capacity of the planetary ecosystem. Total throughputs of the sociosphere (human socioeconomic system) must remain within the regenerative and assimilative biocapacity of the ecosphere (Earth's systems). The epistemological and scientific agenda in the ecosphere/biosphere must focus on the enhancement of ecosystem resilience and integrity, maintenance of ecological processes and ecosystem health, restoration and redevelopment of degraded ecosystems, protection of tropical and Amazonian forests, natural resources accounting, integration of ecological services into economic valuation and ecoregional/bioregional landscape management, etc. The *Anthropocene* faces monumental challenges of maintaining biodiversity and vital ecological services, productive capacity, and resilience of Nature, which are essential for the sustainable living of human and nonhuman beings. Therefore, the goal of Earth's Stewardship must not be understood as the protection of Nature for itself; rather, it must be understood as the protection necessary for the continuation and sustainable living of all beings, including *Homo sapiens*.

13.6.2 Revisiting Sustainable Development

The phrase "*sustainable development*" has been so profusely used in development literature by professionals, planners, conservationists, and even politicians as if it has become a new "*development mantra.*" Due to the lack of a shared language and universally agreed definition based on the science of Earth's systems and its ecology, the phrase has become very ambiguous. Everybody uses this phrase conveniently to convey their own idea about development as they perceive it to be sustainable. For a development to be sustainable, total throughputs (production, consumption, and waste generation) of such development must remain within the regenerative and assimilative biocapacity of the planet Earth so that Earth can sustain and maintain such development. Total throughputs of the current growth-driven "*sustainable development*" model have far exceeded the regenerative biocapacity of the planet Earth's systems, and, therefore, this concept has become an *oxymoron*. Now we need two planets Earth to sustain the current growth model's production, consumption, and waste generation, which we so fondly and mistakenly call "*sustainable development.*" It is necessary to reconceptualize sustainable development and living that is possible within planet Earth's biocapacity. Ecosociocentrism embodies two fundamental components: ecological integrity in the *ecosphere* and social integrity in the *sociosphere*. Ecological integrity entails interconnectedness, ecosystem processes, and the self-organizing complexity of living system. Social integrity entails social and economic components based on the assumption that social and economic justices are the foundation of social integrity. Therefore, from the Ecosociocentric Paradigm, sustainable development may be defined as "*the development that satisfies human needs of present and future generation while maintaining the resilience and the biocapacity of Earth' systems in such a way that human socio-economic throughputs remain within the regenerative biocapacity of the planetary ecosystem.*" Only such development can sustain the social and ecological integrity for the fulfillment and actualization of human potential and the protection of the living systems on planet Earth.

We need to understand that humans have yet to reach the stage and, I doubt, will ever reach that stage to override Nature's ecological and physical laws. Today's environmental and climate crisis problems have emanated from men's attempts to defy natural laws and their limits. What we proudly call human success or progress is the root cause of today's problems. Environmental degradation, overpopulation, hunger, and the destruction of biodiversity and wilderness are inseparably connected to our rapidly increased extractive ability through agricultural, mining, medical, military, and industrial technologies to manipulate the natural forces and processes for anthropocentric interest. Theoretically, there is no limit to the human capacity to destroy ecological and physical processes. However, there is a limit to the amount of total waste throughput (entropy disorder), Earth can assimilate without adverse reactionary feedback that could be detrimental to humans. It is essential to realize that humans do not have the choice to act and behave in *ecologically hostile* manner and ways because they cannot escape the catastrophic consequences of Nature's backlash.

Humans are the only creatures on Earth who can choose to behave in certain ways because they possess the consciousness and mental capacity to think, reason, and predict the future. For the sustainable future of humankind, humans must not override the natural and ecological laws through the use of science and technology and learn to live within the carrying capacity of the biosphere (planetary ecosystem). Humans should direct the development of science and technology that can, instead, increase the resilience and regenerative biocapacity of the land, soil, water bodies, terrestrial ecosystem, and planetary ecosystem. From an epistemological perspective, we do not need the science and technology that can disconnect us from Nature, not the one that disrupts and destroys the ecological integrity and processes but rather the ones that can strengthen the interconnectedness and interdependencies of all the components of the biosphere or the planetary ecosystem. Only such science and technology can build the foundation of sustainable living for humankind and other living beings.

Until now, human behavior has been primarily shaped by the desires and motivation of excessive possession of material goods, services, and insatiable consumerism through extractive science and technologies that disrupted and destroyed the regenerative biocapacity of the planet Earth. This type of development and its *modus operandi* cannot and must not continue. It must be replaced with science and technology that can restore the resilience and regenerative capacity of the land, water, oceans, terrestrial ecosystems, and the biosphere. It falls on the scientific community's responsibility to change their conventional mindsets that science and technology are neutral to values, and they are *value-free*. Einstein's theory of relativity and quantum mechanics have dismantled the notion of a neutral observer. The observer participates in the act of measurement, and the universe changes by the act of measurement. Is it possible to direct and develop value-based science and technology that can provide solutions to environmental problems and restore the resilience and biocapacity of Earth's systems? The impact effects of interaction between the social and natural world have been beneficial to humans and detrimental to Nature. Short-sighted, excessive exploitation of Nature has given rise to a set of conflicts between society and Nature. Economic and socio-cultural activities under the current development paradigm have massively depleted and degraded planetary resources. The current conflict that arose from the interaction between the social and natural worlds can be called an eco-social conflict. Estimates show that if present rates of growth-driven development remain unchanged, the depletion and degradation of the biosphere will reach the point of irreversible instability and is expected to occur in the second half of this century (Chapin et al., 2011; Panayotou, 1990; Costanzaa et al., 2014).

The human-induced disruption and destruction of ecological systems that gave rise to ecological crises warrant different patterns of human behavior concerning the environment. This value judgment must be critically examined in light of the current ecological crisis. It calls our attention to the moral necessity of behavioral changes from a hyper-anthropocentric worldview that regards values in Nature as only *instrumental*, not *intrinsic*. Nature embodies some intrinsic values and must not be treated as only instrumental. For example, how should we treat the biotic pyramid that describes the movement of life from the soil and the microorganisms therein through vegetation, herbivores, and to carnivores, and primates? The value contained in a pyramid is correlated with the richness of the base, the number of levels, the diversity of the forms, and the complexity of the living forms at the top. The biotic pyramid does not provide an anti-anthropocentric view of value, and this pyramid, the *web of interconnectedness*, deserves to be treated as the *intrinsic value* in itself and for itself. Buddhism provides a profoundly deeper reverence for life in all forms to symbolize and identify with creation and revolt against human excesses in all forms. From a Buddhist *Eco-Dharma* perspective, there must be a sense of sanctity for life and life processes on the Earth that provides a moral imperative for sustainable development.

13.6.3 Pragmatic Approach to Environmental Ethics

Hyper ecocentrists argue that the preservation of the wilderness in Nature is necessary for itself. They argue that we have a moral obligation to the biosphere, which means our moral obligation should transcend the biosphere from our current obligation to our fellow human beings who are alive now. Suppose this moral push for the preservation of the biosphere had stemmed from our wanting to preserve reasonable conditions of life for our descendants and other biotic communities in Nature. This is a fair deal between the present and future generations and is morally a sustainable argument. Human rationality acknowledges the necessity for the preservation of the ecosphere/biosphere. However, on the contrary, it becomes challenging to argue that the preservation of the ecosphere/biosphere has intrinsic value independent of human interests. Hyper anthropocentrism that results from human arrogance, which regard humans as the lord of creation and master of Nature, should be rejected. This attitude will certainly not help advance human happiness and prosperity and the existential worth of other living beings. It will instead bring suffering and misery to humans and all other living beings in the long run. *Homo sapiens* evolved in Nature along with other members of the biotic community through million years of evolutionary process. The survival and well-being of the human species are more important than the immediate satisfaction of needless needs of certain human groups (business conglomerates, and ecologically hostile consumers) of people. This is certainly not a matter of enlightened human self-interest but rather an egocentric and ecologically hostile self-aggrandizement that can endanger the continuation of the human species and other living beings. Therefore, protecting and preserving Earth's ecosystem, wilderness areas with animals (whales, tigers, pandas, etc.), tropical and Amazon rainforests, and satisfying immediate human needs is imperative for ensuring a livable future for humans and other nonhuman beings.

From the perspective of ethical pragmatism, it is crucial to have an environmental ethics that integrates some aspects of ecocentrism/biocentrism and still be compatible with giving human beings a unique position in Nature through their capacity for self-awareness, self-reflection, and imagination. So long as humans do not pursue their egocentric needs, purposes, and interests to satisfy their self-aggrandizement and recognize the worth/rights of nonhuman lives to exist in Nature, environmental ethics may have some chance to impact positively. Environmental ethics must be informed by prudence and enlightened human interests that consider the needs of future generations to keep Earth's systems and natural processes in proper functioning so that nonhuman biotic and human communities can coexist and flourish together. From this perspective, it is worth pondering the underlying ethical implication of Aldo Leopold's Land Ethics (1949). If I were to rephrase Aldo Leopold's famous statement, I would add one phrase that is "*satisfaction of the immediate needs of humanity*" to his original statement, which would read as: "*A thing is right when it tends to promote the integrity, stability, beauty of the biotic community and the satisfaction of the immediate needs of humanity. It is wrong when it tends otherwise.*" The argument from ecocentric or biocentric camp that advocates the preservation of wilderness areas at the expense of the survival of human beings is difficult to justify. From the perspective of *ethical pragmatism*, a compromise solution can be worked out by the sustainable use of wilderness areas to meet basic human needs without compromising the regenerative ecological processes in the wilderness. What is not acceptable is the argument that the preservation of some species of biotic communities or wilderness areas must take precedence over the present and future interests of human beings. On the contrary, equally important is the fact that there is a necessity to preserve the ecological integrity of Earth's ecosystem, which is a necessary condition for the survival and well-being of the human species and other species of the biotic community. The biggest challenge is whether we can construct an ethical paradigm that recognizes the needs of both human and nonhuman biotic communities and creates conditions conducive to the flourishing of both.

Ethical pragmatism, in my view, integrates the *sociosphere* (human socioeconomic sphere) and *ecosphere* (biosphere) and may be given a name which I would like to call "*Ecosociocentrism: The Earth First Paradigm.*" The ethical view of "*Ecosociocentrism*," an alternative paradigm proposed here in this book, embodies the concept that human response to Nature must be for the collective needs and interests of humans and other beings, maintaining self-organizing creative processes (autopoiesis) which are intrinsically intertwined in Nature. The *sociosphere* and *ecosphere* are inseparable and always intertwined and entangled in a *dialectical* interactive nexus. A correct understanding of the Nature of this interaction is necessary to realize sustainable human development in the sociosphere and to prolong and maintain environmental sustainability in the ecosphere. A healthy and sustainable relationship between the *sociosphere* and *ecosphere* is a prerequisite for sustainable human development and the regenerative biocapacity of the planet Earth. If the present generation fails to seize this exceptional historical moment, the future generation will have no viable choice except to condemn its parent generation for their wrongdoings. Here, it is suggested that a new ethical paradigm is required in order to reconcile instrumental and intrinsic values found in Nature, promote eco-civilization globally, and build the groundwork for sustainable development. Ethics undoubtedly plays a significant role in challenging and changing the prevalent social development paradigm. It is crucial to comprehend how our worldviews, which serve as the lens through which we see and interpret the external world, influence and form our social and ecological values. Our perspective and comprehension of the role of ethics in relation to the problems of development, the environment, and conservation have been shaped by our worldview. When comprehended in their right perspective, facts can transform into values and vice versa. For instance, if we accept the ecological processes and planetary ecosystem's integrity as facts based on the ecological and scientific knowledge, we have so far amassed, we must unquestionably value these ecological processes and planetary ecosystem's integrity and, as a result, do our best to preserve or improve these ecological processes and the integrity of the planetary ecosystem. Ecological processes and the integrity of the planet's ecosystems here serve to describe both facts and values, and they cannot be distinguished from one another.

Stanley and Loy (2016) argue that there is a link between our affinity with Nature, our love of Nature, our profound spiritual experience, and our moral senses. We cannot ignore our biological evolutionary lineage, which has established that we are *human primates* and the dominant animal on planet Earth. We have created a human *superstructure* (human empire) that has altered the planet Earth so much that the *entropy* effect that has resulted from ecological overshoots will pose an existential threat. A question that troubles me is: How can *human superstructures* continue to exist when the *ecological infrastructures* (biophysical ecosystem components) are being ruptured, eroded, and destroyed? No superstructure can exist when its infrastructures are destroyed. Human social superstructures exist on natural infrastructures (ecosystem structures, processes, and services), and it is inconceivable to think that this human superstructure can exist on Nature's broken and dysfunctional infrastructures. So long as human social, economic, and cultural superstructures cultivate a harmonious and healthy relationship with the natural infrastructures (ecosystem structures, processes, and services), humanity can prolong its existence on Earth.

The biophysical system, the biome, and the ecosystems that exist in their manifold facets constitute the complex nexus and web of interdependence and interconnectedness that is so central to the existence of living systems, including humans must receive a non-anthropocentric interest that can recognize the *intrinsic value* of this web of interconnectedness. Is not humanity better off with the recognition that human beings are integral parts of Nature just like any other beings and that the nexus of interdependence and interconnectedness is what essentially sustains everything, including us, and that the breakdown of this nexus will inevitably endanger our existence, the existence of *Homo sapiens* as a species? What we need today is the subversion of the hyper-anthropocentric view that Nature belongs to man, and man is the master of Nature. This needs to be replaced with the view that man is an integral part of Nature just like any other entity and should learn to live cultivating a harmonious relationship with Nature. Ecological consciousness is a continuously evolving process that can liberate us from our ignorance, delusions, and misunderstanding of our relationship with Nature. The convergence of devastating crises from the ecosphere and the sociosphere demands nothing less than a bold vision of a new development paradigm guided by *ecological wisdom consciousness* that envisages a sustainable, healthy, and balanced relationship between the ecosphere and the sociosphere, laying the foundation for eco-civilization.

13.7 Collective Consciousness of Interdependence

Interdependence and interconnectedness are so pervasive that they exist on all levels of Nature (inanimate, animate, vegetative, and human). It is inconceivable for humans to be separated from this interconnectedness and interdependence. Human society is the reflection of this pattern since we are the products of Nature. As science advances, we may discover laws guiding our own development, showing how intimately we are interconnected and interdependent with all levels of Nature. The global network of human society is so interconnected and interdependent that it seems impossible for anyone to make any independent motion in this system. Everything is included in this network that shows us the interdependence requiring us to act accordingly. Modern physics tells us the story that an observer of an event taking place either in outer space or within the microcosm brings the observer into that event, which, thus, changes the event itself. Such a phenomenon also testifies to the dependence between various levels, in this case, between humans and the skill levels of Nature. We have discovered this interconnected and interdependent network gradually over the generations, and its very discovery is part of our own development. We have witnessed a gradual degradation and breakdown of this interdependence in Nature over the last few decades, but we have never critically looked into the resolution of this systemic breakdown in Nature. Instead, aggressive anthropocentric activities in the world accelerated this imbalance between humans and Nature. This is the cause of the global crisis taking place in every field of human experience today: personally, socially, globally, economically, and ecologically.

For the first time in history, we have reached a multiscale global crisis, which has intensified over the last few decades. Around 50 years ago, members of the Club of Rome began perceiving an oncoming crisis upon which they published various articles. The problem is that we have entered into a new, interconnected, and interdependent system, yet we continue to act as the same self-centered beings we were in the past, thus not changing our attitude and behavior with the new conditions Nature presents to us. We are living in an era of a great dilemma: on the one hand, we discover our global interconnectedness and interdependence more and more, yet on the other hand, we fail to understand the implications of this interdependence and interconnectedness and continue our business-as-usual as if nothing has happened. The question thus arises, how can we begin to match the degree of interconnectedness and interdependence we are revealing in Nature with our attitudinal and behavioral changes? Michael Laitman (2006) expresses the view that according to our current developmental point as humans in this system, we are on the animal level of existence, and Nature is pushing us to realize the human level of existence. At the animal level of existence, we have concerns solely for our own individual selves and bodies. He further states: "*If we could shift our concern away from our bodies to concern for society as a whole, then we would discover a whole new dimension of collective wisdom, intelligence, attainment, and experience, a harmonious and happy life in balance with Nature. We stand before a consciousness revolution, whether we want it or not. As Consciousness changes, we will perceive/attain the world in a new way.*"

The term *symbiosis* was first used in biology by Heinrich Anton de Barry, a German mycologist, in the year 1879, who described it as *the living together of, unlike organisms*. Ever since, this term has been used to describe the relationship between two or more species that depend on each other for their survival. The concept of symbiotic relationship has been extended to include the definition which states: "*The relationship between two or more different species of organisms who are interdependent on each other for their benefits is known as a symbiotic relationship.*" Most of the symbiotic relationship activities are associated with food, protection, and reproduction, which form the basis for the survival of the species. As mentioned earlier, *symbiosis* is the relationship between two or more taxonomic groups of organisms that show dependency on each other. Symbiotic relationships in Nature have evolved over a long evolutionary time through the interaction of the organisms among themselves and with their environment in their constant struggle to survive. In this process of interaction with their environments, species not only modified their environments but also brought modifications in themselves, and their survival strategies resulted from their interaction with the environment (Fig. 13.1).

How can we build and maintain "*social*" and "*ecological symbiotic niches*" that nurture *cooperation* and *compassion*, and *wisdom* that can hold the planetary ecosystem intact and functioning upon which the biotic community continues to exist and flourish? The collective emergence of *ecological Consciousness*, perhaps, offers this optimism. *Enlightened minds* like Buddha, ancient Rishis, and Saints have established rich traditions which teach us that it is possible for every human being to elevate their consciousness to a higher level, which I would like to call, an *ecological wisdom consciousness* that makes it possible to change human behavior to oneself and to the rest in the web of interrelatedness so that we can realize the essence of the universal well-being *mantra* coded in the Sanskrit "*Basudaiva Kutumbakam*," meaning the "*Earth is my Family*." Unless human beings are collectively guided by an *ecological wisdom consciousness* at the global level, the protection, preservation, and sustainability of the Earth's systems (planetary ecosystem) become a "*Sisyphean Myth*."

Fig. 13.1 Collective Consciousness can materialize The Earth First Paradigm. If we do not embrace The Earth First Paradigm, we will be sailing in a leaky boat on a Stormy Sea

If humanity fails in this mission, its fate cannot be different from the fate of a leaky boat that is sailing on a stormy sea.

Collective *ecological wisdom consciousness* of the people can change the direction and behavior of the political institutions and power centers. This collective *ecological awakening* must be expressed and translated into desirable political outcomes. Unless the direction of the political economy of the nation-states can be oriented to respecting the fundamental laws of ecology, the sustainability of ecosystem and ecosystem services cannot be maintained and prolonged. Unless a new breed of informed and environment-friendly politicians and managers take over the powerful decision-making institutions with a vision of creating a sustainable society in which human behavior is shaped and guided by the rational and conscious efforts of meeting the essential needs of all rather than satisfying the greed and self-aggrandizement of a few, the planetary ecosystem cannot be protected, and the regenerative capacity of the Earth cannot be maintained let alone talk about the healthy planetary ecosystem.

The *ecological awakening* or consciousness at the grassroots level is absolutely necessary to change the current scenario of ecological hostility to ecological complaisance. It is pathetically unfortunate to hear environmentally hostile hyperbolic rhetoric from the most powerful political leader (former president Trump) challenging the conclusions of the scientific community that anthropogenic activities have accelerated global warming, impacting climate change all over the world. President Trump's decision to withdraw from Paris Accord and subsequently relax domestic environmental laws and regulations is not only deplorable but also extremely dangerous in that such hostile and myopic actions push environmental destruction to the extreme. Humanity has the power to alter present ecologically destructive activities and direct them to a more sustainable mode of production, consumption, and uses. The new breed of politicians and managers must cultivate a worldview based on the understanding of the natural and social ecology of interdependence and interconnectedness. For this to happen, the transition from the current egocentric consumer culture to a culture guided by ecological wisdom consciousness is a must as envisioned *by Ecosociocentrism: The Earth First Paradigm.*

13.8 Ecosociocentrism: The Earth First Paradigm

It is clear from the preceding discussion that reconciling the conflicts between the *sociosphere* (social world) and *ecosphere* (natural world) through the appropriate mode of interaction that maintains social and ecological integrity is necessary and helpful to conceive sustainable living and sustainable development. The overriding necessity is to develop a new global development ethics that seeks to preserve and enhance the resilience and integrity of planetary ecosystems and processes in the ecosphere and social justice and human prosperity in the sociosphere. Such an ethical system not only recognizes the intrinsic values of the diverse life forms and self-organizing complexity in Nature but also affirms that humanity must live within the biocapacity of the planet Earth to ensure the perpetuation of all species, including *Homo sapiens*. *Ecosociocentrism*, the proposed development paradigm, inspires humanity to maintain the web of interconnectedness and interdependence in the ecosphere and ensures and enables the actualization of human potential in the sociosphere. As the logical fusion of ecocentrism and sociocentrism, Ecosociocentrism rejects the exclusive hyper-ecocentric or hyper-sociocentric views. *Ecosociocentrism* asserts that anthropocentric development activities in the sociosphere must remain within the regenerative biocapacity of the ecosphere (planet Earth).

13.8.1 Ecosociocentrism: A Synthesis

Ecosociocentrism entails two interactive dimensions, namely, *ecosphere* and *sociosphere*, respectively. The social dimension, or "*sociosphere*," refers to the human social, economic, and cultural system (Arizpe, 1991). Any ethical approach that regards society as its center of consideration and emphasizes social goods over individuals can be termed *sociocentric* ethics, and such an ethical system can be called *sociocentrism*. A brief discussion of *sociocentrism* is necessary to clarify some conceptual underpinnings before elaborating on *ecosociocentrism*.

The *sociocentric* approach, by definition, is the opposite of the individual *egocentric* approach. The sociocentric ethical perspective provides knowledge and understanding about societies' relationship with natural resources, how and why they are being used, and for what human purposes. Hence, sociocentric ethics is grounded in society's utilitarian principles that a society ought to act in such a way as to ensure the greatest good for the greatest number of people. According to this ethics, the social good must be maximized, the social evil be minimized, and the human actions directed to augment the happiness of society must be considered good. The understanding of human behavior and social interaction heavily relies on the recognition of human needs as the most influential driving forces. Coate and Rosati (1988) emphasize the crucial role of individual needs in facilitating and necessitating societal existence, asserting that human beings must interact to fulfill their needs. In this context, all forms of politics, including global politics, are intrinsically intertwined with processes and outcomes related to the satisfaction or deprivation of human needs. Therefore, politics derive their significance and purpose from the interactions of individuals as they pursue their needs within the sociosphere, forming the foundation and essence of political dynamics. The notion of development can only hold social significance when it is directed toward meeting human needs. Within the realm of human needs, it is vital to differentiate between fundamental human needs and those created by culture in a broader sense. Max-Neef et al. (1989) emphasize that fundamental human needs remain consistent across cultures and historical periods. They emphasize that human needs should be viewed as an interconnected and interactive system, with the exception of the primordial need for subsistence to sustain life. Within this system, there are no hierarchical arrangements; instead, simultaneous fulfillment, complementarity, and trade-offs characterize the process of satisfying needs. For an individual to achieve self-actualization, the satisfaction of human needs is a prerequisite. Consequently, development should primarily focus on the satisfaction of those needs and the quality of life of individuals rather than material possessions. It should be regarded as a transformative process that brings about substantial improvements in people's quality of life. The paramount objectives of any development endeavor should revolve around enhancing human conditions and fulfilling basic human needs. It is crucial to understand that development should not be limited to mere economic growth; it must encompass the betterment of life quality, social justice, equity, and the realization of human potential. When development programs and policies of nation-states do not entail equity, social justice, and equitable access to resources, it becomes impossible for the people, especially in poor developing countries, not to engage in activities that degrade their own physical and natural environments. Hence, the sociocentric ethical approach assumes that social justice, equity, and human well-beings cannot be ignored because the root of environmental degradation lies in the failure to address these human societal issues.

The ecological dimension or *ecosphere* (biosphere) refers to the whole exospheric/biospheric planetary system that provides the material basis for the existence of all forms of life, including humankind, on planet Earth. The ethical approach that regards the biospheric planetary ecosystem as the center of values can be termed *ecocentrism*, which the proponents of deep ecology have advocated for decades and has been discussed above. The stability and integrity of the *ecosphere* and *sociosphere* are a highly interconnected, interdependent, and intertwined phenomenon that determines the fate of humanity and the living system on planet Earth. Humanity's biggest challenge in the *Anthropocene* is organizing its anthropogenic activities in the sociosphere and living in harmony and stewardship with the ecosphere. The *sociosphere*, as a subsystem of the *ecosphere*, must remain within the limits of the biocapacity of the *ecosphere* without destroying its integrity and resilience. This is only possible through a development paradigm that is informed and guided by values and ecological wisdom consciousness that recognizes and respects ecological laws and processes that operate in the *ecosphere* and brings necessary changes in the *sociosphere* (men's socioeconomic and cultural system). The *dialectics* of the interaction between the *ecosphere* and the *sociosphere* does not need to be antagonistic; instead, they should be symbiotic or mutualistic, if not synergistic. The fundamental premise of *Ecosociocentrism: The Earth First Paradigm* is to reverse the *dialectics* of this antagonistic interaction between *ecosphere* and *sociosphere* and maintain ecological and social integrity in ecosphere and sociosphere to ensure sustainable living for all. "*Save Earth First to Save All*" is the "*mantra*" of this paradigm.

As pointed out, ecosociocentrism embodies two fundamental components: ecological integrity in the ecosphere and social integrity in the sociosphere. Ecological integrity entails interconnectedness, ecosystem processes, and the self-organizing complexity of living system. Social integrity entails social and economic components based on the assumption that social and economic justices are the foundation of social integrity. Therefore, from the *ecosociocentric paradigm*, sustainable development may be defined as "*the development that satisfies human needs of present and future generation while maintaining the resilience and the biocapacity of Earth' systems in such a way that human socio-economic throughputs remain within the regenerative biocapacity of the planetary ecosystem*". Only such development can sustain the social and ecological integrity for the fulfillment and actualization of human potential and the protection of the living systems on planet Earth. The current infinite growth model of neoliberal capitalism is the very antithesis of social and environmental sustainability unless its fallacious assumptions are reformed, and its *modus operandi* changed to value the Nature and its creative and regenerative processes that form the basis for continuation of human civilization and that of living system in the biosphere.

The insurmountable challenge that sustainable development faces today is integrating social and ecological integrity with the recognition of *instrumental* and *intrinsic* values in Nature and placing them at the heart of normative discourse on development. *Ecosociocentrims* seeks to integrate social and ecological integrity with an ethics-based development approach that embodies both instrumental and some intrinsic values in Nature.

The alarming environmental challenges that originated from the current development paradigm, with its reductionistic approach and egocentric consumerism, are horrifying and deeply concerning. Should the current trajectory of planetary environmental destruction and degradation persist, the biocapacity and resilience of the planetary ecosystem will cross the critical threshold, imperiling the very existence of humanity and the intricate *web of life*. The ecological consequences would be unimaginably catastrophic. The loss would extend beyond the breathtaking natural landscapes and diverse ecosystems, which have sustained biotic communities for millions of years through the process of evolution, to encompass the fabric of human civilization itself. What awaits is an uncertain and potentially irreversible chapter of evolution. Regrettably, the current egocentric neoliberal corporate market capitalism has undermined not only the prospects of future generations but also the survival of *Homo sapiens* as a species, as *Noam Chomsky (2020), the great visionary thinker of our time has thoughtfully articulated in his inaugural speech at STAR Scholars Network Conference of 2020.*

The relationship between the current development model and Nature is analogous to a parasite-host model in which humans are acting like the *parasite* and the biosphere, the *host*. This parasite-host model must be turned around from destroying the Earth to caring for the Earth, going from Nature-domination-destruction to stewardship of Nature. Perhaps, because of technological achievements, human arrogance remains parasitic on the biosphere, but the survival of a parasite, as scientists point out, depends on reducing the virulence and establishing reward feedback that benefits the host (Alexander, 1981; Anderson & May, 1981, 1982; Levin & Pimentel, 1981; Washburn et al., 1991). The fact is that the less *virulent parasite* can survive longer and also helps the *host* to live longer. Similar relationships hold for herbivory ad predation (Dyer et al., 1986; Lewin, 1989; Owen & Weigert, 1976). In terms of human affairs, this concept involves reducing waste throughputs and destroying natural ecosystem processes to reduce human virulence (impact throughputs), prolong ecosystem processes and biocapacity, and invest more in Earth care. The time has come to take care of the Earth First, which, in turn, will take care of everything in it, including us, the *Homo sapiens*. That is what the "Ecosociocentrism: The Earth First Paradigm" envisages.

13.8.2 Ecosociocentrism and Values in Nature

Natural systems are characterized by certain dynamic processes, attributes, or properties that evolved through millions of years of evolutionary processes. These attributes or the properties of the natural system can be considered to have certain values (instrumental and intrinsic), as depicted in Fig. 13.2 (conceptualization of instrumental and intrinsic values in Nature). *Self-organizational complexity, resilience, diversity*, and *interconnectedness* are the attributes of the natural system, which can be regarded as both *instrumental* and *intrinsic* values. Recognition and protection of these values in the natural system on planet Earth are essential for sustaining the future of humankind and the living system.

Protection and conservation of biological diversity, habitats, ecosystems, and ecosystems processes is not a policy option but a necessity for maintaining a healthy and productive planetary ecosystem for humankind's survival. Ecosystem processes and biological diversity must be considered as a system of interacting components of planetary ecosystems rather than the list of individual isolated entities, organisms, and elements. The value of the ecosystem and biological diversity should be based not only on their contribution to the maintenance of ecological processes and ecosystem health (instrumental values) upon which human existence is directly

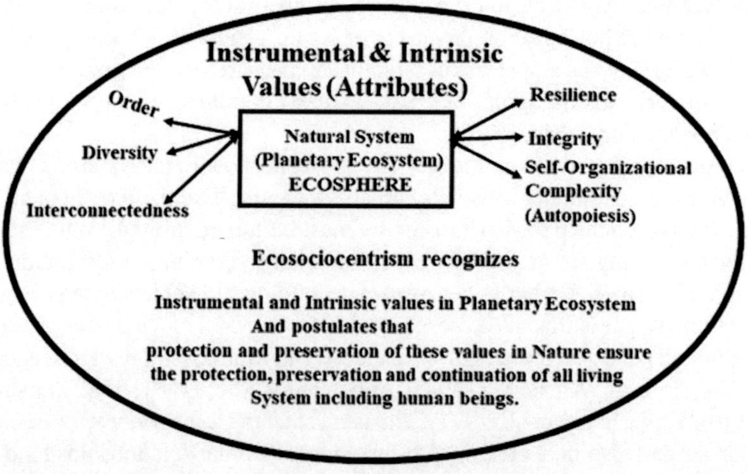

Fig. 13.2 Conceptualization of Instrumental and Intrinsic Values in Nature

dependent but also on their very existence for the flourishing of the biotic community (intrinsic values) and the continuation of the biological evolutionary process. This requires a shift from the prevailing social paradigm to a new paradigm characterized by a gentler and holistic approach deeply embedded in justice and social and ecosystemic well-being. This ethical paradigm, proposed to be called "*Ecosociocentrism*" postulates that *diversity, autopoiesis* (self-organizing complexity), *interconnectedness, integrity*, and *coevolution* are the *intrinsic* properties of the ecosphere and sociosphere. These attributes have values and must be allowed to flourish, both in the ecosphere and sociosphere. This paradigm asserts that actions taken by humans that foster the stability, integrity, and diversity on planet Earth are deemed morally righteous and fair, and actions that undermine the health, stability, integrity, and diversity of social and ecological systems on planet Earth are morally incorrect and unjust. The ecosociocentric paradigm advocates for the harmonization of anthropocentric activities with principles of social justice and ecological laws and processes. This harmonization and integration are necessary to ensure the prosperous coexistence and fulfillment of both human and biotic communities on planet Earth.

13.8.3 Conceptual Framework of Ecosociocentrism

The *sociosphere* (socioeconomic and cultural system) is the subsystem of the *ecosphere*, and the nature of their interaction is *dialectical*, which manifests not in linear but rather in cyclical progression propelled and maintained by positive and negative feedback mechanisms. The nature of the current interaction between the sociosphere (human socioeconomic system) and the ecosphere has become intensely antagonistic. Consequently, the impacts of such antagonism have manifested in various forms and formats, including the mass extinctions of diverse species, global warming, adverse climate changes, degradation and desertification of once productive ecosystems, shrinkage of fresh water, marine life, ocean acidification, the rapid increase of greenhouse gases and breakdown of Earth's systems, etc. This antagonistic interaction must be reversed through appropriate human cultural and behavioral changes in the sociosphere because men are the main driver of this antagonistic interaction. Without changing the cultural patterns of resource extraction, production of goods and services, consumptions, and waste generation (total throughputs) in the sociosphere, degradation in exospheric integrity and its biocapacity cannot be reversed and stopped. Scientific *epistemology*, human intellect, and ingenuity, and ecological wisdom consciousness should prevail over greed, insatiable consumerism, human arrogance, and the lust for power and domination. Only the cultural change informed by ecological wisdom consciousness can free us from the illusion of excessive consumerism and self-aggrandizement and help us devise the pathways for the transformation of the current growth-oriented unsustainable development model to a more sustainable and just development model that can help us live in stewardship with planet Earth.

The conceptual framework of *ecosociocentrism*, as depicted in Fig. 13.3, stipulates that human social and economic system must remain within the regenerative biocapacity of the ecosphere/biosphere for a sustainable future of humanity and the perpetuation of a living system in Nature. The conceptual framework delineates that sustainable development is inconceivable when the rate of resource extraction, consumption, and waste throughput production from the *sociosphere* exceeds the regenerative and assimilative biocapacity of the *ecosphere/biosphere*. Since the integrity and sustainability of the human social and economic system in the *sociosphere* are interconnected and dependent upon the integrity and sustainability of the planetary ecosystem or *ecosphere*, resource extraction, production, processing, consumption, and waste throughput must remain within the biocapacity of the Planet Earth to realize sustainable development, intergenerational equity, and flourishing of all living entities in Nature.

Ecosociocentrism holds that the autonomous self-organizing complexity of life, diversity, resilience, interconnectedness, and coevolution are the intrinsic properties of the planetary ecosystem, which have values in themselves and must be allowed to flourish in the ecosphere and sociosphere. *Hence, human actions that promote social and ecosystemic health, resilience, and diversity are morally right and just, and human actions that degrade social ecology, resilience, ecosystem health, and diversity are morally wrong and unjust.* The ecosociocentric paradigm demands that anthropogenic activities must be reconciled with social and ecological integrity for the fulfillment of both human and biotic communities. The ten principles presented below provide the foundational basis and embody the developmental and ethical imperatives of *Ecosociocentrism*.

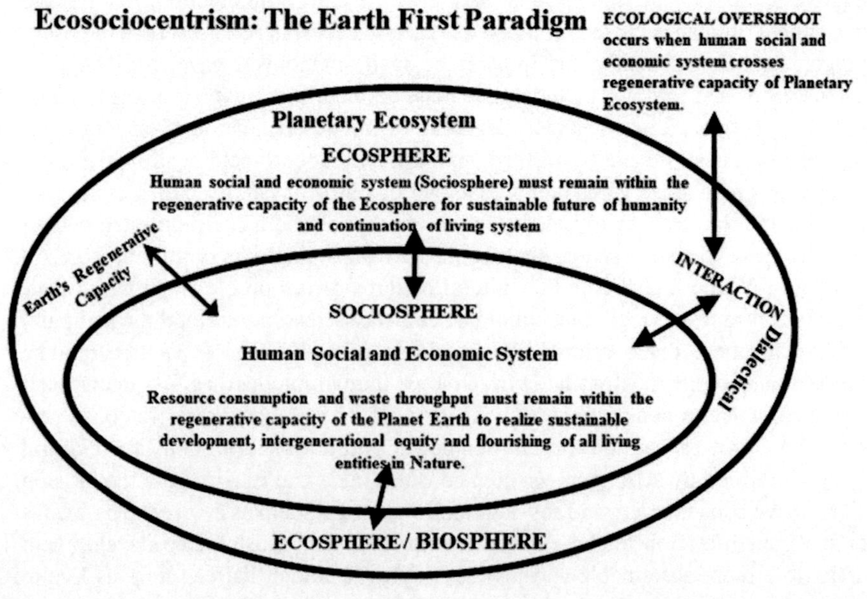

Fig. 13.3 Conceptual Framework of Ecosociocentrism: The Earth First Paradigm

Ecosociocentrism is a harmonious synthesis of the two seemingly opposite but complementary tendencies, *ecocentrism,* and *sociocentrism.* It integrates the ecosphere (biosphere) and the sociosphere and provides a more balanced and comprehensive perspective on ecospheric and sociospheric phenomena that ultimately regulate the behavior of the biosphere and sociosphere. This paradigm advocates value-based development that regards ethical values and justice as the ultimate reference point for human behavior in the sociosphere or ecosphere for achieving the health and well-being of all beings, including humanity and the planetary ecosystem. Sustainable development will be a *Sisyphus's Myth* without realizing ethical values and justice in the use and management of manmade and natural systems, both in the sociosphere and ecosphere, because both spheres interact dialectically with each other and the nature of the outcome of this interaction determines each other's state of health and wellbeing. The following assumptions and principles have been advanced here to provide a theoretical foundation for ecosociocentrism:

13.8.4 Assumptions of Ecosociocentrism

1. Sociosphere is a subsystem of the ecosphere (planetary ecosystem).
2. Infinite economic growth is impossible as it is constrained by entropy law and the biocapacity of planet Earth's systems (planetary ecosystem).
3. Human economic subsystem cannot function beyond the boundary of the biocapacity of the planetary ecosystem. A planetary ecosystem can sustain the human economic subsystem to satisfy human needs only if it remains in its regenerative biocapacity.
4. The sound economic system internalizes the environmental externalities and natural resource accounting into the economic matrix and does not discount Nature.
5. Technological innovation can modify and improve the efficiency of natural resource uses but cannot substitute natural life support systems and services on which human existence depends.
6. The fundamental attributes of a sustainable economic system are characterized by regenerative and distributive nature.
7. The distributive and regenerative nature of the economy ensures integrity in both the sociosphere and the ecosphere, laying the foundation for ecocivilization and sustainability.

13.8.5 Directive Principles of Ecosociocentrism

Undoubtedly, numerous technological advancements and innovations have emerged, holding the potential to protect the environment and preserve Earth's habitability for humanity and the living system. Examples include geoengineering methods to capture and store carbon for future use as a fuel source, green energy (hydrogen fuel), implementing white roofs in cities to reflect light into space, employing reflective materials to cover deserts, stimulating phytoplankton growth in oceans through iron seeding to absorb atmospheric carbon, incorporating solar glass windows and rooftops in residential and skyscraper architecture, and harnessing wind power and solar energy on a large scale. While these technologies contribute to mitigating certain adverse effects, they fall far short of resolving today's environmental crises since they address only the surface-level manifestations rather than tackling the underlying drivers of this crisis. Stubblefield (2018) astutely observes that Anthropocenologists tend to sidestep the root causes of present-day crises and confine themselves to addressing their surface-level manifestations. To effectively respond to the current crisis of the *Anthropocene*, we must confront its fundamental cause: *the ecologically hostile mode of production and consumption patterns*. Unless we are willing to transform the prevailing models of producing and consuming goods and services, and unless science and technology are deliberately designed and directed toward this objective, the planetary environmental and ecological crisis will worsen, surpassing a point of no return. Such a stage in history would be profoundly unfortunate for *Homo sapiens* during the *Anthropocene* epoch.

Building upon scientific and epistemological studies, discoveries, innovations, and a comprehensive understanding of human social-cultural dynamics, physical and bio-ecological processes, and evolutionary principles, Ecosociocentrism: The Earth First Paradigm has been conceived with the following ten principles that serve as a rational foundation for achieving environmental sustainability, sustainable human development, intergenerational equity, and the flourishing of living systems on planet Earth. These principles presuppose and provide the basis for developing pragmatic environmental and developmental ethics, guiding human behavior towards sustainable living and responsible stewardship of Planet Earth heralding a new era of ecocivilization:

1. Human socioeconomic system (sociosphere) is a subsystem of the larger ecosphere (biophysical system) or planetary ecosystem and cannot exist independently. The sustainability of the socioeconomic system (sociosphere) is invariably interconnected with and dependent upon the integrity and sustainability of the ecosphere (biophysical system).
2. Sustainable development is inconceivable when the rate of resource extraction, consumption, and throughput production from the human socioeconomic system exceeds the regenerative and assimilative biocapacity of the Earth's systems (ecosphere). Human socioeconomic subsystems must operate within the regenerative biocapacity of Planet Earth's ecosphere or biosphere.
3. The nature of the interaction between the sociosphere (socioeconomic system) and the ecosphere is *dialectical*, both causing changes and impacts in each other, which gives rise to a new relational state or equilibrium that may be less or neutral or more detrimental to the wellbeing of human beings and other life forms in Nature.

4. Human rationality, intellect, and wisdom can change the trajectory of the environmental crisis and detrimental changes (global warming, climate change, and destruction of planetary ecosystem) and its consequences toward environmental and social sustainability in which actualization of human potential and flourishing of other life forms is possible.
5. Human-caused destruction and degradation of the planetary ecosystem that generates life-sustaining environmental goods and services undermines the security and survival of all life forms, including human beings. Humanity cannot survive by destroying its *niche*, the Planet Earth. Hence, we need to save "The Earth First", to save humanity and the rest of the biotic community.
6. The life-sustaining environmental services and interconnectedness, self-organizing complexity on which depend the very existence of human beings and other life forms in Nature must be considered to have both *instrumental* and *intrinsic* values.
7. Protection and preservation of biological diversity, ecosystem resilience and the web of interconnectedness, self-organizing complexity of life, and life-sustaining environmental services are the fundamental basis for building social and environmental sustainability.
8. Humanity's development endeavor and behavioral conducts must be guided by *pragmatic* environmental and development ethics that embodies both *instrumental* and *intrinsic* values in Nature and cultivates and nurtures humanity to live sustainably within the means of the planet Earth's systems.
9. Promotion of equity and opportunity for all to satisfy their basic needs and realize their human potential in the *sociosphere* ennobles humanity to expand its moral capacity to incorporate the nonhuman biotic community into one moral plane or one single biotic community of inseparable *interconnectedness*. Socially and ecologically sustainable society can emerge only from realizing distributive economic justice in the sociosphere and regenerative natural economy in the ecosphere.
10. Ecosociocentrism: The Earth First Paradigm affirms that humanity's actions that protect and preserve the integrity, resilience, web of interconnectedness, health, and the functioning of the ecosphere and sociosphere are right and just and morally wrong if they do otherwise.

Hence, ecosociocentrism attempts to resolve human and environmental problems in a way that leads to realizing a just and ecologically sustainable society that functions as a harmoniously interacting component of the larger biospheric system on planet Earth. The fundamental premise of ecosociocentrism is that a socially and ecologically sustainable society can only emerge from realizing economic justice in the sociosphere and a regenerative natural economy in the ecosphere. Distributive economic justice and regenerative natural economy can promote and maximize the wellbeing of the maximum number of people in the sociosphere and flourishing of biotic community in the ecosphere/biosphere. This paradigm requires balancing sociocentric and ecocentric (biocentric) approaches to solving today's human and environmental problems. Too much emphasis on the *sociosphere* without taking account of what is happening in the *ecosphere/biosphere* and vice versa would take humanity nowhere except to its own misery and self-destruction because *sociosphere* and *ecosphere/biosphere* are so inseparably and intricately interlinked that both must be adequately represented in development ethical and management consideration of social and natural capital systems. Ecosociocentrism, the Earth First Paradigm envisages achieving social and ecological sustainability with development strategies that can sustain health, diversity, and resilience of natural ecosystem and social integrity, diversity, and resilience in sociosphere.

13.8.6 Ethical View of Ecosociocentrism

Ecosociocentrism: The Earth First Paradigm postulates that anthropocentric activities that promote ecospheric (biospheric) and sociospheric health and well-being must be considered good, just, and right and the activities that endanger unjust, and morally wrong. This view integrates humanity with Nature and claims that all living systems including humans depend on planetary ecosystems and ecosystem services for their existence and continuation. This elevates humanity to a higher dimension of development where reverence for life, diversity, a web of interconnectedness and interdependence, symbiosis and coevolution, cooperation, and sustainable living become the guiding principles and ideals for human behavior in contrast to the prevailing egocentric paradigm, which encourages excessive consumerism, unbridled growth, and profit, unhealthy competition for accumulating power, wealth, and domination and control which ultimately accelerates the destruction of planetary ecosystem and the very existence of human civilization. A striking parallel exists between the diversity of ecosystem landscapes and species in Nature and the human cultural landscapes. The origin of the diverse form of human cultures can be found in the diverse forms of Nature that existed in the historical past. The diversity of human cultures originated from the interaction and adaptation of human beings to the diverse ecological systems (Gadgil, 1987) suggesting the inseparable interconnectedness of the sociosphere and ecosphere. Ecosociocentrism provides a holistic worldview that postulates that everything is connected to everything, and that the system's well-being depends on the harmonious interaction and the functional integrity and stability of the interacting components, of which the human system is just one (Table 13.1).

The *ecosociocentric paradigm* suggests that the human social system (*sociosphere*) is dependent on the *ecosphere* or *biosphere* for its material resources and services. Therefore, the sociosphere is considered a subsystem of the planetary ecosystem. To ensure the sustainability of the human socioeconomic system, it is crucial for humans to recognize their place in Nature and conduct their activities in a way that does not harm the health and functioning of the planetary ecosystem. This recognition is essential because a healthy planetary ecosystem contributes to the well-being of its subsystems the socioeconomic system. A famous quote by Mahatma Gandhi captures the idea that there are sufficient resources in the world to

Table 13.1 Distinguishing features and positions of dominant and the Earth First Paradigm

Current Dominant Social Paradigm	Ecosociocentrism: The Earth First Paradigm
Assumptions *Human being*: Because human beings are fundamentally different from all creatures on Earth, they must exercise control and rule over other organisms and Nature. They are the master of Nature. *Human society*: The world provides unlimited opportunities for human beings, and it is proper for human society to control and dominate rest of Nature. Social and cultural environments are the crucial context for human affairs, and biophysical environment is essentially irrelevant. *Human behavior*: For every problem there is a solution, progress is a never-ending process. Human beings can choose their goals and can learn to do whatever is necessary. All social problems are ultimately soluble.	**Assumptions** *Human being*: However exceptional qualities human may have, they still remain one of many other species, coevolved, interconnected, and interdependent in the planetary ecosystem (ecosphere). Greater happiness emanates from maintaining the web of interconnectedness and misery from its destruction. *Human society*: Human society (sociosphere) is just a subsystem of the ecosphere (planetary ecosystem) that imposes physical and ecological restrains on human affairs. Social and cultural environment should be compatible with the ecological laws. Integration of sociosphere with ecosphere brings greater happiness and fulfillment and well-being of all beings including humans. *Human behavior*: No matter how inventive human may be, they have to live within the context of physical and ecological laws (entropy law). Human behaviors consistent with natural laws will ensure a greater happiness and stability both in sociosphere and ecosphere.
Responsibilities and duties: Maximization of individual self-interest, what is good for individual is good for society. Those who cannot compete for resources are unfit and will be eliminated and those who are fit will take control of resources and enjoy.	*Responsibilities and duties*: All living beings, human, and nonhumans have values in themselves and are intricately involved in maintaining the integrity and stability of biospheric ecosystem essential for flourishing of both human and nonhuman beings. Our duty as a conscious being is to ensure and strengthen this stability and integrity, not to undermine it. Respect for life, social justice, and cooperation and coexistence rather than individual domination, control, and ego satisfaction are the key values essential for ensuring sustainable future for all.
Metaphysical Base: A mechanistic paradigm based on following reductionistic assumptions: The whole is equal to the sum of the parts. Causation is a matter of external action on inert and inactive parts. Quantitative change is more important than qualitative change.	*Metaphysical Base*: A holistic paradigm based on the assumptions: Everything is connected to everything else. The whole is greater than the sum of its parts. The world is active and alive and possesses internal causation. The change and the ongoing processes are dialectical in nature always driving to a new relational complexity.

(continued)

Table 13.1 (continued)

Current Dominant Social Paradigm	Ecosociocentrism: The Earth First Paradigm
End goals: The knowledge of innate ideas, pleasure, control, and domination over Nature and other beings and individual self-aggrandizement.	*End goals*: Unity, stability, and diversity in Nature and human society. Self-sustaining natural and human system (sustainable system) with social justice and opportunity for self-actualization for all. The web of interconnectedness and interdependence that has evolved over millions of years of evolutionary process in Nature is the basis for unity, stability, and diversity.

meet everyone's needs but not their greed. It highlights the fact that the Earth can satisfy human needs, but it is incapable of satisfying human greed. Time may be running out for us to realize this truth. Therefore, it is necessary to establish a moral value value system rooted in ecological facts, principles, reasoning, and social justice. Such value system should promote goal-directed creativity, cooperation, mutualism, pluralism, and synergistic interactions while discouraging unhealthy competition, domination, aggression, control, and greed. Nature inherently exhibits characteristics like *autopoiesis, symbiosis, diversity*, and *interconnectedness.* These values are intrinsically valuable and must be acknowledged and protected in Nature. They are also manifested in the *sociosphere* through *social autopoiesis*, *social symbiosis*, *cultural diversity*, and *interconnectedness.* These attributes serve as the foundation for the development of human socioeconomic and cultural superstructures that are environmentally enabling. In essence, what drives and sustains phenomena in the ecosphere/biosphere also drives and sustains phenomena in the sociosphere (human socioeconomic system). Therefore, these attributes deserve recognition in the sociosphere the ecosphere.

The evolution of cooperation, group selection, symbiosis, organizational complexity (autopoiesis), and interactive synergism in Nature validates the importance of these attributes. To foster diversity, cooperation, symbiosis, interconnectedness, and interdependence in both ecosphere and sociosphere, an ethical paradigm is necessary. This paradigm aims for development that promotes these values. Thus, the moral values of the Ecosociocentrism can be summarized in the following statement: *human actions that foster diversity, symbiosis,resilience, interconnectedness, and sustainable living in the ecosphere/biosphere and sociosphere (human social system) should be considered morally good and right. Conversely, actions that undermine these values should be deemed unjust and morally wrong.*

13.9 Policy Imperatives of Ecosociocentrism

The principles of Ecosociocentrism—the Earth First Paradigm provide foundation for developing environmental policies and strategies that accurately perceive and interpret the social and planetary phenomena surrounding us. What is necessary is a framework that offers balanced and comprehensive perspectives on the social and ecological dimensions, highlighting the importance and value of biological diversity and ecosystems. It is essential to understand the social and economic causes contributing to the loss of biodiversity and ecosystem destruction and identify policy measures and strategies for reversing this trend and restoring functional integrity to natural systems. To develop appropriate policy instruments and measures, it is crucial to recognize the value of life support systems and services generated by natural systems. This recognition stems from an understanding of how social and economic forces drive the destruction of natural systems and the impact of such destruction on the well-being of both human and biotic communities. Faulty development paradigms, population growth, ecologically hostile production and consumption patterns, poverty, inequitable development, and a lack of knowledge and understanding of ecosystem processes and environmental services all contribute to the loss of biological diversity and planetary ecosystem destruction. Pearce et al. (1990) argue that the multifunctionality of environmental resources, including their role in generating life support systems and services, should be acknowledged and accounted in national development strategy. This is important due to our limited understanding of the life support functions of natural environments and our inability to substitute for these vital functions. Additionally, losses of environmental resources and biodiversity are often irreversible. Although we know that natural ecosystems and environmental resources provide life support systems and services, we still need to understand the exact mechanisms by which they do so. Nevertheless, our lack of complete understanding should not serve as a justification for their destruction and irreversible loss, as this only perpetuates humanity's misery and suffering.

Prominent environmentalists argue that policies promoting the continuity of all life forms, ecosystem processes, natural capital, human dignity, and social capital, rather than focusing solely on private capital, lead to greater environmental harmony and social benefits, including human happiness and social security (Martinez-Alier et al., 2016; Kelemen et al., 2015; Baveye et al., 2013). What is needed today are development policies that prioritize ecological processes, ecosystem health, and biological diversity, ensuring the sustainable use of ecosystems while promoting human dignity, equity, cooperation, and a pluralistic social system. These policies should move away from encouraging the Spencerian notion of "individual competition" and the "survival of the fittest" in resource allocation.

To formulate appropriate policy measures, it is crucial to have a comprehensive understanding of various aspects, including the importance of ecosystem health and services, the valuation of biodiversity, the global environmental and climate crisis, and the social and economic forces driving the destruction of Earth's systems and the consequential impacts. The forces responsible for the environmental and climate crisis, loss of biological diversity, and ecosystem destruction can be summarized as ecologically hostile production and consumption patterns, population growth, poverty, inequitable development patterns, and the lack of valuation of natural capital and the environmental services within the current dominant paradigm (Aargau et al., 2016; Martinez-Alier et al., 2016; Kelemen et al., 2015; Baveye et al., 2013; Upreti, 1994; MacNeill, 1989). In the sociosphere, ecosociocentrism emphasizes the need to fulfill basic human needs, improve the quality of human life culturally and materially, and promote social autopoiesis, equity, justice, and social symbiosis. It stands in contrast to egocentric individualism and dominant, exploitative, and monolithic cultural superstructures. In the biosphere, ecosociocentrism stresses the importance of maintaining ecological processes, ecosystem health, autopoiesis, diversity, and natural capital not only for the existence of the living system but also for the survival of humanity itself.

Based on the analysis of the issues discussed above, this paradigm recommends six policy strategies that must be integrated into the development planning process of nation-states and effectively implemented. However, these policy strategy measures hold little meaning and significance if nation-states show no political commitment. Implementation of these strategy measures with firm political determination by nation-states is necessary for achieving the social goals associated with environmental protection and sustainable development as envisaged by ecosociocentrism.

13.9.1 Cultural Adaptation: An Imperative for Survival

The rapid and extensive cultural evolution of humanity has surpassed its biological evolution, resulting in a precarious situation where the survival of the entire living system, including *Homo sapiens*, is questioned by rational thinkers. Technological advancements since the industrial revolution have led to profound environmental impacts, referred to as *technometabolism*, by altering the input and output of materials and energy through various technological processes (Boyden & Dovers, 1992). These impacts have been overwhelmingly negative, endangering the well-being of humanity and the functional health of Earth's systems. The consumption rates, waste production, and greenhouse gas emissions by developed countries starkly illustrate the magnitude of the environmental impact (Aargau et al., 2016; Martinez-Alier et al., 2016; Kelemen et al., 2015; Baveye et al., 2013; MacNeill, 1989). The natural system, with limited assimilative capacity for waste products and toxins generated by this technometabolism, also possesses very little resilience and adaptability. Suppose the present patterns of ecologically unsustainable resource, energy consumption, and waste production in Western high-energy societies are not controlled and altered through well-meaning and deliberate societal actions. In that case, the most likely scenario is that it will end as a consequence of resource depletion and the damage and possibally, the irreversible breakdown of the planetary ecosystem caused by the *entropic effects* (waste products) of this technometabolism. The integrity and the stability of the biosphere and biological system will collapse; the survival of *Homo Sapiens becomes a dubious hypotheis?*. Such a scenario would result from the collapse of the integrity and stability of the biosphere and the biological system, jeopardizing the survival of *Homo sapiens* (Heinen & Low, 1992). Human history demonstrates that *Homo sapiens* have thrived and evolved by adopting and adapting different survival strategies in adverse circumstances.

Cultural diversity, born out of the need for survival, represents the essence of human versatility and flexibility. However, the current dominant consumption culture in high-energy societies must now undergo a transformation, necessitating fundamental changes in production patterns, resource extraction and consumption patterns, the waste throughput management, and the underlying assumptions, beliefs, and value systems that shape our consumption culture and worldview. If this cultural shift occurs in a timely manner, it has the potential to create a new biosensitive/ecosensitive society, representing a fifth ecological phase in human existence. This society would harmoniously coexist with Nature, maintaining the functional health of both the biospheric ecosystems and the human population (Boyden & Dovers, 1992).

The current excessive consumption-oriented culture prevalent in Western countries fosters consumption patterns, assumptions, expectations, and values that are antagonistic to ecological sustainability and, consequently, the survival of humanity. It is evident that it is not Nature itself but rather the human culture that requires direction, control, and transformation to ensure the survival of *Homo sapiens*. Overcoming this cultural impediment and facilitating cultural change necessitates our capacity and flexibility for adaptation and the invention of ecologically enabling culture (Caldwell, 1987). The primary challenges faced by industrial consumer society today are twofold: raising awareness among individuals about the ecological impacts of their consumption patterns and providing viable alternatives to break free from the addiction to ecologically hostile consumption patterns (Durning, 1989; Boyden & Dovers, 1992; Upreti, 1994). Achieving cultural adaptation toward an ecologically sustainable society is contingent upon collective efforts by government agencies, and social, cultural, educational, scientific, and political institutions. These efforts should entail raising societal awareness and consciousness while concurrently designing and restructuring the political-economy, socio-cultural landscape, and institutions in a manner that fosters ecological sustainability, equity, and social justice. Cultural restructuring necessitates simultaneous socioeconomic and technological transformations (Durning, 1989; Boyden & Dovers, 1992; Upreti, 1994). The crucial question remains: can governments and various social, cultural, and political and educational institutions worldwide, particularly in the Western world, comprehend the necessity and urgency of ecologically and socially sustainable cultural adaptation? Should human cultural evolution lead to the destruction of the very natural foundation on which its existence depends? The answer lies within our rich evolutionary history, but the time to seek that change is running out (Boyden & Dovers, 1992; Upreti, 1994).

13.9.2 Poverty Eradication and Debt Abrogation: A Moral Imperative

Persistent poverty and inequitable development patterns pose significant obstacles to achieving ecological and social sustainability. In many developing countries, these patterns have perpetuated a poverty trap, upheld by social, cultural, and political institutions controlled by the elites and vested interest groups. The marginalized majority, living in economic deprivation, and lacking access to education find themselves excluded from the decision-making process, unable to exercise their rights and interests. As a result, development programs primarily serve the interests of urban elites and business groups, perpetuating a cycle of corruption and inequality. It is the moral responsibility of both global and national elites, who control resources and powerful institutions, to dismantle this poverty trap and empower the impoverished to realize their human potential. Effective strategies that promote equitable distribution of development outcomes and prioritize the needs of the poorest must be implemented by national governments and supported by international governing bodies and development agencies. Drawing lessons from successful examples like China, South Korea, Kerala in India, and Sri Lanka, it is evident that poverty eradication requires political determination and programs focused on equity and opportunities. However, poverty elimination in developing countries is not achievable without the assistance of wealthy nations and complementary reforms at the national and international level.

The burden of debt carried by the global South remains a grave concern. The staggering debt levels and the net transfer of capital from poor Southern countries to rich Northern countries are unjustifiable. This debt burden obstructed meaningful recovery and exacerbated the suffering of developing nations. The COVID-19 pandemic has further highlighted the dire situation, with developing countries struggling to access vaccines while still prioritizing debt service payments over public health expenditures. The current debt trap must be dismantled, and support should be provided to enable the diversification of developing economies, rather than subjecting them to the policies that perpetuate their role as suppliers of raw materials and the consumers of goods produced elsewhere. The question remains: Do wealthy nations possess the moral imperative and courage to address this unfortunate debt trap and assist developing countries in eradicating poverty? It is crucial for global Northern countries to recognize their moral obligation and work toward a more equitable and sustainable global economic system. By addressing poverty and debt relief, they can contribute to the well-being of humanity as a whole and prevent the further destruction of ecosystems and the degradation of our planet.

The daunting debt burden carried by the global South remains a grave concern that impedes their progress. In 1988, developing countries' debt exceeded one trillion dollars, with an annual interest payment of 60 billion dollars. Shockingly, there was a net flow of over 43 billion dollars from developing to developed countries, a situation that is morally unjustifiable and detrimental to a significant portion of humanity (MacNeill, 1989). Consequently, industrialized nations share responsibility for the creation and perpetuation of poverty in developing countries. The current plight of the global South, compounded by the ongoing debt crisis, hampers their ability to achieve meaningful recovery. According to Daniel Munevar's report (2021), between 2010 and 2020, the public debt of developing countries increased from an average of 40.2% to 62.3% of their GDP. In 2020 alone, more than one-third of this increase amounted to a staggering 1.9 trillion dollars. Developing countries allocate over 20% of their government revenues to debt servicing, leaving them struggling to acquire resources for essential needs such as accessing COVID-19 vaccines. In 2020, these countries continued to pay their external creditors over 372 billion dollars in debt service, a figure 1.6 times larger than the resources allocated to public health expenditure during the same year. Emerging markets and developing countries face an overwhelming external debt of around 11 trillion dollars, with 3.9 trillion dollars due in debt service in 2020. Of this amount, approximately 3.5 trillion dollars are principal repayments, with about 1 trillion dollars due on medium and long-term debt. For the poorest countries (those eligible for IDA support), the 2020 medium and long-term debt (MLT) debt service was about $36 billion (Homi Khras, 2020). All developing country regions are seriously affected. This predicament extends beyond low-income or African countries, and there are increasing calls for debt standstills to alleviate the burden on developing countries. Debt poses a global development emergency that parallels the global health emergency created by the pandemic.

As pointed out by Miller (1991), 5% of the Earth's population residing in North America consumes one-third of the world's resources. If only 15% of the global population consumed at the rate of North Americans, there would be nothing left for the remaining 85% in the world. This disparity highlights the socio-cultural and economic system's detrimental effects, pushing a significant portion of humanity toward social disaster and dehumanization while fostering an unsustainable society rooted in ecologically detrimental consumerism in the West. The persistence of ecologically damaging consumption patterns in the developed world, coupled with poverty and debt servicing, inequitable development, and population pressure in developing world, along with existing economic arrangements between developed and developing nations, not only further destroy ecosystems and biodiversity on Earth but also lead to an inescapable downward spiral of unprecedented human suffering intertwined with ecological degradation on a global scale. It is incumbent upon the global North to acknowledge its moral responsibility in dismantling the global poverty trap and supporting developing nations in diversifying their economies. Rather than subjecting these countries to policies that perpetuate the debt trap and relegate them to suppliers of raw materials, the rich Northern nations should extend a helping hand. This necessitates a departure from exerting pressure through institutions such as the World Bank, IMF, and WTO and embracing a global economic policy that empowers developing nations through debpt abrogation. The pivotal question remains: do industrialized, wealthy nations possess the moral imperative and courage to dismantle the unfortunate debt trap they created and assist developing countries in eradicating poverty?

13.9.3 Optimum Population

The population of our planet has experienced a remarkable increase, surpassing threefold growth, while the world economy has expanded twentyfold since 1900. Fossil fuel consumption and industrial production have risen by 30% and 50%, respectively, with four-fifths of this surge occurring since 1950 (MacNeill, 1989).

MacNeill (1989) contended that meeting the needs and aspirations of then five billion people can be achieved without jeopardizing the well-being of the projected eight to ten billion individuals in the future. However, this requires fundamental changes in how nations manage the global economy. To accomplish this, various ecological, social, institutional, and political challenges must be overcome, as economic and ecological sustainability are intricately intertwined. Environmental crises including climate change, biodiversity loss, global warming, ozone depletion, acid rain, and soil degradation serve as feedback from Earth's ecological system to the world's socioeconomic system (Repetto, 1992), reinforcing an inseparable interconnection and interdependence between the ecosphere and the sociosphere.

Developed countries stabilized their population growth with rapid economic growth and improvement in basic necessities of life. As income levels rise, education improves, urbanization progresses, and women are empowered, negative population growth rates have emerged in the developed world. Conversely, developing countries find themselves entangled in the complexities of rapid population growth needing to extricate themselves from this unfortunate paradox. In developing countries, an effective family planning program hinges upon the understanding that the children born will have a chance to thrive. Ensuring child survival necessitates the availability of robust healthcare systems, access to safe drinking water, a balanced diet, and educational opportunities for women. These basic needs are often taken for granted by people in developed industrialized nations, while only privileged ones in developing countries can access such essential provisions even today. If government leaders in developing countries are genuinely committed to optimum population growth, appropriate policy measures must be implemented, guaranteeing the availability of essential healthcare facilities, potable water, schools, and employment and educational opportunities in rural areas. This calls for substantial political and financial reforms in developing countries and also support from developed countries.

In summary, achieving optimum population growth on planet Earth demands a global collaborative effort. By addressing the underlying challenges through meaningful changes in how nations manage the global economy, we can strike a balance that satisfies the needs of both present and future generations. Developed countries, with their successful approaches to population management, should extend their support to developing nations by advocating for and assisting in the provision of fundamental necessities such as healthcare, clean water, education, and employment opportunities. Through these collective actions, we can foster a sustainable future for all with optimum population.

13.9.4 Landscape Ecosystem and Ecoregionalism: A Basis for Conservation and Sustainability

The valuation and preservation of biological diversity and planetary ecosystems have sparked intense debate and confusion regarding their appropriate forms, scales, and methods of conservation, management, and protection (Perring et al., 1992; Norton &

Ulanowicz, 1992; Munashinghe, 1992; Upreti, 1994). These discussions significantly influence the formulation of policies aimed at conserving and managing biological diversity. In this context, the primary objective of such policies should be to uphold the overall health of ecosystem dynamics that sustain and nurture diversity on a larger geographical scale. The values attributed to species and ecosystems must primarily consider their contribution to these larger dynamics, which are responsible for the productivity, stability, and integrity of planetary ecosystems (ecosphere). Consequently, policies need to focus on safeguarding total diversity at the landscape level of ecological organization. Comprehensive conservation and management plans at the ecosystem level aim to protect the health and integrity of broader ecological systems. These plans also involve public engagement and education to foster a better understanding of ecological management, empowering individuals to articulate and express their values in conservation and management strategies. As eloquently elaborated by Norton and Ulanowicz (1992), ecosystem health serves as the pivotal policy concept guiding environmentally conscious management. Furthermore, one can argue that public values such as asthetic, economic, and moral—all depend on protecting the processes that support the health of larger ecological systems. Consequently, preserving ecosystems' capacity to react creatively and productively to disturbances should takes precedence over short-term objectives pursued by individuals and economic interest groups. Resource management strategies must be compatible with maintaining a healthy functional ecosystem. We cannot afford not to have a healthy functional ecosystem with diversity, and resilience, all of which are the determinants of the system's overall productivity.

Resource management strategies must align with the preservation of a healthy ecosystem. The integrity, diversity, and resilience of an ecosystem are the essential determinants of its overall productivity. It is crucial to view biological diversity as a system rather than merely a list of individual organisms and species (Gee, 1992). This self-organizing system sustains and maintains itself through homeostatic responses and adaptive mechanisms in the face of changing environmental conditions. One significant attribute of this system is its ability to creatively adapt to new conditions, enabling natural systems to rebound from extensive human intervention and exploitation, as well as absorb and assimilate the wastes generated by human economic subsystems. This capacity for creativity and waste assimilation constitutes the ecosystem's functional health. This capacity, encompassing creativity and resilience, is intrinsically linked to the available biological diversity within the system. Consequently, the concept of ecosystem health, defined as the system's capacity for creative self-organization and waste assimilation, underscores the importance of maintaining autopoietic activity on a scale that spans multiple human generations. When formulating public policies, the preservation of this *autopoietic capacity* and ecosystem stability must take precedence.

Economists often attempt to assign value to species by quantifying their potential direct contributions and aggregating them to assess the overall value of ecosystems. However, quantifying the direct potential contributions of individual species to humankind is an immensely challenging task, as it requires accounting for the yet-to-be-discovered values of species and ecosystems. Our lack of knowledge regarding their values may inadvertently lead to the elimination of diverse organisms and ecosystems, as a priority may be given to species whose values are already known for conservation and protection. Assessing the contribution of an individual species to ecosystem services defies quantification due to the complex and highly interactive processes that occur at the ecosystem level. In other words, phenomena occurring at higher levels or within an ecosystem cannot be adequately explained by phenomena at lower levels. More precisely, the behavior of an organism or species cannot fully elucidate the behavior of a community or an ecosystem.

One potential approach to valuing biological diversity is through the assessment of ecosystem services. Ecological services are generated through ecosystem processes in which each organism and species play a role. This approach involves considering the valuation of biological diversity as a system that generates ecological services, in addition to valuing individual species for their direct use. Valuing and quantifying the ecological services provided to humans within a particular landscape ecosystem may be more feasible and tangible. This approach accurately reflects the true value of biological diversity as a system and underscores the importance of protecting ecosystem processes and designing appropriate ecosystem management principles and practices, thereby safeguarding as many species as possible. This valuation approach emphasizes the development of appropriate interface *modus operandi* between ecology and economics and underscores the need to preserve the self-perpetuating features of ecological systems in order to achieve ecological and social objectives of species and ecosystem preservation and sustainable development.

Environmental problems can manifest at the local, regional, and global levels. The magnitude of environmental challenges faced in parts of Africa, Asia, South America, and developed industrialized nations indicates that these issues are transnational and necessitate regional and global solutions. Countries within a specific *ecological region* (bioregion) should develop unified, integrated, and cooperative approaches to address ecological and environmental problems and concurrently resolve socioeconomic and political issues faced by the people living in the *ecoregion*. This approach recognizes that environmental problems within an ecological region (ecoregion) ultimately undermine the security and livelihoods of all its inhabitants in the ecoregion. While environmentally sound development programs and policies in one country may have a limited impact on the environmental health of the *ecoregion*, reinforcing similar policies across all countries in the region is essential. An ecoregion can be defined as a landscape ecosystem characterized by distinct boundaries, incorporating essential elements of a dynamic system that holds significant socioeconomic value for the region. Scientific understanding of the internal functioning of ecological systems within a landscape guides the establishment of boundaries and identification of key compartments within the system. Certain species, known as *keystone species*, play critical roles in regulating and maintaining ecosystem stability, while specific regions within an ecoregion, referred to as *keystone regions*, may be instrumental in preserving the integrity and stability of the larger ecoregion. Consequently, the conservation of biological diversity and habitats within these *keystone regions* should be prioritized. For example, in a *Himalayan ecoregion*, the upper bioregion encompassing major river systems and watershed areas may be considered a *keystone region* due to the potential ecological catastrophes that could occur if these areas were destroyed. Such catastrophes would not only affect the people living in that specific ecoregion/bioregion but also those residing in downstream bioregions. The destruction of watershed areas and deforestation in the high Himalayan ecoregion/bioregion of Nepal, for instance, would result in environmental problems for both the people of Nepal and those in India and Bangladesh along the Gangetic plain (Upreti, 1994; Erickholm, 1975). Policymakers in industrialized and developing countries must mutually agree that solutions to environmental and human problems, conservation and protection of biodiversity and habitat, restoration, and ecosystem development must be pursued within the framework of well-connected landscape ecosystem processes on an ecoregional basis.

13.9.5 Restoration of Degraded Ecosystems: An Ecological Urgency

Ecosystems are dynamic communities of plants, animals, and microorganisms interacting with their physical environment as a functional unit. Ecological restoration aims to initiate or expedite the recovery of ecosystems that have suffered damage, degradation, or destruction. Restoration ecology, a scientific field, focuses on the repair and re-establishment of disturbed ecosystems through human intervention. The restoration of degraded ecosystems and the preservation of species' natural habitats are pivotal tasks from the standpoint of ecosystem management and redevelopment. Restoration initiatives may involve replicating a predisturbance ecosystem or creating a new ecosystem in locations where it did not previously exist. These activities draw upon various principles from the realm of landscape ecology.

Ecological restoration endeavors to return a degraded ecosystem to its historical trajectory, rather than restoring it to its exact historical state. The ecosystem may not fully recover to its former condition due to contemporary ecological dynamics, including the influence of global climate change, which can cause it to evolve along an altered trajectory. Similar ecological factors might have also impacted nearby undisturbed ecosystems, diverting their original trajectory. While historical considerations are important in restoration efforts, current conditions must also be considered. The objective of ecological restoration is to reestablish a self-organizing ecosystem that follows a trajectory toward complete recovery. While initial restoration activities can often set a degraded ecosystem on the path to recovery relatively quickly, achieving full recovery may take years, decades, or even centuries. For instance, while tree planting could initiate a forest restoration process, it would take hundreds of years to attain full recovery if the destroyed forest had 500-year-old trees. Throughout this recovery period, unanticipated barriers to restoration may arise, or additional restoration activities may become feasible during later stages of development. Thus, even though individual restoration activities may reach completion, in most cases, the overall restoration process continues as the ecosystem gradually recovers and matures. Numerous natural areas have already been altered, depleted, eroded, or contaminated by hazardous substances, with many of these areas having transformed into barren deserts. However, these degraded ecosystems can be transformed from ecological liabilities to valuable ecological assets through efforts to redevelop them into functioning ecosystems (Brinck et al., 1988; Cairns, 1988; Upreti, 1994). Restored ecosystems have the potential to enhance biodiversity and safeguard natural systems in diverse ways. They can serve as ecological buffers between waste treatment systems and natural environments, offering protection. Additionally, they can provide refuge for rare, endangered, or threatened species. A damaged ecosystem, acting as an ecological barrier, can be converted into a "bridge" connecting ecosystems on either side. Finally, restored ecosystems can be utilized to replenish and purify groundwater aquifers.

The subsidy-stress gradient model as proposed by Odum et al. (1979) serves as a theoretical justification for ecosystem redevelopment. This model asserts that a mildly perturbed natural system often exhibits increased productivity, with perturbation functioning as a subsidy. However, as perturbation continues and intensifies, the ecosystem progresses through several stages: *a stress phase*, where the system begins to deteriorate; *a replacement phase*, where organisms are replaced by those better adapted to the new conditions; and a *terminal phase*, where the perturbation becomes so severe that life becomes unsustainable. Five stressors, as Brinck et al. (1988) point out, encompass both human impacts and extreme natural events affecting ecosystems: *harvesting of renewable resources*, *discharge of pollutants into the environment*, *physical alteration of the environment*, *the introduction of exotic species*, and *extreme natural events*. Ecosystems possess resilience, resisting changes imposed by these stressors and tending to return to a state of equilibrium. On the other hand, ecosystems that have evolved in unstable environments demonstrate greater resilience, maintaining their structure and patterns despite disturbances (Rapport et al., 1985). Resilient ecosystems exhibit a higher capacity for adapting to change and substantial variability. It is noteworthy that the majority of the world's natural systems demonstrate resilience (Brinck et al., 1988). With proper knowledge of ecosystem structure and functioning, it becomes possible to redevelop degraded or damaged ecosystems that are approaching the critical, unsustainable phase. The redevelopment of degraded ecosystems with afforestation/reforestation holds great significance from the perspectives of environmental conservation and sustainable development. Some of these can be summarized as follows:

- First, by restoring these ecosystems, we contribute to the conservation and preservation of biodiversity, ensuring the survival of various plant and animal species that depend on these habitats. This helps to maintain ecological balance and prevent further loss of valuable flora and fauna.

- Second, redeveloped ecosystems serve as crucial natural resources, providing essential ecosystem services such as clean air and water, soil fertility, and climate regulation. These services are vital for human well-being and sustainable development, supporting agriculture, tourism, and other economic activities.
- Moreover, the restoration of degraded ecosystems enhances the resilience and adaptive capacity of natural systems in the face of environmental challenges, including climate change. Restored ecosystems can mitigate the impacts of climate change by sequestering carbon, reducing soil erosion, and improving water retention.
- Additionally, redeveloped ecosystems act as green infrastructure, offering natural solutions for flood control, coastal protection, and water purification. This reduces the reliance on costly and energy-intensive engineering interventions, promoting sustainability and resilience in the face of natural disasters.
- Lastly, the redevelopment of degraded ecosystems contributes to social well-being by providing recreational spaces, promoting cultural and educational opportunities, and fostering a sense of connection with the natural world. These benefits have positive impacts on human health, quality of life, and community cohesion.
- Overall, the redevelopment of degraded ecosystems is a crucial aspect of environmental conservation and sustainable development, as it supports biodiversity conservation, provides essential ecosystem services, enhances resilience to environmental challenges, and improves the well-being of both humans and the natural world.

In the wake of the European Commission's proposal in 2022, the European Parliament has recently advanced its Nature Restoration Law aimed at rehabilitating and restoring degraded ecosystems throughout Europe, despite significant resistance from some political factions. The scientific community has stood firmly behind this legislation, urging lawmakers to endorse it. Under the proposed law, a mandate would be issued to all member states to roll out restoration initiatives for 20% of Europe's terrestrial and marine regions by 2030. This expands to include the restoration of all degraded ecosystems that are on the brink by 2050. The legislation is strategic, setting definitive targets for individual habitats and species. For example, it aims to turn around the decline of crucial pollinating insects by 2030 and revitalize seagrass beds and other marine ecosystems, according to Rabesandratana's report of July 10, 2023. This proactive initiative by the European Parliament sets an exemplary precedent for the global community. If every nation could emulate this initiative and establish laws to restore and regenerate their degraded ecosystems, it could dramatically enhance the resilience and biocapacity of our planet. In doing so, we would strengthen Earth's natural systems and ensure the sustainable use of our shared resources for future generations.

13.9.6 Integration of Economics and Ecology: Foundation for Sustainable Development

The integration of economics and ecology through natural resource accounting plays a vital role in promoting environmental protection and sustainable development. By incorporating ecological considerations into economic decision-making, natural resource accounting provides a comprehensive framework for evaluating the true costs and benefits of development activities. Traditional economic analysis has overlooked the assignment of economic value to changes in natural resource stocks. This oversight can be attributed to the fact that Keynesian macroeconomic models were developed during the Great Depression, when economists were primarily focused on issues such as consumption, saving, investment, business cycles, and commodity prices, without much concern for the scarcity of natural resources, as highlighted by Repetto (1992). However, it is widely acknowledged that the human economic system operates within the planetary ecosystem (ecosphere), which supplies all the essential resources. Achieving ecologically sustainable development hinges on the prerequisite that a nation's basic stock of ecological capital remains constant or does not decline over time. Sustaining a constant, if not increasing, natural capital stock is crucial to meet the present generation's needs and aspirations, while ensuring fairness and equity for future generations.

Governments have a vital role in designing public policies tailored to minimize deforestation, desertification, habitat destruction, and species loss, as well as declines in air and water quality. In developing countries, these policies must address the challenges faced by landless people and provide viable alternatives to curb marginal land cultivation and deforestation. Furthermore, developed countries must adopt policies that do not incentivize deforestation, species loss, and desertification in developing nations. The activities of American and European transnational corporations operating in these countries have resulted in massive destruction of forests and other natural resources. There is a direct link between deforestation in Amazonia Brazil and the expansion of pasture for dairy and beef industries, with the products being sold in the markets of the United States and Europe. A significant portion of these industries is owned by American and European companies. The question arises as to whether the governments of these industrialized countries can design policies that effectively regulate the behavior of multinational corporations to protect the environment and the ecosystem of global significance. Unfortunately, the world has witnessed the opposite thus far.

Achieving ecological sustainability necessitates the appropriate integration of economics and ecology (environment) in decision-making processes. MacNeill (1989) aptly observes that economic and ecological systems are intertwined in the real world, yet they have been disconnected in the institutions where important decisions are made. The most crucial step toward ecologically sustainable development involves internalizing the external environmental costs associated with the production, consumption, and disposal of goods and services, as argued by Barbier (1987), Daly (1990), Pearce et al. (1990), and Repetto (1992). Currently, the market treats biospheric resources such as the atmosphere, oceans, and other commons as free goods, externalizing or transferring the costs of air and water pollution, land degradation, noise pollution, and resource depletion to society. Society bears the costs of health and property damage, ecosystem degradation, loss of biodiversity, and pollution. Government intervention is imperative to internalize these costs, while decision-makers must base their choices on long-term ecological consequences rather than short-term economic gains. Failing to integrate ecological consequences into economic decision-making will inevitably lead to greater human suffering rather than alleviating it, as warned by Costanzaa et al. (2014) and Jacobs et al. (2015).

The most significant aspect of the integration of economics and ecology through natural resource accounting is the assignment of economic value to changes in natural resource stocks. Traditional economic analysis has totally ignored accounts for the depletion or degradation of natural resources, as it primarily focuses on indicators such as gross domestic product (GDP) and market transactions. However, natural resource accounting recognizes that the planet's ecosystems are the foundation for all economic activity, and their degradation can have significant economic implications. By quantifying and valuing changes in ecological capital, such as forests, soils, and biodiversity, decision-makers gain a more accurate understanding of the trade-offs associated with development choices. This integrated approach enables governments and policymakers to assess the sustainability of economic activities. By considering the changes in ecological capital alongside economic indicators, natural resource accounting reveals the true impacts of development on the environment. This information not only provides a more accurate depiction of economic performance but also offers insights into how economic policies affect ecological systems, as emphasized by Barbier (1987) and Repetto (1992). For example, it can unveil whether reported increases in GDP are achieved at the expense of declining stocks of soils, water, forests, species, habitats, parks, historical sites or fisheries. This information empowers decision-makers to identify unsustainable practices, adjust policies, and implement measures that mitigate environmental degradation and promote sustainable development. The need for resource accounting is particularly urgent and essential in developing countries due to the predominantly biomass-dependent nature of their economies, requiring policies for ecologically sustainable developmentSustainable development.

Furthermore, natural resource accounting facilitates the identification and internalization of external environmental costs. Market systems often externalize the environmental consequences of economic activities, leaving society to bear the burden of pollution, resource depletion, and ecosystem degradation. By incorporating the costs of environmental damage into economic calculations, natural resource accounting highlights the full costs of production, consumption, and disposal. This helps in the formulation of policies that encourage responsible resource management and ensure that economic activities consider the long-term ecological consequences. By internalizing these costs, decision-makers can incentivize sustainable practices, reduce environmental impacts, and promote the transition towards a more environmentally conscious and resilient economy. In the context of sustainable development, natural resource accounting provides a robust framework for balancing economic growth with environmental conservation. It enables governments to design policies that prioritize the preservation of ecological capital, preventing its decline over time. By safeguarding natural resources and ecosystems, sustainable development becomes attainable, meeting the needs of the present generation while preserving resources for future generations. Natural resource accounting also aids in the formulation of strategies to address pressing environmental challenges such as deforestation, habitat destruction, pollution, and land degradation. By accurately assessing the costs and benefits of different development pathways, decision-makers can identify sustainable alternatives, promote green technologies, and create incentives for the protection and restoration of ecosystems. In summary, the integration of economics and ecology through natural resource accounting is essential for environmental protection and sustainable development. It enables decision-makers to make informed choices that consider the ecological consequences of economic activities, internalizes environmental costs, and prioritize the preservation of natural resources. By embracing this integrated approach, societies can strive toward a more sustainable future, where economic growth is harmonized with environmental well-being and the needs of present and future generations are balanced.

Designing appropriate policies and strategies that promote environmental conservation, biological diversity, and preservation of natural ecosystems is not merely an option but an absolute necessity. However, achieving this goal requires a paradigm shift to provide a more relevant perceptual and interpretive framework from which such strategies can emerge. The prevailing dominant social paradigm has neglected critical issues, including the excessive resource consumption in developed industrialized countries, the acute poverty and inequitable development patterns in the developing world, the significant capital flight from the global South to the global North, and the high population growth rates in impoverished developing nations, often due to political or ideological reasons. Unless the current paradigm is restructured or replaced by value-based pragmatic development ethics informed by ecological wisdom consciousness, it will continue to jeopardize Earth's systems and lead to the collapse of the biospheric ecosystem, ultimately pushing humanity into an unfortunate evolutionary stage in the *Anthropocene*. Therefore, I argue that "Ecosociocentrism: The Earth First Paradigm" holds the potential to free humanity from its current unfortunate development dillema.

13.10 Conclusion

The alarming environmental challenges that originated from the current development paradigm, with its reductionistic approach and egocentric consumerism, are horrifying and deeply concerning. Should the current trajectory of planetary environmental destruction and degradation persist, the *biocapacity* and *resilience* of the planetary ecosystem (some subsystems have been suspected to have already crossed) will cross the critical threshold, imperiling the very existence of humanity and the intricate *web of life*. The ecological consequences would be unimaginably catastrophic. The loss would extend beyond the breathtaking natural landscapes and diverse ecosystems, which have sustained living communities for millions of years through evolutionary process, to encompass the fabric of human civilization itself. What awaits is an uncertain and potentially irreversible chapter of evolution. Regrettably, our egocentric and indivisualistic economic pursuits have undermined not only the prospects of future generations but also the survival of *Homo sapiens* as a species.

It appears that *Homo sapiens* is uniquely the only *species* to inflict harm upon its own existence. In Nature, hardly any other species has self-destructively contrived circumstances leading to their own extinction. There is no quick fix for the current environmental catastrophe including climate change. We may be able to perceive the potential resolution to the current situation more clearly if many approaches and viewpoints are combined. Political economy and ecology undoubtedly offer the primary context for the interaction of various methods, but human *spirituality* and *ethical imperative* also have great potential for resolving this dilemma. It requires the combination of strategies consisting of political economy, science, technology, religion, spirituality, and ethics. Perhaps the eclecticism of such strategies and viewpoints has a much larger capacity to address and fix the current dilemma humanity is facing. In my opinion, *ethical environmental pragmatism* as envisioned in this book, focuses more on bringing practical insights from various sources (Buddhism, Gaian, and System Theory, political economy, and ecology, etc.) that can assist political and power elites in changing their fixed mindset and make them realize that humanity cannot survive by destroying the very *niche environment* of planet Earth in which it evolved and flourished. Nevertheless, the successful execution of those insightful strategies would not be possible without concurrent rectification of the systemic shortcomings prevalent in our contemporary consumer market capitalism. It is necessary for neoliberal capitalism to initiate reforms and transformative adjustments in order to become it more congruent with the operation of both the Earth's natural systems and the mechanisms of modern democracy. Such reforms would entail a paradigm shift, transforming capitalism from a potentially destructive force into a more benevolent system that ensures sustainable development and social justice, thereby fostering a more harmonious coexistence with our planet Earth.

With unflinching certainty, we can assert that modern industrial neoliberal capitalism needs transformative changes to align it more harmoniously with the Earth's natural systems and the fundamental tenets of modern democracy. This is not a mere suggestion, but a logical imperative derived from an in-depth analysis of our current predicament created by modern neoliberal capitalism. To begin with, capitalism, as it exists today, operates in stark contrast to Earth's ecological systems. While the planet functions on the principle of cyclic regeneration and balance, modern neoliberal capitalism thrives on a linear model of constant growth with ever-increasing consumption. This incongruity has accelerated environmental destruction, ecosystems breakdown, climate crisis, and Nature's resource exhaustion, which, in turn, threaten our survival and survival of the living system on planet Earth. To rectify this, we must steer modern hyper-consumeristic capitalism toward circular, cyclical, and regenerative economy principles, which ensure that resources are reused, regenerated and waste is minimized, echoing Earth's natural processes of regeneration. Furthermore, neoliberal capitalism, in its current form, undermines the democratic principle of equality and equitable development. It widens the wealth gap, concentrates power and wealth in the hands of a few, and promotes social and economic inequity; the conclusions of Thomas Piketty's seminal work cannot be refuted even by the staunch adherents of neoliberal capitalism. To align capitalism with modern democracy, it must inject in it with ecological, social and economic justice. The democratization of neoliberal capitalism will inevitably pave the way to democratic socialism which will be kindlier and gentler with its fair and equitable development and social inclusion in *sociosphere* and less burdensome in *ecosphere* with its environmentally enabling *modus operandi* and consumption culture.

Therefore, to ensure the survival of both our planet and human civilization, it is paramount that we transform neoliberal capitalism from its current destructive *modus operandi* into environmentally enabling *modus operandi*. This transformation would not reduce capitalism's potential for wealth creation if it can be transformed to incorporate Nature's capital into its analytical and operational framework; this will change the concept of the wealth and make it more ecologically sustainable and socially just. It would enable capitalism to foster sustainable development and social justice, which are cornerstones of a harmonious coexistence with our planet Earth.

While Earth has witnessed numerous crises over geological timescales, the current crisis is uniquely distinctive, as it is the first planetary crisis brought about by human activities. It is incumbent upon *Homo sapiens* to assume responsibility for this crisis, forsaking destructive behaviors in favor of creative and enabling approaches and constructive *modus operandi* that can secure a sustainable future for humanity and the world at large. No deity or supreme being can salvage humanity and the planetary ecosystem from the environmental calamity and climate crisis engendered by human greed and hubris. Only through a transformative process within humanity itself can we secure a sustainable existence for all. This transformation must extend to our way of life, our thoughts, our understanding of worldly phenomena, and, above all, our consciousness. To facilitate this transformation, a paradigm that fosters ecological harmony between *sociosphere* and *ecosphere* is indispensable.

The problem, however, is not insurmountable. The key lies in overcoming the monolithic mindset of these powerful corporate and political elites that drive the global corporate economy in a bid to amplify their greed, utterly disregarding the plight of common people. As they persist in exploiting Earth's resources to increase their wealth and power, people worldwide bear the brunt, precipitating an impending catastrophic environmental collapse. Now, more than ever, we need a seismic shift in our *collective consciousness*. We must rise, unified in purpose, to usher for environmentally sustainable and equitable development and environmental justice that respects the natural rights of other living being to exist on planet Earth. This compelling vision is neither fanciful nor idealistic; it is an imperative that requires us to subvert the current pathological development and demand change. It is time to realize that our survival is intricately linked to the health of our planet (the Earth's systems), and by extension, our actions, and choices. We must, therefore, strive to dismantle the reigning culture of ecologically hostile *consumerism* and influence the political process and attitudes of politicians and corporate elites to guide humanity away from the brink of disaster and toward a future of environmental sustainability and equity. The only way out for our survival is to learn to live within the regenerative biocapacity of the planet Earth with ecological consciousness. If we dream alone, it will be just a dream but if we dream collectively, it will be a reality.

Ecosociocentrism: The Earth First Paradigm underscores the imperative of fulfilling basic human needs and enhancing the quality of human life through self-determination and empowering human communities, thereby attaining social justice and equitable development opportunities in the *sociosphere*. This paradigm affirms that values such as *social symbiosis, autopoiesis, cooperation, equity*, and *cultural diversity*, which foster social interconnectedness, should be promoted, and safeguarded as opposed to *egocentric individualism* and *exploitative, inequitable*, and *homogeneous* cultural constructs. Moreover, this paradigm posits that maintaining *ecological processes, ecosystem health, biodiversity*, and the *natural capital* within the *ecosphere* is essential not only for the existence of humanity in the *sociosphere* but also for the broader biotic community in the *ecosphere*. Sustainable interaction between these two spheres, which mutually benefits both realms, is the key to unlocking and realizing the human potential in the *sociosphere* and the living system in the *ecosphere*. Consequently, the *sociosphere*, as a subsystem of the *ecosphere*, must remain within the Earth systems'biocapacity to ensure its own sustainability.

The proposed paradigm, "Ecosociocentrism: The Earth First Paradigm" seeks to integrate social and ecological integrity with an ethics-based development approach that embodies both instrumental and intrinsic values in Nature. Glenn Albrecht (2016) coined the term "Symbiocene" to describe an anticipated era in Earth's history characterized by human intelligence replicating the symbiotic and mutually reinforcing life-reproducing forms and processes found in living systems. Ecosociocentrism: The Earth First Paradigm embraces the fundamental interconnectedness of symbiotic life and challenges the prevailing Hobbesian, Spencerian, and Cartesian views of Nature as a cut-throat battleground for survival. It aims to redefine humanity's relationship with Nature, catalyzing humanity's profound potential for transitioning from the current Anthropocene to the Symbiocene. Included in the prescription of this paradigm are ten directive principles and six policy strategies to achieve the underlying objectives and goals of sustainable living on planet Earth. It would be fallacious, as aptly described by Morowitz (1991), to entertain the notion that growth-driven neoliberal market capitalism, will rescue humanity from the unfortunate entanglement it has created—a mere *"Sisyphean Myth"*. Planet Earth is the sole abode for all living entities, including human beings. Only a healthy and nourishing Earth can ensure the security of humanity and sustainable living, which is possible only if "*The Earth First Paradigm*" becomes the conscious working algorithm of humanity in the *Anthropocene epoch* of the twenty-first century. This is not merely a hopeful vision of the future; it is a logical and necessary path that we must embark upon if we wish to exist on this planet, our dear home.

Bibliography

Aargau, A., Jacobs, S., & Cliquet, A. (2016). What's law got to do with it? Why environmental justice is essential to ecosystem service valuation. *Ecosystem Services, 22B*, 221–227.

Abatzoglou, J. T., & Barbero, R. (2014). Observed and projected changes in absolute temperature records across the contiguous United States. *Geophysical Research Letters, 2014*, 6501. https://doi.org/10.1002/2014GL061441

Ackerman, F. (2020). Still dead after all these years: Interpreting the failure of general equilibrium. *Journal of Economic Methodology, 9*(2020), 119–139. Regarding its lack of predictive power, refer to every single dynamic stochastic model that somehow missed the Great Recession back in 2008.

Adams, F. (2018). *Earth will survive. We may not.* Downloaded from Medium News, 12/24/18. http://sk.sagepub.com/reference/ethics/n175.xml

Adriana, E., Ford, S., Graham, H., & White, P. C. L. (2015). Integrating human and ecosystem health through ecosystem services frameworks. *Ecohealth, 12*, 660–671. Published online 2015 Sep 24. https://doi.org/10.1007/s10393-015-1041-4

Agrawal, A., Costella, C., Kaur, N., Tenzing, J., Shakya, C., & Norton, A. (2019). *Climate resilience through social protection. Background paper to the 2019 report of the Global Commission on Adaptation.* Rotterdam and Washington, DC. https://gca.org/reports/climateresilience-through-social-protection/

Alexander, M. (1981). Why microbial parasites and predators do not eliminate their prey and hosts. *Annual Reviews in Microbiology, 35*, 113–133.

Allen, C. D. (2009). Climate-induced forest dieback: An escalating global phenomenon. *Unasylva, 231*(232), 60.

Allen, T. F. H., & Starr, T. B. (1982). *Hierarchy: Perspectives for ecological complexity*. University of Chicago Press.

Anderson, R. M., & May, R. M. (1981). The population dynamics of microparasites and their invertebrate hosts. *Philosophical Transactions of the Royal Society of London. B, Biological Sciences, 291*(1054), 451–524.

Anderson, R. M., & May, R. M. (1982). Coevolution of hosts and parasites. *Parasitology, 85*, 411–426.

Arguelles, J. (1996). *The Mayan factor, path beyond technology* (2nd ed.). Bear & Co. 1987.

Arguelles, J. (2010a). *Manifesto for the Noosphere: The next stage in the evolution of human consciousness*. Evolver Editions.

Arguelles, J. A. (2010b). *Planetary whole system design science: A contribution to World Shift.* Retrieved 5 Feb 2010, from http://www.lawoftime.org/planetarywholesystem.html

Arizpe, L. (1991). A global perspective to build sustainable future. *Journal of Society for International Development, 3*(4), 7–9.

G. Upreti, *Ecosociocentrism*, https://doi.org/10.1007/978-3-031-41754-2

Atkinson, B. (1972). *This bright lands*. Doubleday Natural History Press.
Attfield, R. (1993). Sylvan, Fox and deep ecology: a view from the continental shelf. *Environmental Values, 2*(1), 21–32.
Attfield, R. (2018). *Environmental Ethics: A very short introduction*. Oxford University Press. https://www.veryshortintroductions.com/view/10.1093/actrade/9780198797166.001.0001/actrade-9780198797166-chapter-8
Aubert, M., et al. (2018). Palaeolithic cave art in Borneo. *Nature, 564*(7735), 254–257.
Axelrod, R. (1980). Effective choice in the prisoner's dilemma. *Journal of Conflict Resolution, 24*(1), 3–25.
Axelrod, R. (1984). *Evolution of cooperation*. Basic Books.
Axelrod, R., & Hamilton, W. D. (1981). The evolution of cooperation. *Science, 211*, 1390–1396.
Bahg, C. G. (1990). Major system theories throughout the world. *Behavioral Science, 35*, 79–107.
Baldwin, J. H. (1985). Environmental planning and management, 1985, Westview Press, Inc, 5500 Central Avenue, Boulder, Colorado 80391.
Balvanera, P., Daily, G. C., Ehrlich, P. R., Ricketts, T. H., Bailey, S.-A., Kark, S., Kremen, C., & Pereira, H. (2001). Conserving biodiversity and ecosystem services. *Science, 291*(5511), 2047.
Balvanera, P., Siddique, I., Dee, L., Paquette, A., Isbell, F., Gonzalez, A., Byrnes, J., O'Connor, M. I., Hungate, B. A., & Griffin, J. N. (2014). Linking biodiversity and ecosystem services: Current uncertainties and the necessary next steps. *BioScience, 64*(1), 49–57. https://doi.org/10.1093/biosci/bit003
Barbier, E. W. (1987). The concept of sustainable economic development. *Environmental Conservation, 14*(2), 101–110.
Bardi, U. (2017). *The Seneca Effect. Why growth is slow, but collapse is rapid* (Vol. XII, p. 203). Springer.
Bark, R. H., Coloff, M. J., MacDonald, D. H., Pollino, C. A., Jackson, S., & Crossman, N. D. (2016). Integrated valuation of ecosystem services obtained from restoring water to the environment in a major regulated river basin. *Ecosystem Services, 22B*, 381–391.
Barnett, H. J., & Morse, C. (1963). *Scarcity and growth: The economics of natural resource availability*. Johns Hopkins University Press.
Barnhill, D. L. (1997). Great earth Sangha: Gary Snyder's view of nature as community. In M. E. Tucker & D. R. Williams (Eds.), *Buddhism and ecology: The interconnection of dharma and deeds*. Harvard University Press. Buddhism and Ecology by Donald Swearer.
Barnosky, A. D., Hadly, E. A., Bascompte, J., Berlow, E. L., Brown, J. H., Fortelius, M., et al. (2012). Approaching a state shift in earth's biosphere. *Nature, 486*(7401), 52–58.
Bartelmus, P. (2013). *Sustainability economics*. Routledge.
Barton, D. N., Andersen, T., Bergland, O., Engebretsen, A., Moe, S. J., Orderud, G. I., Tominaga, K., Romstad, E., & Vogt, R. D. (2016). Eutropia: Integrated valuation of lake eutrophication abatement decisions using a Bayesian belief network. Chap. 14. In Z. P. Niel (Ed.), *Handbook of applied systems science*. Routledge. https://doi.org/10.4324/9781315748771.ch14
Bar-Yam, Y. (1997). *Dynamics of complex systems*. Perseus Books.
Bates, M. (1960). *The Forest and the Sea: The Economy of Nature and Ecology*. Alfred A. Knopf Inc. and Random House Inc.
Bateson, G. (1972). *Steps to an ecology of mind: Collected essays in anthropology, psychiatry, evolution and epistemology*. Jason Aronson.
Baveye, P. C., Baveye, J., & Gowdy, J. (2013). Monetary valuation of ecosystem services: It matters to get the timeline right. *Ecological Economics, 95*, 231–235. https://doi.org/10.1016/j.ecolecon.2013.09.009
Baxter, W. (1974). *People or penguins: The case for optimal pollution* (p. 1974). Columbia University Press.
Beck, D., & Cowan, C. (2005). *Spiral dynamics: Mastering values, leadership, and change* (42829th ed.). Wiley-Blackwell.
Beck, D. E., Cowan, C. C., & Dynamics, S. (1996). *Mastering values leadership, and change*. Blackwell. "The Hierarchy of Service is but the fulfillment of the Higher Will.".

Berkes, F., & Folke, C. (1992). A system perspective on the interrelations between natural, human-made and cultural capital. *Ecological Modelling, 56*(1–8), 1992.

Bernie, S. (2020). *Our revolution: A future to believe in*. Thomas Dunne Books.

Berry, T. (1990a). *The Dream of the Earth*. Sierra Club.

Berry, T. (1990b). *The great work: Our way into the future* (Vol. 4, pp. 104–105). Bell Tower, 1999.

Berry, T. (1999). *The great work: Our way into the future*. New York Harmony/Bell Tower, 1999. https://thomasberry.org/quote/the-great-work-our-way-into-the-future/

Berryman, A. A., & Millstein, J. A. (1989). Are ecological systems chaotic—and if not, why not? *Trends in Ecology & Evolution, 4*(1), 26–28.

Bertalanffy, L. V. (1968). General systems theory as integrating factor in contemporary science. *Akten des XIV. Internationalen Kongresses für Philosophie, 2*, 335–340.

Bertalanffy, L. V. (1972). The history and status of general system theory. In G. J. Klir (Ed.), *Trends in general system theory* (pp. 21–41).

Bhandari, M. P. (2019). Sustainable development: Is this paradigm the remedy of all challenges? Does its goals capture the essence of real development and sustainability? *With Reference to Discourses, Creativeness, Boundaries and Institutional Architecture, SocioEconomic Challenges, 3*(4), 2019. https://doi.org/10.21272/sec.3(4).97-128.2019

Bhikkhu, B. (1985). Heartwood from the Boo Tree. Bangkok, United States Overseas Mission Foundation, 1985. Buddhasasanik Kap Kan Anurak Thamachat, pp. 34–45.

Bishop, R. (1978). Endangered species and uncertainty: The economics of safe minimum standards. *American Journal of Agricultural Economics, 60*, 1978.

Block, N. (1995). On a confusion about a function of consciousness. *Behavioral and Brain Sciences, 18*(2), 227–247.

Blumm, M. C., & Guthrie, R. D. (2012). Internationalizing the public trust doctrine: Natural law and constitutional and statutory approaches to fulfilling the Saxion vision. *UC Davis Law Review, 45*, 741–808.

Boden, T., Marland, G., & Andres, R. J. (2017). *Global, regional, and national fossil-fuel CO2 emissions (1751–2014) (V. 2017)*. Carbon Dioxide Information Analysis Center (CDIAC), Oak Ridge National Laboratory (ORNL). https://doi.org/10.3334/CDIAC/00001_V2017

Bonan, G. B., Pollard, D., & Thompson, S. L. (1992). Effects of boreal forest vegetation on global climate. *Nature, 359*, 716–718.

Bonner, J. T. (1980). *The evolution of culture in animals*. Princeton University Press.

Bonnet, M. (2017). Environmental consciousness, sustainability, and the character of philosophy of education. *Studies in Philosophy and Education, 36*(3), 333–347.

Bormann, F. H. (1976). An inseparable linkage: Conservation of natural ecosystems and conservation of fossil energy. *Bioscience, 26*, 759.

Boucher, D. S., James, S., & Keeler, K. H. (1982). The ecology of mutualism. *Annual Review of Ecology and Systematics, 13*, 315–347.

Boulding, K. E. (1993). The economics of the coming spaceship earth. In H. E. Daly & N. Kenneth (Eds.), *Townsend edited valuing the earth: Economics, ecology, ethics*. The MIT Press.

Boyden, S. (1987). *Western civilization in biological perspective: Patterns in biohistory*. Oxford University Press.

Boyden, S., & Dovers, S. (1992). Natural resource consumption and its environmental impacts in western world: Impacts of increasing per capita consumption. *Ambio, 21*(1), 63–69.

Brennan, A., & Lo, Y. S. (2020). Environmental Ethics. In E. N. Zalta (Ed.) *The Stanford encyclopedia of philosophy* (Winter 2020 Edition), Retrieved from https://plato.stanford.edu/archives/win2020/entries/ethics-environmental/

Brinck, P., Nilson, L. M., & Sevedin, U. (1988). Ecosystem redevelopment. *Ambio, 17*(2), 84–89.

Brown, L. (2008). *Plan B 3.0*. Norton.

Brown, G., Kunin, G. M., Kunin, W. E., & Swierbinski, J. E. (Eds.). (n.d.). University of Washington Press, Seattle.

Brundtland, G. H. (1987). Report of the world commission on environment and development: Our common future.

Bryner, J. (2012, April 4). *The reality of climate change, live science managing editor.*
Buis, A. (2011). *Climate change may bring big ecosystem changes.* https://climate.nasa.gov/news/645/climate-change-may-bring-big-ecosystem-changes/
Burton, B. (2016). *Many species now going extinct may vanish without a fossil trace.* https://today.uic.edu/many-species-now-going-extinct-may-vanish-without-a-fossil-trace
Butler, R. A. (2020). *Rainforest Information.* https://rainforests.mongabay.com/
Cadotte, M. W., Carscadden, K., & Mirotchnick, N. (2011). Beyond species: Functional diversity and the maintenance of ecological processes and services. *Journal of Applied Ecology, 48*(5), 1079–1087. https://doi.org/10.1111/j.1365-2664.2011.02048.x
Cahen, H. (1988a). Against the moral considerability of ecosystems. *Environmental Ethics, 10,* 196–216.
Cahen, H. (1988b). Against the moral considerability of ecosystems. *Environmental Ethics, 10,* 203–220.
Cairns, J., Jr. (1988). Increasing diversity by restoring damaged ecosystem. In E. O. Wilson (Ed.), *Biodiversity* (pp. 230–251). National Academy Press, xiii + 521 pp., illustr., 1988.
Calaprice, A. (2000). *Albert Einstein, the expanded quotable Einstein* (p. 316). Princeton University Press.
Caldwell, L. N. (1987). Moral thrust: The attainable ideal in human and ecological terms, pp.? In N. Polunin & J. H. Brucenett (Eds.), *Maintenance of the biosphere: Proceeding of the third international conference on environmental future.* St. Martin's Press, vi + 228 pp. illustr., 1987.
Callicott, B. (1989). *In defense of the land ethic: Essays in environmental philosophy.* State University Press of New York Press.
Callicott, J. B. (1992). Can a theory of moral sentiments support a genuinely normative environmental ethic? *Inquiry, 35*(2), 183–198.
Callicott, J. B. (1995). The value of ecosystem health. *Environmental Values, 4*(4), 345–361.
Callicott, B. (2006). Explicit and implicit values. In J. Scott, D. Goble, & F. Davis (Eds.), *The endangered species act at thirty: Conserving biodiversity in human-dominated landscapes* (Vol. II, pp. 36–48). Island Press.
Callicott, J. B. (2013). *Thinking like a planet: The land ethic and the earth ethic.* Oxford University Press.
Callicott, J. B., Crowder, L. B., & Mumford, K. (1999). Current normative concepts in conservation. *Conservation Biology, 13*(1), 22–35.
Camazine, S., Deneubourg, J. L., Franks, N. R., Sneyd, J., Theraulaz, G., & Bonabeau, E. (2001). *Self-organization in biological systems.* Princeton University Press.
Campbell, J. (1986). *The inner reaches of outer space: Metaphor as myth and religion* (p. 17). Alfred van der Mark Editions/St. James's Press Ltd.
Campero, A. (2018). *Genes vs cultures vs consciousness: A brief story of our computational minds,* Middletown, DE.
Campero, A. (2020). *Genes vs cultures vs consciousness: A brief story of our computational minds.* Massachusetts Institute of Technology (MIT).
Capra, F. (1990). The crisis of perception. *The Futurist, 24*(1), 64.
Capra, F. (1991). *The turning point: Science, society and the rising culture.* Bantam Books.
Capra, F. (1993). A systems approach to the emerging paradigm. In *The new paradigm in business: Emerging strategies for leadership and organizational change* (pp. 230–237). Tarcher Books.
Capra, F. (1996). *The web of life.* Anchor Books (A Division of Bantam Dell Publishing Group, Inc).
Capra, F. (1999). *Systems theory and the new paradigm in Carolyn Merchant edited key concepts in critical theory: Ecology.* Humanity Books.
Capra, F. (2002). *Hidden connections.* Doubleday.
Capra, F. (2014). *Il Tao della fisica.* Adelphi Edizioni spa.
Capra, F., & Luisi, P. L. (2014). *The systems view of life: A unifying vision.* Cambridge University Press.
Capra, F., & Steindl-Rast, D. (1991). Bookend: Belonging to the universe. *Business Ethics: The Magazine of Corporate Responsibility, 5*(6), 38–38.

Carleton, T., & Greenstone, M. (2021). *Updating the United states government's social cost of carbon*. University of Chicago, Becker Friedman Institute for Economics Working Paper, (2021-04).

Carmody, J., & Carmody, D. L. (1993). *Native American religions: An introduction*. Paulist Press.

Carpenter, S. R., Mooney, H. A., Agard, J., Capistrano, D., DeFries, R. S., Díaz, S., Dietz, T., Duraiappah, A. K., Oteng-Yeboah, A., Miguel, H., & Sterling, S. R. (1990). Towards an ecological worldview. In J. R. Engel & J. G. Engel (Eds.), *Ethics of environment and development*. University of Arizona Press.

Carpenter, S. R., DeFries, R., Dietz, T., Mooney, H. A., Polasky, S., Reid, W. V., & Scholes, R. J. (2006). Millennium ecosystem assessment: Research needs. *Science, 314*(5797), 257–258.

Carpenter, S. R., Mooney, H. A., Agard, J., Capistrano, D., Defries, R. S., Díaz, S., Dietz, T., Duraiappah, A. K., Oteng-Yeboah, A., Pereira, H. M., Perrings, C., Reid, W. V., Sarukhan, J., Scholes, R. J., & Whyte, A. (2009). Science for managing ecosystem services: Beyond the millennium ecosystem assessment. *Proceedings of the National Academy of Sciences of the United States of America, 106*(5), 1305–1312. https://doi.org/10.1073/pnas.0808772106. Epub 2009 Jan 28. PMID: 19179280; PMCID: PMC2635788.

Carrington, D. (2018, December 5). Brutal News: Global Carbon Emissions Jump to All-Time High in 2018, *The Guardian*.

Carrington, D. (2019). Environment editor: Mon 4 Feb 2019 06.45 EST Last modified on Mon 4 Feb 2019 13.15 EST.

Carrington, D. (2020a). Coronavirus: "Nature is sending us a message", says UN environment chief. In *The Guardian* [Internet]. The Guardian; 25 Mar 2020. Available: https://www.theguardian.com/world/2020/mar/25/coronavirus-nature-is-sending-us-a-message-says-un-environment-chief

Carrington, D. (2020b). Pandemics result from destruction of nature, say UN and WHO. In *The Guardian Guardian*; 17 June 2020. Available: https://www.theguardian.com/world/2020/jun/17/pandemics-destruction-nature-un-who-legislation-trade-green-recovery

Carson, R. (1962). *Silent spring*. Houghton, Mifflin.

Carson, H. W. (1990). *The global ecology hand book*. Beacon Press.

Carter, J. (2017a). The awkward relationship between Homo sapiens and Planet Earth. *Colorado Native Plant Society newsletter Aquilegea, 41*(5) Fall 2017.

Carter, J. (2017b). *Colorado Native Plant Society newsletter Aquilegea* (vol. 41(5) Fall 2017).

Cashford, J., Gaia, & Mundi, A. (2017). In L. Vaughan-Lee (Ed.), *Spiritual ecology – The cry of earth (second edition)*. The Golden Sufi Center. www.goldensufi.org

Catton, W. R. (1987). The world's most polymorphic species: Carrying capacity transgressed two ways. *Bioscience, 37*, 413–419.

CBD. (2010). *CBD report on convention on biological diversity, key international instrument for sustainable development*.

CDC. (2013). *Centers for Disease Control and Prevention (2013) Health-Related Quality of Life (HRQOL): Well-Being Concepts*. http://www.cdc.gov/hrqol/wellbeing.htm. Accessed 4 May 2015.

Chan, K. M. A., Balvanera, P., Benessaiah, K., Chapman, M., Díaz, S., Gómez-Baggethun, E., Gould, R. K., Hannahs, N., Jax, K., Klain, S. C., Luck, G., Martín-López, B., Muraca, B., Norton, B., Ott, K., Pascual, U., Satterfield, T., Tadaki, M., Taggart, J., & Turner, N. J. (2016). Why protect nature? Rethinking values and the environment. *PNAS, 113*, 1462–1465. https://doi.org/10.1073/pnas.1525002113

Chapin, F. S., Power, M. E., Pickett, S. T. A., Freitag, A., Reynolds, J. A., Jackson, R. B., Lodge, D. M., Duke, C., Collins, S. L., Power, A. G., & Bartuska, A. (2011). Earth stewardship: Science for action to sustain the human-earth system. *Ecosphere, 2*(8), 1–20. https://doi.org/10.1890/ES11-00166.1

Chapman, A. D. (2009). *Number of living species in Australia and the world* (2nd ed.). Australian Biodiversity Information Service. https://www.dcceew.gov.au/sites/default/files/env/pages/2ee3f4a1-f130-465b-9c7a-79373680a067/files/nlsaw-2nd-complete.pdf

Chawla, L. (2006). Learning to love the natural world enough to protect it. *Barn–forskning om barn og barndom i Norden, 2*(2006), 57–78.

Cherrett, J. M. (1989). The contribution of ecology to an understanding of the natural world. In J. M. Cherrett (Ed.), *Ecological concepts*. Blackwell Scientific Publ.

Chew, G. (1999). In L. Brink, R. C. Brower, & K. K. Phua (Eds.), *Architect of bootstrap*. World Scientific Publishing Pvt. Ltd.

Chomsky, N., & Pollin, R. (2020). *Climate crisis and the global green new deal: The political economy of saving the planet*. Verso Books.

Ciriarcy-Wantrup, S. V. (1959). *Resource conservation: Economics and politics*. University of California Division of Agricultural Services.

Clark, T. W., & Harvey, A. H. (1988). *Management of the Greater Yellow Stone Ecosystem: An annotated bibliography*. Northern Rockies Conservation Cooperative.

Clark, T. W., & Zaunbrecher, D. (1987). The greater yellow stone ecosystem: The ecosystem concept in natural resource policy and management. *Renewable Resource Journal, Summer*, 8–16.

Clements, F. E. (1916). *Plant succession: An analysis of the development of vegetation*. Carnegie Institution of Washington.

Cliff. (1990). Historical perspectives on sustainable development. *Environment, 32*(9), 4–9.

Cloud, P. E. (1988). Gaia modified. *Science, 240*, 1716.

Club of Rome. (n.d.). Springer International Publishing.

Coate, A. R., & Rosati, J. A. (1988). *The power of human needs in world society*. Lunne Reinner Publisher. xi + 283 pp., illustr. 1988.

Coate, A. R., & Rosati, J. A. (1988). The basic human needs. In Coate and Rosai (Ed.), *The power of human needs in world society*. Lunne Reinner Publisher.

Cohen, P., & Polunin, N. (1990). The viable culture. *Environmental Conservation, 17*(1), 3–6.

Cole, D. C. (1998). Human health and agroecosystem: Making the links. New directions in animal production systems. In *Proceeding of the Annual Meeting of Canadian Society of Animal Science*, July 5–8, 1998, Vancouver, British Columbia, Canada.

Collins, S. (1982). *Selfless persons: Imagery and thought in Theravada Buddhism*. Cambridge University Press.

Connell, J. H., & Slatyer, R. O. (1977). Mechanisms of succession in natural communities and their role in community stability and organization. *The American Naturalist, 111*(982), 1119–1144.

Cook, F. (1989). The jewel net of Indra. In J. B. Callicott & R. T. Ames (Eds.), *Nature in Asian traditions of thought: Essays in environmental philosophy* (pp. 213–229). SUNY Press.

Cook, W. A. (2008). *Issues in bioethics and concepts of scale*. Peter Lang.

Cook, J. (2010). *2009 2nd hottest year on record while Sun is coolest in a century, Skeptical Science*. https://skepticalscience.com/2009-2nd-hottest-year-on-record-sun-coolest-in-a-century.html

Corbett, J. (1981). Refugees from terror. *Fellowship, 47*(9), 14.

Corzilius, D. B. (1992). Conserving biological diversity in agricultural/forestry systems. *Bioscience, 42*(5), 354–362.

Costa, M. H., & Pires, G. F. (2010). Effects of Amazon and Central Brazil deforestation scenarios on the duration of the dry season in the arc of deforestation. *International Journal of Climatology: A Journal of the Royal Meteorological Society, 30*(13), 1970–1979.

Costanza, R., & Mageau, M. (1999). What is a healthy ecosystem? *Aquatic Ecology, 33*, 105–115.

Costanza, R., d'Arge, R., de Groot, R., Farber, S., Grasso, M., Hannon, B., Limburg, K., Naeem, S., O'Neil, R. V., Paruelo, J., Raskin, R. G., Sutton, P., & van den Belt, M. (1997). The value of the world's ecosystem services and natural capital. *Nature, 387*, 253–260.

Costanza, R., Graumlich, L. J., & Steffen, W. (2007). *Sustainability or collapse? an integrated history and future of people on earth*. MIT Press.

Costanza, R., Kubiszewski, I., Ervin, D., Bluffstone, R., Boyd, J., Brown, D., et al. (2011). Valuing ecological systems and services. *F1000 Biology Reports, 3*, 14.

Costanza, D. P., Blacksmith, N., Coats, M. R., Severt, J. B., & DeCostanza, A. H. (2016). The effect of adaptive organizational culture on long-term survival. *Journal of Business and Psychology, 31*, 361–381.

Costanzaa, R., Groot, R., Sutton, P., Ploeg, S. d., Anderson, S. J., Kubiszewski, I., Ferber, S., & Turner, R. K. (2014). Changes in the global value of ecosystem services. *Global Environmental Change, 26*, 152–158. https://www.sciencedirect.com/science/article/pii/S0959378014000685

Cotgrove, S. F. (1982). *Catastrophe or cornucopia: The environmental, politics and the future*. Wiley, USA, xi + 154 pp., illustr., 1982.

Crick, F., & Koch, C. (1992). The problem of consciousness. *Scientific American, 1992*, 153–159.

Daily, G. C., & Matson, P. A. (2008). Ecosystem services: From theory to implementation. *Proceedings of the National Academy of Sciences USA, 105*, 9455–9456.

Daly, E. H. (1990). Towards some operational principles of sustainable development. *Ecological Economics, 2*, 1–6.

Daly, H. E. (1993). Sustainable growth: An impossibility theorem. In H. E. Daly & N. Kenneth (Eds.), *Townsend edited valuing the earth: Economics, ecology, and ethics*. The MIT Press.

Daly, H. E., & Cobb, B. J. (1989). *For the common good: Redirecting the economy towards community, the environment, and a sustainable future*. Beacon Press.

Daniel, J. G., & Kulasingam, A. (1974). Problems arising from large scale forest clearing for agricultural use. *Malaysian Forester, 37*, 152–160.

Davies, J. C. (1977). The priority of human needs and the stages of political development. *American Society for Political and Legal Philosophy, 17*(156-196), 1977.

Dawkins, R. (1976). *The selfish gene*. Oxford University Press.

De Groot, R. S., Wilson, M. A., & Boumans, R. M. J. (2002). A typology for the classification, description, and evaluation of ecosystem functions, goods and services. *Ecological Economics, 4*, 393–408.

Den Boer, P. J. (1986). The present status of competition exclusion principle. *Trends in Ecology & Evolution, 1*, 25–28.

Descartes, R. (1955). *The philosophical works of Descartes. [2 Vols.]*. Dover Publications.

Descartes, R. (1992). *Mind and body*. American Psychological Association.

Devall, B. (1992). Deep ecology and radical environmentalism. *Society and Natural Resources, 4*, 247–258.

Devall, B. (2001). The deep, long-range ecology movement 1960–2000—a review. *Ethics and the Environment, 6*(1), 18–41. http://www.jstor.org/stable/40339002

Díaz, S., Demissew, S., Joly, C., Lonsdale, W. M., & Larigauderie, A. (2015). A Rosetta stone for nature's benefits to people. *PLoS Biology, 13*(1), e1002040.

Dietrich, J. (1933). *Is universe friendly or unfriendly?* The First Unitarian Society 1526 Harmon Place Minneapolis, Minn. The Humanist Pulpit Series X IV No. 2.

Dirzo, R., Young, H. S., Galetti, M., Ceballos, G., Isaac, N., & Collen, B. (2014). Defaunation in the Anthropocene. *Science, 345*(6195), 401–406.

Dobzhansky, T. (1970). *Genetics of evolutionary processes*. Columbia University Press.

Dombois, D. M., Kartawinata, K., & Handley, L. L. (1983). Conservation of species and habitats. In R. A. Carpenter (Ed.), *Natural system for development: What planners need to know* (pp. 260–281). Macmillan Publishing Company. xii + 491 pp., illustr., 1983.

Donohue, I., Hillebrand, H., Montoya, J. M., Petchey, O. L., Pimm, S. L., Fowler, M. S., et al. (2016). Navigating the complexity of ecological stability. *Ecology Letters, 19*(9), 1172–1185.

Dunning, H. C. (1989). The public trust: A fundamental doctrine of American property law. *Environmental Law, 19*, 515.

Durning, A. B. (1989). Poverty and the environment: Reversing the downward spiral. *Worldwatch Paper, 92*, 1–83.

Durning, A. T. (1992). *How much is enough? The consumer society and the future of the earth. The Worldwatch environmental alert series*. W. W. Norton & Company.

Dussault, A. C. (2014). *Can Autopoiesis ground a response to the Selectionist critique of Ecocentrism?* Centre interuniversitaire de recherche sur la science et la technologie (CIRST).

Dyer, M. I., DeAngelis, D. L., & Post, W. M. (1986). A model of herbivore feedback in plant productivity. *Mathematical Biosciences, 79*, 171–184.

E360 Digest. (2018). *Nearly every ecosystem on the planet will be transformed by climate change*. https://e360.yale.edu/digest/nearly-every-ecosystem-on-the-planet-will-be-transformed-by-climate-change

Earth Charter International. (2000). *The Earth Charter*. (link).

Ebeling, W. (1989). *Chaos DOUBLEHYPHENOrdnungDOUBLEHYPHENinformation: Selbstorganisation in der Natur*. Frankfurt (M.).

Ebi, K. L., Balbus, J. M., Luber, G., Bole, A., Crimmins, A., Glass, G., Saha, S., Shimamoto, M. M., Trtanj, J., & White-Newsome, J. L. (2018). Human health. In D. R. Reidmiller, C. W. Avery, D. R. Easterling, K. E. Kunkel, K. L. M. Lewis, T. K. Maycock, & B. C. Stewart (Eds.), *Impacts, risks, and adaptation in the USGCRP (2018). United States: Fourth national climate assessment* (Vol. II, p. 544, 551–552). U.S. Global Change Research Program. https://doi.org/10.7930/NCA4.2018.CH14

Eckel, M. D. (1997). Is there a Buddhist philosophy of nature? In M. E. Tucker, D. Ryuken, & Williams (Eds.), *Buddhism and ecology – The interconnection of dharma and deeds* (p. 1997). Harvard University Press.

Eckersley, R. (1999). The failed promise of critical theory. In C. Merchant (Ed.), *Key concepts in critical theory: Ecology*. Humanity Books.

Ehrenfeld, D. (1978). *The arrogance of humanism* (p. 1978). Oxford University Press.

Ehrlich, A. (1992). *Healing the planet: Strategies for resolving the environmental crisis*. Surrey Beatty, and Sons.

Ehrlich, P. R., & Ehrlich, A. H. (1992). The value of biodiversity. *Ambio, 21*(3), 219–226.

Ehrlich, P. R., & Ehrlich, A. H. (1993). Why is not everybody as scared as we are? In H. E. Daly & K. N. Townsend (Eds.), *Valuing the earth: Economics, ecology, ethics*. The MIT Press.

Ehrlich, P. R., & Kennedy, D. (2005). Millennium assessment of human behavior. *Science, 300*(2005), 562–563.

Ehrlich, P. R., & Mooney, H. A. (1983). Extinction, substitution, and ecosystem services. *Bioscience, 33*, 248–254.

Ehrlich Paul, R., & Ehrlich, A. H. (1993). Why is not everyone as scared as we are? In H. E. Daly & K. N. Townsend (Eds.), *Valuing the earth: Economics, ecology, ethics*. The MIT Press.

Ehrlich Paul, R., & Wilson, E. O. (1991). Biodiversity studies: Science and policy. *Science, 253*, 758–762.

Ehrlich, P. R., Ehrlich, A. H., & Daily, G. C. (1993). Food security, population and environment. *Population and Development Review*, 1–32.

Einstein, A. (2000). In A. Calaprice (Ed.), *The expanded quotable Einstein* (p. 316). Princeton University Press.

Ellen McArthur Foundation. (2012). *Towards the Circular Economy*. Ellen McArthur Foundation. HTTPS://ARCHIVE.ELLENMACARTHURFOUNDATION.ORG/EXPLORE/THE-CIRCULAR-ECONOMY-IN-DETAIL. http://www.ellenmcarthurfoundation.org/assets/downloads/publications/Ellen-McArthur-Foundation-Towards-the-Circular-Economy-vol.1.pdf

Elliot, R. (1992). Intrinsic value, environmental obligation, and naturalness. *The Monist, 75*, 138–160.

Emmons, R. A., & Paloutzian, R. F. (2003). The psychology of religion. *Annual Review of Psychology, 54*(1), 377–402.

Engel, J. R. (1990). The ethics of sustainable development. In J. R. Engel & J. G. Engel (Eds.), *Ethics of environment and development, global challenge, international response*. University of Arizona Press.

Erickholm, E. P. (1975). The deterioration of mountain environments. *Science, 189*, 764–770.

Fann, N., Brennan, T., Dolwick, P., Gamble, J. L., Ilacqua, V., Kolb, L., Nolte, C. G., Spero, T. L., & Ziska, L. (2016). Air quality impacts. In *The impacts of climate change on human health in the United States: A scientific assessment* (pp. 69–98). U.S. Global Change Research Program. https://doi.org/10.7930/J0GQ6VP6

FAO. (1989). In FAO (Ed.), *Food and Agriculture Organization of the United Nations (FAO)*. The State of Food and Agriculture.

Field, T. (1985). Attachment as psychobiological attunement: Being on the same wavelength. In M. Reite & T. Field (Eds.), *Psychobiology of attachment and separation*. Academic Press.

Field, R., & Parrott, L. (2017). Multi-ecosystem services networks: A new perspective for assessing landscape connectivity and resilience. *Ecological Complexity, 32*, 31–41. https://doi.org/10.1016/j.ecocom.2017.08.004

Fischbach, G. D. (1992). Mind and brain. *Scientific American, 1992*, 48–57.

Fisher, B., Turner, R., & Morling, P. (2009). Defining and classifying ecosystem services for decision making. *Ecological Economics, 68*, 643–653. https://doi.org/10.1016/j.ecolecon.2008.09.014

Fleming, E., Payne, J., Sweet, W., Craghan, M., Haines, J., Hart, J. F., Stiller, H., & Sutton-Grier, A. (2018). Coastal effects. In D. R. Reidmiller, C. W. Avery, D. R. Easterling, K. E. Kunkel, K. L. M. Lewis, T. K. Maycock, & B. C. Stewart (Eds.), *Impacts, risks, and adaptation in the United States: Fourth national climate assessment* (Vol. II, pp. 322–352). U.S. Global Change Research Program. https://doi.org/10.7930/NCA4.2018.CH8

Folke, C., & Kaberger, T. (1991). *Linking the natural environment and economy: Essays from Eco-Eco Group*. Kluvier Academic Publ.

Folke, C., Maler, K. G., & Perring, C. (1992). Biodiversity loss: An introduction. *Ambio, 21*(3), 200.

Folke, C., Polasky, S., Rockstro¨m, J., Galaz, V., Westley, F., Lamont, M., Scheffer, M., O¨ Sterblom, H., Carpenter, S. R., Stuart Chapin, F., III, Seto, K. C., Weber, E. U., Crona, B. I., Daily, G. C., Dasgupta, P., Gaffney, O., Gordon, L. J., Hoff, H., Levin, S. A., Lubchenco, J., Steffen, W., & Walker, B. H. (2021). Our future in the anthropocene biosphere. *Ambio, 50*(2021), 834–869. https://doi.org/10.1007/s13280-021-01544-8

Ford, D. (1981). Is nature really motherly? *The Coevolution Quarterly, Spring, 198*, 58–63.

Foss, G. (2020). *COVID-19 Curves and Carrying Capacity*. https://medium.com/theparallel/covid-19-curves-carrying-capacity-3af204d72b1e

Fox, W. (1990). *Toward a transpersonal ecology: Developing new foundations for environmentalism*. Shambhala Publications, 1990; US reprint edition: New York: The State University of New York Press, 1995.

Fox, W. (1995). *Toward a transpersonal ecology: Developing new foundations for environmentalism*. SUNY Press.

Fox, W. (2006). *A theory of general ethics: Human relationships, nature, and the built environment*. The MIT Press.

Frechette, K. S. (1985). Environmental ethics and global imperatives. In R. Repetto (Ed.), *The global possible: Resources, development, and the new century* (pp. 97–127). Yale University Press. xv + 538 pp., illustr., 1985.

Fromm, E. (1977). *To have or to be* (Vol. 8, p. 137). Continuum.

Furtado, B. A., Fuentes, M. A., & Tessone, C. J. (2019). Modeling and applications: State-of-the-art and perspectives. *Complexity, 2019*, Article ID 5041681. https://doi.org/10.1155/2019/5041681

Gadgil, M. (1987). Diversity: Cultural and biological. *TREE, 2*(12), 369–373.

Galston, A. W. (1992). Photosynthesis as a basis for life support on earth and in space. *Bioscience, 42*(7), 490–493.

Gao, B. C., & Goetz, A. F. (1990). Column atmospheric water vapor and vegetation liquid water retrievals from airborne imaging spectrometer data. *Journal of Geophysical Research: Atmospheres, 95*(D4), 3549–3564.

Gardiner, S. M. (2004). Ethics and global climate change. *Ethics, 114*(3), 555–600.

Gee, H. (1992). The objective case for conservation. *Nature, 357*, 639–639.

Gell-Mann, M. (1994). *The quark and the jaguar: Adventures in the simple and the complex* (Vol. 47, p. 89). W. H. Freeman.

GEN. (2020). Geneva environment network: Geneva environment dialogues – The impact of COVID-19 on climate science. In *Geneva Environment Network. Geneva Environment Network*; 02 July 2020. Available: https://www.genevaenvironmentnetwork.org/events/geneva-environment-dialogues-the-impact-of-covid-19-on-climate-science/

Georgescu-Roegen, N. (1993). The entropy law and the economic problem. In H. E. Daly & N. Kenneth (Eds.), *Townsend edited valuing the earth: Economics, ecology, ethics*. The MIT Press.

Getty Images. https://www.gettyimages.com/detail/photo/textured-cracked-mud-landscape-iceland-royalty-free-image/1176594622?adppopup=true

Glenn, A. (2016). Exiting the Anthropocene and Entering the Symbiocene. *Minding Nature, 9*(2).

Gifford, R., & Nilsson, A. (2014). Personal and social factors that influence pro-environmental concern and behaviour: A review. *International Journal of Psychology, 49*(3), 141–157.

Glick, P. C. (1980). Remarriage: Some recent changes and variations. *Journal of Family Issues, 1*(4), 455–478.

Global Footprint Network (GFN). (2022). https://www.footprintnetwork.org/biodiversity/

Goerner, S., et al. (2009). Quantifying economic sustainability: Implications for free enterprise theory, policy, and practice. *Ecological Economics, 69*, 79.

Gómez-Baggethun, E., & Martín-López, B. (2015). Ecological economics perspective in ecosystem services valuation. In J. Martínez-Alier & R. Muradian (Eds.), *Handbook of ecological economics* (pp. 260–282). Edward Elgar.

Gómez-Baggethun, E., De Groot, R., Lomas, P. L., & Montes, C. (2010). The history of ecosystem services in economic theory and practice from early notions to markets and payment schemes. *Ecological Economics, 69*, 1209–1218. https://doi.org/10.1016/j.ecolecon.2009.11.007

Gómez-Baggethun, E., Martín-López, M., Barton, D., Braat, L., Saarikoski, H., Kelemen, M., et al. (2014). In E. Gómez-Baggethun, D. Barton, P. Berry, R. Dunford, & P. Harrison (Eds.), *EU FP7 OpenNESS Project Deliverable 4.1, State-of-the-art report on integrated valuation of ecosystem services*. European Commission.

Goodland, R., & Ledec, G. (1987). Neoclassical economics and principles of sustainable development. *Ecological Modelling, 38*, 19–46.

Goreau, T. J. (1990). Balancing atmospheric carbon dioxide. *Ambio, 19*(5), 230–235.

Gorney, J. E. (1979). The field of illusion in literature and the psychoanalytic situation. *Psychoanalysis and Contemp. Thought, 2*, 527.

Gould, S. J. (1982). Darwinism and the expansion of evolutionary theory. *Science, 216*, 380–387.

Gould, J. L. (1988). Timing of landmark learning by honey bees. *Journal of Insect Behavior, 1*, 373–377.

Gow, D. D. (1992). Poverty and natural resources: Principles for environmental management and sustainable development. *Environmental Impact Assessment Review, 12*, 49–65.

Gowdy, J. M. (1992). Economic growth versus the environment. *Environmental Conservation, 19*(2), 102–104.

Green, L. W., Richard, L., & Potvin, L. (1996). Ecological foundations of health promotion. *American Journal of Health Promotion, 10*(4), 270–281.

Greenfield, P. (2021, May 27). Investing 0.1% of global GDP could avoid breakdown of ecosystems, says UN report, *The Guardian*.

Griffin, D. R. (1976). *The question of animal awareness: Evolutionary continuity of mental experience*. Rockefeller University Press.

Gross, R. M. (1997). Buddhist resources for issues of population, consumption and the environment. In M. E. Tucker & D. R. Williams (Eds.), *Buddhism and ecology, the interconnection of Dharma and Deeds*. Harvard University Press.

Gross, N., Bagousse-Pinguet, Y. L., Liancourt, P., Berdugo, M., Gotelli, N. J., & Maestre, F. T. (2022). Biodiversity and ecosystem functioning relations are modified by environmental harshness. *Science, 371*(6530), 589–593.

Grossman, G., & Krueger, A. (1995). Economic growth and environment. *Quarterly Journal of Economics, 110*(2), 353–377.

Guerry, A. D., Polasky, S., Lubchenco, J., & Vira, B. (2015). Natural capital and ecosystem services informing decisions: From promise to practice. *PNAS, 112*(24), 7348–7355. https://doi.org/10.1073/pnas.150375111

Guha, R. (1989). Radical American environmentalism and wilderness preservation: A third world critique. *Environmental Ethics, 11*, 71–85.

Hall, C. A. S. (1990). *Quantifying sustainable development: The future of tropical economies*. Academic Press.

Hall, C. A. S. (2004). In W. E. Gibson (Ed.), *Sanctioning resource depletion: Economic development and neoclassical economics in eco-justice – The unfinished journey*. State University of New York Press.

Hansen, J., Sato, M., Kharecha, P., & von Schuckmann, K. (2011). Earth's energy imbalance and implications. *Atmospheric Chemistry and Physics, 11*, 13421–13449. https://doi.org/10.5194/acp-11-13421-2011

Harting, J. H., & Vallentyne, J. R. (1990). Use of an ecosystem approach to restore degraded areas of the Great Lakes. *Ambio, 19*(4), 47–56.

Haskell, B. D., Norton, B. G., & Costanza, R. (1992). *Ecosystem health: New goals for environmental management*. Island Press.

Hausfather, Z. (2017, November 13). *Analysis: Global CO2 emissions set to rise 2% in 2017 after three-year plateau, carbon brief*.

Havel, V. (1990). *Disturbing the peace*. Faber and Faber.

Havel, V. (1998). *Spirit of the earth*. Resurgence.

Hawking, S. (2017). *A brief history of time*. Bantam Books and House.

Hayhoe, K., Wuebbles, D. J., Easterling, D. R., Fahey, D. W., Doherty, S., Kossin, J., Sweet, W., Vose, R., & Wehner, M. (2018). Our changing climate. In D. R. Reidmiller, C. W. Avery, D. R. Easterling, K. E. Kunkel, K. L. M. Lewis, T. K. Maycock, & B. C. Stewart (Eds.), *Impacts, risks, and adaptation in the United States: Fourth national climate assessment* (Vol. II, p. 76). U.S. Global Change Research Program. https://doi.org/10.7930/NCA4.2018.CH2

Hegerl, G. C., F. W. Zwiers, P. Braconnot, N. P. Gillett, Y. Luo, J. A. Marengo Orsini, N. Nicholls, J. E. Penner, & P. A. Stott. (2007). Understanding and attributing climate change. In: *Chapter 9 in Climate change 2007: The physical science basis. Contribution of Working Group I to the Fourth Assessment Report of the Intergovernmental Panel on Climate Change*.

Heinen, J. T., & Low, R. S. (1992). Human behavioral ecology and environmental conservation. *Environmental Conservation, 19*(2), 105–116.

Heisenberg, W. (1958). *Physics and philosophy: The revolution in modern physics*. Harper & Row Publishers.

Heisenberg, W. (1971). *Physics and beyond: Encounters and conversations*. Harper Tochbooks.

Herring, S. C., Christidis, N., Hoell, A., Kossin, J. P., Schreck, C. J., III, & Stott, P. A. (2018). Explaining extreme events of 2016 from a climate perspective. *Bulletin of the American Meteorological Society, 99*, S1–S157.

Hertsgaard, M. (1993). *Earth odyssey*. Broadway Books.

Hertsgaard, M. (1999). *Earth odyssey: Around the world in search of our environmental future*. Broadway Books.

Hickel, J. (2020). *Less is more: How degrowth will save the world*. Windmill Books/Penguin Random House.

Hill, R., Dyer, G. A., Lozada-Ellison, L.-M., Gimona, A., Martin-Ortega, J., Munoz-Rojas, J., & Gordon, I. J. (2020). A social–ecological systems analysis of impediments to delivery of the Aichi 2020 Targets and potentially more effective pathways to the conservation of biodiversity. *Global Environmental Change, 34*(22-34), 2015. https://doi.org/10.1073/pnas.08087721

Hoffmann, D. L., et al. (2018). U-Th dating of carbonate crusts reveals Neandertal origin of Iberian cave art. *Science, 359*(6378), 912–915.

Holden, C. (1990). Multidisciplinary look at a finite world. *Science, 249*, 18–19.

Holdgate, M. (1984). UNEP: Some personal thoughts. *Mazingira, 17*, 20.

Holing, C. S. (1986). The resilience of terrestrial ecosystems: Local surprise and global change. In W. C. Clark & R. E. Munn (Eds.), *Sustainable development of biosphere*. Cambridge University Press.

Holland, J. H. (1998). *Emergence: From chaos to order*. Perseus Books.

Holling, C. S. (1973). Resilience and stability of ecological systems. *Annual Review of Ecology and Systematics, 4*(1), 1–23.

Hooper, D. U., Chapin, F. S., III, Ewel, J. J., Hector, A., Inchausti, P., Lavorel, S., Lawton, J. H., Lodge, D. M., Loreau, M., Naeem, S., Schmid, B., Setälä, H., Symstad, A. J., Vandermeer, J., & Wardle, D. A. (2005). Effects of biodiversity on ecosystem functioning: Consesus of current knowledge. *Ecological Monographs*. https://doi.org/10.1890/04-0922

HOPE. (2015). *Health of populations and ecosystems glossary of terms*. http://www.york.ac.uk/healthsciences/research/public-health/projects/hope/glossary/. Accessed 15 Jan 2015.

Horowitz, M. J. (1988). *Introduction to psychodynamics: A new synthesis*. Basic books.

Houghton, R. A. (1990). The future role of tropical forests in affecting the carbon dioxide concentration of the atmosphere. *Ambio, 19*(5), 204–208.

Howard, B. R. (2002). Of roaches, rats, and rattlesnakes, imaging the world, the third voice. In R. B. Tapp (Ed.), *Eco-humanism*. Prometheus Books.

Hsiang, S., & Kopp, R. E. (2018). An economist's guide to climate change science. *Journal of Economic Perspectives, 32*, 3–32.

Hsiang, S., Kopp, R., Jina, A., Rising, J., Delgado, M., Mohan, S., Rasmussen, D. J., Muir-Wood, R., Wilson, P., Oppenheimer, M., Larsen, K., & Houser, T. (2017). Estimating economic damage from climate change in the United States. *Science, 356*, 1362–1369.

Hsiang, S., Oliva, P., & Walker, R. (2019). The distribution of environmental damages. *Review of Environmental Economics and Policy, 13*, 83.

Hubbert, M. K. (1993). Exponential growth as transient phenomenon in human history. In H. E. Daly & N. Kenneth (Eds.), *Townsend edited valuing the earth: Economics, ecology, ethics*. The MIT Press.

Huntingford, C., & Mercado, L. M. (2016). High chance that current atmospheric greenhouse concentrations commit to warmings greater than 1.5 °C over land. *Scientific Reports, 6*, 30294.

ICIMOD. (2020). *Ministerial declaration on the HKH call to action*. https://www.icimod.org/wp-content/uploads/2020/11/20201015_Declaration_Signed_MinisterialMountainSummit_ICIMOD.pdf

ICIMOD. (2021). *HKH2Glasgow: An urgent call for climate action for the Hindu Kush Himalaya*. file:///C:/Users/Microcenter/Downloads/HimalDoc2021_CollectiveHKH2Glasgow.pdf

ICVA. (1988). International Council of Voluntary Agencies and Rome Development Forum, Report.

IDRC. (1998, January 14). *Ecosystem approaches to human health*. International Development Research Center. ecohealth@idrc.ca

IEA. (2016, March 16). *Decoupling of global emissions and economic growth confirmed*. International Energy Agency.

IFPRI. (1995). *International Food Policy Research Institute (IFPRI) 1995. A 2020 vision for food, agriculture, and the environment. The vision, the challenge, and recommended action*. IFPRI.

IISE. (2011). *International Institute for Species Exploration (IISE)*. https://www.coursehero.com/file/p785drg/1-International-Institute-for-Species-Exploration-IISE-2011-State-of-Observed/

IPBES. (2015). The Intergovernmental Platform on Biodiversity and Ecosystem Services (IPBES). Conceptual Framework — connecting nature and people. *Current Opinion in Environmental Sustainability, 2015*(14), 1–16.

IPBES. (2020). Workshop report on biodiversity and pandemics of the intergovernmental platform on biodiversity and ecosystem services. In P. Daszak, C. das Neves, J. Amuasi, D. Hayman, T. Kuiken, B. Roche, C. Zambrana-Torrelio, P. Buss, H. Dundarova, Y. Feferholtz, G. Foldvari, E. Igbinosa, S. Junglen, Q. Liu, G. Suzan, M. Uhart, C. Wannous, K. Woolaston, P. Mosig Reidl, K. O'Brien, U. Pascual, P. Stoett, H. Li, & H. T. Ngo (Eds.), *IPBES secretariat*. https://doi.org/10.5281/zenodo.4147317

IPCC. (2007). Climate change, IPCC fourth assessment report. *The Physical Science Basis, 2*, 580–595.

IPCC. (2014a). Summary for policymakers. In C. B. Field, V. R. Barros, D. J. Dokken, K. J. Mach, M. D. Mastrandrea, T. E. Bilir, M. Chatterjee, K. L. Ebi, Y. O. Estrada, R. C. Genova, B. Girma, E. S. Kissel, A. N. Levy, S. MacCracken, P. R. Mastrandrea, & L. L. White (Eds.), *Climate change 2014: Impacts, adaptation, and vulnerability. Part A: Global and sectoral aspects.*

Contribution of working group II to the fifth assessment report of the intergovernmental panel on climate change. Cambridge University Press.

IPCC. (2014b). Climate change 2014: Synthesis report. Contribution of Working Groups I, II and III to the Fifth Assessment Report of the Intergovernmental Panel on Climate Change.

IPCC. (2018). *Global Warming of 1.5°C. Special Report.*

Isbell, F., Craven, D., Connolly, J., Loreau, M., Schmid, B., Beierkuhnlein, C., et al. (2017). Biodiversity increases the resistance of ecosystem productivity to climate extremes. *Nature, 526*(7574), 574–577.

IUCN and WWF. (1980). *World conservation strategy: Living resource conservation for sustainable development.*

Jackson, T. (2017). *Prosperity without growth: Foundations for the economy of tomorrow, second edition*. Routledge.

Jackson, R. B., Canadell, J. G., Le Quéré, C., Andrew, R. M., Korsbakken, J. I., Peters, G. P., & Nakicenovic, N. (2016). Reaching peak emissions. *Nature Climatic Change, 6*, 7–10.

Jackson, R. B., Le Quéré, C., Andrew, R. M., Canadell, J. G., Peters, G. P., Roy, J., & Wu, L. (2017). Warning signs for stabilizing global CO2 emissions environ. *Research Letters, 12*, 110–202.

Jacobs, S., Dendoncker, N., Barton, D., & Gomez-Baggethun, E. (2015). Integrated valuation of ecosystem services in science-policy practice. In *Proceedings of the 8th conference of the ecosystem services partnership, 9-13th November 2015, Stellenbosch, South Africa*. http://www.espconference.org/espconference2015

Jacobs, S., Dendonckerb, N., Martín-Lópezc, B., Bartone, D. N., Gomez-Baggethund, E., Boeraevef, F., McGrathg, F. L., Vierikkoh, K., Genelettii, D., Seveckej, K. J., Pipartb, N., Primmerk, E., Mederlyl, P., Schmidtm, S., Aragãoo, A., Baralp, H., Barkq, R. H., Bricenor, T., Brognab, D., Cabrals, P., De Vreeset, R., Liqueteu, C., Muellerv, H., PehKelvin, S.-H., Phelany, A., Rincónz, A. R., Rogersaa, S. H., Turkelbooma, F., ReethWouter, V., van ZantenBoris, T., Wamad, H. K., & Washbournae, C.-L. (2016). Ecosystem services: A new valuation school: Integrating diverse values of nature in resource and land use decisions. *Ecosystem Services, 22*(2016), 213–220.

Janssens-Maenhout, G., Crippa, M., Guizzardi, D., Muntean, M., Schaaf, E., Olivier, J. G., et al. (2017). *Fossil CO2 & GHG emissions of all world countries* (Vol. 107877). Publications Office of the European Union.

Johnson, L., & Morally, A. (1991). *Deep world.* Cambridge University Press.

Jones, A. K. (1990). Social symbiosis: A gaian critique of contemporary social theory. *The Ecologist, 20*(3), 108–111.

Kabilsingh, C. (1990). *Buddhist monks and forest conservation in radical conservation: Buddhism in contemporary world* (pp. 301–311). Sathirakoses-Nagapradipa Foundation.

Kai, M., Chan, A., Rebecca Shaw, M., Cameron, D. R., Underwood, E. C., & Daily, G. C. (2006). *Planning for conservation of ecosystem services*, Published: October 31, 2006. https://doi.org/10.1371/journal.pbio.0040379

Katz, E. (1992). The call of the wild. *Environmental Ethics, 14*, 265–273.

Kauffman, S. (1990). Spontaneous order, evolution, and life. *Science, 247*, 1543–1544.

Kauffmann, G. B. (1991). Biosphere pioneer. *The World & I, 6*, 316–323.

Kay, J. J., & Schneider, E. (1994). Embracing complexity: The challenge of the ecosystem approach. *Alternatives, 20*(3), 32–39.

Kelemen, E., Barton, D., Jacobs, S., Martín-López, B., Saarikoski, H., Termansen, G., Bela, L., Braat, R., Demeyer, M., García-Llorente, E., Gómez-Baggethun, J., Hauck, H., Keune, S., Luque, I., Palomo, G., Pataki, M., Potschin, C., Schleyer, P., Tenerilli, F., & Turkelboom. (2015). In S. C. Klain, T. A. Satterfield, & K. M. Chan (Eds.), *Preliminary guidelines for integrated assessment and valuation of ecosystem services in specific policy contexts EU FP7 OpenNESS Project Deliverable 4.3.* (p. 2014). European Commission FP7.

Kellert, S. R., & Speth, J. G. (2009). The coming transformation: Values to sustain human and natural communities. *Yale School of Forestry & Environmental Studies.*

Kerr, R. A. (1988). Nolonger willful, Gaia becomes respectable. *Science, 240*, 393–395.

Kettel, B. (1996). Women, health and environment. *Social Science Medicine, 42*, 1367–1379.

Keune, H., Kretsch, C., de Blust, G., Gilbert, M., Flandroy, L., Van den Berge, K., Versteirt, V., Hartig, T., de Keersmaecker, L., Eggermont, H., Brosens, D., Dessein, J., Vanwambeke, S., Prieur-Richard, A. H., Wittmer, H., Van Herzele, A., Linard, C., Martens, P., Mathijs, E., Simoens, I., Van Damme, P., Volckaert, F., Heyman, P., & Bauler, T. (2013). Science–policy challenges for biodiversity, public health and urbanization: Examples from Belgium. *Environmental Research Letters, 8*, 025015. https://doi.org/10.1088/1748-9326/8/2/025015

Khras, H. (2020). *What to do about the coming debt crisis in developing countries*. https://www.brookings.edu/blog/future-development/2020/04/13/what-to-do-about-the-coming-debt-crisis-in-developing-countries/#:~:text=Emerging%20markets%20and%20developing%20countries,debt%20service%20due%20in%202020.&text=Around%20%241%20trillion%20is%20debt,which%20is%20normal%20trade%20finance

Kinsley, D. (1995). *Ecology and religion: Ecological spirituality in cross-cultural perspective*. Prentice Hall.

Klöckner, C. A. (2013). A comprehensive model of the psychology of environmental behaviour—A meta-analysis. *Global Environmental Change, 23*(5), 1028–1038.

Koch, C. (2018). What is consciousness? *Scientific American, 318*(6), 60–64. https://doi.org/10.1038/scientificamerican0618-60

Kolasi, E. (2017). *Energy, economic growth, and ecological crisis by Erald Kolasi*. https://monthlyreview.org/2019/06/01/energy-economic-growth-and-ecological-crisis/

Kolasi, E. (2018). The physics of capitalism. *Monthly Review, 70*(1), 29–43.

Kolodziej, E. A. (1991). The cold war as cooperation. *Bulletin of the American Academy of Arts and Science, 44*(7), 9–39.

Korten, D. (2006). The great turning yes! *A Journal of Positive Future, Summer, 2006*, 16.

Korten, D. (2007). *The great turning: From empire to earth community*. Berrett-Koehler Publishers.

Kropotkin, P. A. (1902). *Organised Vengeance Called'Justice* (Vol. 14). "Freedom" Office.

Krugman, P. (2009, June 28). Betraying the planet, *New York Times*.

Krugman. (2022). *Credible irresponsibility revisited*. https://www.gc.cuny.edu/sites/default/files/2022-03/Krugman-Credible-Irresponsibility-Revisited.pdf

Kuhn, T. S. (1962). *The structure of scientific revolutions*. University of Chicago Press.

Kuznets, S. (1955). Economic growth and income inequality. *American Economic Review, 45*(1), 1–28.

Kuznets, S. (1973). Modern economic growth: Findings and reflections. *The American Economic Review, 63*(3), 247–258.

Laitman, M. (2006). *From Chaos to Harmony: The solution to the global crisis according to the wisdom of Kabbalah*. Laitman Kabbalah Publishers.

Lall, U., Johnson, T., Colohan, P., Aghakouchak, A., Brown, C., McCabe, G., Pulwarty, R., & Sankarasubramanian, A. (2018). Water. In D. R. Reidmiller, C. W. Avery, D. R. Easterling, K. E. Kunkel, K. L. M. Lewis, T. K. Maycock, & B. C. Stewart (Eds.), *Impacts, risks, and adaptation in the United States: Fourth national climate assessment* (Vol. II, pp. 145–173). U.S. Global Change Research Program. https://doi.org/10.7930/NCA4.2018.CH3

Lappe, F. M. (2009). *Liberation ecology*. Resurgence.

Laszlo, E. (1978). Evolution und Invarianz in der Sicht der Allgemeinen Systemtheorie. In H. Lenk & G. Rohpohl (Eds.), *Systemtheorie als Wissenschaftsprogramm* (pp. 221–238). Konigstein.

Lawrence, E. J. (1992). Towards a moral considerability of species and ecosystems. *Environmental Ethics, 14*(2), 145–157.

Le Quéré, C., Jackson, R. B., Jones, M. W., Smith, A. J. P., Abernethy, S., Andrew, R. M., et al. (2020). Temporary reduction in daily global CO2 emissions during the COVID-19 forced confinement. *Nature Climate Change, 10*(7), 647–653.

Lenton, T. M. (2010). Tipping elements in the Earth's climate system. *Proceedings of the National Academy of Sciences, 105*(6), 1786–1793.

Leopold, A. (1949). *A Sand County Almanac and sketches here and there*. Oxford University Press.

Leopold, A. (1966). *A Sand County Almanac: With essays on conservation from Round River: The land ethic* (pp. 237–264). Ballantine Books.

Leopold, A. (1990). *Thinking like a mountain*. Lone Goose Press.

Levchenko, V. F., Kazansky, A. B., & Sabirov, M. A. (2017). Development of the biosphere in the context of some fundamental inventions of biological evolution. In *Evolutionary physiology and biochemistry - advances and perspectives*. https://www.intechopen.com/chapters/59044. https://doi.org/10.5772/intechopen.73297

Levin, S. A. (2000). Multiple scales and the maintenance of biodiversity. *Ecosystems, 3*(6), 498–506.

Levin, S. S., & Pimentel, D. (1981). Selection of intermediate rates of increase in parasites host systems. *The American Naturalist, 117*, 308–315.

Levin, S. A., Barrett, S., Aniyar, S., Baumol, W., Bliss, C., Bolin, B., et al. (1998). Resilience in natural and socioeconomic systems. *Environment and Development Economics, 3*(2), 221–262.

Levins, R. (1974). Discussion paper: The qualitative analysis of partially specified systems. *Annals of the New York Academy of Sciences, 231*(1), 123–138.

Lewin, R. (1989). Sources and sink complicate ecology. *Science, 243*, 477–478.

Lin, D., Wambersie, L., & Wackernagel, M. (2021). *Estimating the Date of Earth Overshoot Day 2021*. Global Footprint Network. https://www.overshootday.org/content/uploads/2021/07/Earth-Overshoot-Day-2021-Nowcast-Report.pdf

Lipton, D., Rubenstein, M. A., Weiskopf, S. R., Carter, S., Peterson, J., Crozier, L., Fogarty, M., Gaichas, S., Hyde, K. J. W., Morelli, T. L., Morisette, J., Moustahfid, H., Muñoz, R., Poudel, R., Staudinger, M. D., Stock, C., Thompson, L., Waples, R., & Weltzin, J. F. (2018). Ecosystems, ecosystem services, and biodiversity. In D. R. Reidmiller, C. W. Avery, D. R. Easterling, K. E. Kunkel, K. L. M. Lewis, T. K. Maycock, & B. C. Stewart (Eds.), *Impacts, risks, and adaptation in the United States: Fourth national climate assessment* (Vol. II, pp. 268–321). U.S. Global Change Research Program. https://doi.org/10.7930/NCA4.2018.CH7

Liu, S., Costanza, R., Farber, S., & Troy, A. (2010). Valuing ecosystem services. Theory, practice, and the need for a transdisciplinary synthesis. *Annals of the New York Academy of Sciences, 1185*, 54–78. https://doi.org/10.1111/j.1749-6632.2009.05167.x

Long, S. (2017). *The connectivity principle: Healing the wounds of separation paperback*. https://www.amazon.com/Connectivity-Principle-Healing-Wounds-Separation/dp/1506904734

Lorenz, E. N. (1963). Deterministic nonperiodic flow. *Journal of the Atmospheric Sciences, 20*(2), 130–141.

Lovejoy, T. E., & Nobre, C. (2018). Amazon tipping point: Last chance for action. *Science Advances, 4*(12), 23–40.

Lovelock, J. E. (1979). *Gaia: A new look at life on earth*. Oxford University Press.

Lovelock, J. E. (1986). Geophysiology: A new look at earth science. *Bulletin of the American Meteorological Society, 67*(4), 392–397.

Lovelock, J. E. (1988). Self-regulation of the earth as a Hoving organism. In *PRONADISA: Reactivos de calidad internacional made in Spain* (p. 141).

Lovelock, J. E. (1991). Gaia: A planetary emergent phenomenon. In W. I. Thompson (Ed.), *Gaia 2: EmergenceDOUBLEHYPHENThe science of becoming*. Lindisfarne Press.

MacArthur, R. H., Diamond, J. M., & Karr, J. R. (1972). Density compensation in island faunas. *Ecology, 53*(2), 330–342.

MacElroy, R. D., Klein, H. P., & Averner, M. M. (1987). The evolution of CELSS for lunar bases: Controlled ecological life support systems. In W. Mendell (Ed.), *Lunar bases and space activities of the 21st century* (pp. 623–633). Lunar and Planetary Institute.

Maclaurin, J., & Sterelny, K. (2008). *What is biodiversity?* University of Chicago Press.

MacNeil, R. C. (2022). *Hopeful realism: A climate manifesto*. https://read.amazon.com/?asin=B0B52F18CX&ref_=kwl_kr_iv_rec_1

MacNeill, J. (1989). Strategies for sustainable economic development. *Scientific American, 261*(3), 155–165.

Macy, J. (2016). The greening of the self. In L. V. Lee (Ed.), *Spiritual ecology* (2nd ed.). Golden Sufi Center.

Macy, J. (2017). The greening of the self. In L. Vaughan-Lee (Ed.), *Spiritual ecology – The cry of the earth*. The Golden Sufi Center.

Madhur, A., Gonzalez, A., Guichard, F., Kolasa, J., & Parrott, L. (2010). Ecological systems as complex systems: Challenges for an emerging science. *Diversity, 2*, 395–410. https://doi.org/10.3390/d2030395

Madore, B. F., & Freedman, W. L. (1987). Self-organizing structures. *American Scientist, 75*, 252.

Malhi, Y., Franklin, J., Seddon, N., Solan, M., Turner, M. G., Field, C. B., & Knowlton, N. (2020). Climate change and ecosystems: Threats, opportunities, and solutions. *Philosophical Transactions of the Royal Society B, 375*, 20190104. https://doi.org/10.1098/rstb.2019.0104

Malhotra, R. (2014). *Indra's Net – Defending Hinduism's Philosophical Unity*. Harper Collins Publisher.

Mann, M. (2023). The fragility of truth in the existential climate crisis. *The Humanist: A Quarterly of Ideas and Actions*, Summer 2023. https://en.wikipedia.org/wiki/Hockey_stick_graph_(global_temperature)

Margulis, L., & Dorion, S. (1997). *Slanted truths: Essays on Gaia, symbiosis and evolution*. Springer. https://doi.org/10.1007/978-1-4612-2284-2

Margulis, L., & Fester, R. (Eds.). (1991). *Symbiosis as a source of evolutionary innovation: Speciation and morphogenesis*. MIT press.

Martínez-Alier, J. (2002). *The environmentalism of the poor*. Edward Elgar Publishing.

Martínez-Alier, J., Munda, G., & O'Neill, J. (1998). Weak comparability of values as a foundation for ecological economics. *Ecological Economics, 26*(3), 277–286.

Martinez-Alier, J., Temper, L., Del Bene, D., & Scheidel, A. (2016). Is there a global environmental justice movement? *Journal of Peasant Studies, 43*, 731. https://doi.org/10.1080/03066150.2016.1141198

Martín-López, B., Gómez-Baggethun, E., García-Llorente, M., & Montes, C. (2014). Trade-offs across value-domains in ecosystem services assessment. *Ecological Indicators, 37*, 220–228.

Marx, K. (1973). *Grundrisse*, Trans. Martin Nicolaus. Penguin.

Marx, K. (1978). Economic and philosophic manuscript of 1844. In R. C. Tucker (Ed.), *The Marx-Engels Reader* (2nd ed.).

Marx, K. (1987). *Capital: A critique of political economy, volume I: The process of production of capital*. Progress Publishers.

Maslow, A. H. (1943). A theory of human motivation. *Psychological Review, 50*(4), 370–396. https://doi.org/10.1037/h0054346

Mathews, J. T. (1989). Redefining security. *Foreign Affairs, 68*(2), 162–177.

Matthews, E. (1989). In N. Dover (Ed.), *The metaphysics of environmentalism in ethics and environmental responsibility*. Gower Publishing Company.

Matthews, B. (2016). Evolution modifies the consequences of biodiversity in ecological communities. *Ecology Letters, 19*(7), 792–802.

Maturana, H. R. (1980). Autopoiesis: Reproduction, heredity and evolution. In *Autopoiesis, dissipative structures and spontaneous social orders, AAAS Selected Symposium 55 (AAAS National Annual Meeting, Houston TX, 3–8 January 1979)* (pp. 45–79). Westview Press.

Maturana, H. R. (1981). Autopoiesis. In M. Zeleny (Ed.), *Autopoiesis: A theory of living organization*. Elsevier-North Holland.

Maturana, H. R., & Varela, F. G. (1980a). *Autopoiesis and cognition: The realization of the living*. Reidel.

Maturana, H., & Varela, F. (1980b). *Autopoiesis: The organization of the living system in Maturana and Valera: autopoiesis and cognition*. D. Reidel.

Max-Neef, M., Elizalde, A., & Hopenheynn, M. (1989). Human scale development: An option for the future. *Development Dialogue, 1*, 3–80.

May, R. M. (2019). *Stability and complexity in model ecosystems* (Vol. 1). Princeton University Press.

Mayr, E. (1954). Change of genetic environment and evolution. In J. Huxley (Ed.), *Evolution as a process*. Allen and Unwin.
McBrien, J. (2016). Accumulation extinction: Planetary catastrophism in the necrocene. In J. Moore (Ed.), *Anthropocene or capitalocene?: Nature, history, and the crisis of capitalism* (pp. 116–137). PM Press.
McCann, K. S. (2000). The diversity–stability debate. *Nature, 405*(6783), 228–233.
McMichael, A. J. (2013). Globalization, climate change, and human health. *The New England Journal of Medicine, 368*, 1335–1343. https://doi.org/10.1056/NEJMra1109341
McNeely, J. A., Miller, K. R., Reid, W. V., Mittermeier, R. A., & Werner, T. B. (1990). *Conserving the world's biodiversity*. World Bank, World Resources Institute, IUCN, Conservation International and World Wildlife Fund.
MEA. (2005). Millennium ecosystem assessment. In *Ecosystems and human well-being: Synthesis*. Island Press.
Meadows, D. (2008). Thinking in systems: A primer. In D. Wright (Ed.), *White River junction*. Chelsea Green.
Meadows, D. H., Randers, D. L., & Behrens III, W. W. (1972). *The limits of growth. A report for the Club of Rome's project on the predicament of mankind*. Retrieved September 20, 2015 from http://collections.dartmouth.edu/published-derivatives/meadows/pdf/meadows_ltg-001.pdf
Meier, G. M. (1976). *Leading issues in economic development* (3rd ed.). Oxford University Press.
Menand, L. (2016). Karl Marx, yesterday and today, a critic at large, *New Yorker*, October 10, 2016. https://www.newyorker.com/magazine/2016/10/10/karl-marx-yesterday-and-today
Menhinick, E. F. (1964). A comparison of some species-individuals diversity indices applied to samples of field insects. *Ecology, 45*, 859–861. https://doi.org/10.2307/1934933
Merchant, C. (1990). Environmental ethics and political conflict: A view from California. *Environmental Ethics, 12*, 45–68.
Merino, E. (1992). *Self-organization in stylolites*. https://earth.indiana.edu/directory/emeriti-faculty/publication-pdf/1992Stylolites.pdf
Messerschmidt, D. A. (1985). Local participation in park resource planning and management. In S. R. Chalise (Ed.), *People and protected areas in the Hindu Kush-Himalayas, Jeffrey A., McNeely, Thorsell, IUCN*. King Mahendra Trust for Nature Conservation and ICIMOD.
Milbrath, L. W. (1989). *Envisioning a sustainable society, learning our way out*. State University of New York Press: xiv + 403 pp., illustr, 1989.
Milgram, S. (1992). In J. Sabini & M. Silver (Eds.), *The individual in a social world: Essays and experiments* (2nd ed.). Mcgraw-Hill Book Company.
Miller, J. G. (1978). *Living systems*. Wiley.
Miller, R. I. (1979). Conserving the genetic identity of faunal population and communities. *Environmental Conservation, 16*(4), 297–303.
Miller, G. T. (1985). *Living in the environment*. Wadsworth Publishing company.
Miller Alan, S. (1991). *Gaia connections: An introduction to ecology, ecoethics, and economics*. Rowman & Littlefield Publishers, Inc.
Miller, K. R., Furtado, J., Deklemm, C., Mcneely, J. A., Myers, N., Soule, M. E., & Trexler, M. C. (1985). Issues on the preservation of biological diversity. In R. Repetto (Ed.), *The global possible: Resources, development, and the new century*. Yale University Press.
Mingers, J. (1980). Towards an appropriate social theory for applied systems thinking: Critical theory and soft systems methodology. *Journal of Applied Systems Analysis, 7*, 41–50.
Mingers, J. (1981). Toward an appropriate social theory for applied systems analysis. *Journal of Applied Systems Analysis, 7*, 141–149.
Mingers, J. (1989). An introduction to autopoiesis-implications and applications. *System Practices, 2*, 159–180.
Mingers, J. (1991). The cognitive theories of Maturana and Varela. *Systems Practice, 4*(4), 1991.
Minteer, B. (2001). Intrinsic value for pragmatists. *Environmental Ethics, 23*(57–75), 57.
Mitchell, M. (2009). *Complexity: A guided tour*. Oxford University Press.

Mithen, S. (1996). *The prehistory of the mind: The cognitive origins of art, religion and science*. Thames and Hudson.

Mohr, M., Laemmel, T., Maier, M., & Schindler, D. (2016). Analysis of air pressure fluctuations and topsoil gas concentrations within a Scots pine forest. *Atmosphere, 7*(10), 125.

Monserrate, M. A., Zambrano, M. A., Ruano, V. O.-C., & Sanchez-Loor, D. A. (2020). Global ecological footprint and spatial dependence between countries. *Journal of Environmental Management, 272*, 111069.

Moore, D. R., & Keddy, P. A. (1988). The relationship between species richness and standing crop in wetlands: The importance of scale. *Vegetatio, 79*, 99–106.

Morand, S., & Lajaunie, C. (2018). *Biodiversity and health: Linking life, ecosystems and societies*. ISTE Press Ltd and Elsevier Ltd.

Moreno, A., & Mossio, M. (2015). *The autonomy of living systems: A philosophical enquiry into biological organization*. Springer.

Morowitz, H. J. (1991). Balancing species preservation and economic consideration. *Science, 253*, 752–754.

Morowitz, H. (1995). The emergence of complexity. *Complexity, 1*(1), 4–5.

Mossio, M., Saborido, C., & Moreno, A. (2009). An organizational account of biological functions. *British Journal for the Philosophy of Science, 60*(4), 813–841.

Mulholland, J. (2019). *Why the climate crisis is the most crucial story we cover in America?*. https://www.theguardian.com/environment/2019/oct/16/climate-crisis-america-guardian-editor-john-mulholland?CMP=gu_com

Muller, F. (1992). Hierarchical approaches to ecosystem theory. *Ecological Modelling, 63*, 215–242.

Munashinghe, M. (1992). Biodiversity protection policy: Environmental valuation and distribution issue. *Ambio, 21*(3), 227–236.

Munevar, D. (2021). A debt pandemic is engulfing the Global South. *Economy and Ecology*. https://www.ips-journal.eu/topics/economy-and-ecology/a-debt-pandemic-is-engulfing-the-global-south-5114/

Munevar, D., & Mariotti, C. (2021). *The 3 trillion dollar question: What different will the IMF's new SDR allocation make for the world's Poorest?* European Network on Debt and Development.

Myers, N. (1984a). Genetic resources in Jeopardy. *Ambio, 13*(3), 171–174.

Myers, N. (1984b). *Gaia: An atlas of plant management, 146, 154–59*. Anchor Books.

Myers, S. S., & Patz, J. A. (2009). Emerging threats to human health from global environmental change. *Annual Review of Environment and Resources, 34*, 223–252. https://doi.org/10.1146/annurev.environ.033108.102650

Myers, N., & Sayensu, E. (1983). Reduction of Biological Diversity and Species Loss. *Ambio, 12*(2), 72–74.

Myrdal, G. (1968). *Asian Drama: An inquiry into the poverty of nations*. Oxford Academy Press.

Naess, A. (1973). The shallow and the deep, long-range ecology movement. *Inquiry, 16*, 95–100.

Naess, A. (1990a). Man apart and deep ecology: A reply to Reed. *Environmental Ethics, 12*(2), 185–192.

Naess, A. (1990b). *Ecology, community, and lifestyle: Outline of an ecosophy*. Cambridge University Press.

Naidoo, R., Balmford, A., Costanza, R., Fisher, B., Green, R. E., Lehner, B., Malcolm, T. R., & Ricketts, T. H. (2007). Global mapping of ecosystem services and conservation priorities. *PNAS*, published ahead of print July 9, 2008. https://doi.org/10.1073/pnas.0707823105

Nash, R. F. (1989). *The rights of nature* (p. 75). University of Wisconsin Press.

Nielsen, N. O. (1998). *Management for agroecosystem health: The new paradigm for agriculture. New directions in animal production systems*. In Proceeding of the Annual Meeting of Canadian Society of Animal Science, July 5–8, 1998, Vancouver, British Columbia, Canada.

Nisbet, E. K., & Zelenski, J. M. (2013). The NR-6: A new brief measure of nature relatedness. *Frontiers in Psychology, 4*, 813.

Nobre, C. A., Sellers, P. J., & Shukla, J. (1991). *Journal of Climatology, 4*, 957–988.

Norton, B. (1991). *Toward unity among environmentalists*. Oxford University Press.
Norton, B. (1995). Why I am not a non-anthropocentrist. *Environmental Ethics, 17*, 341–358.
Norton, B. G., & Ulanowicz, R. E. (1992). Scale and biodiversity policy: A hierarchical approach. *Ambio, 21*(3), 244–249.
O'Malley, B. (2020). Three crises threaten human survival, Chomsky warns, University World News, The Global Window on Higher Education, 12 December 2020. https://www.universityworldnews.com/post.php?story=20201212053831736
O'Neil, R. V. (1988). Hierarchy theory and global change. In T. Rosswall & R. G. Woodmansee (Eds.), *Scale and global change*. John Wiley and Sons, Ltd.
O'Neill, J. (1993). Ecology, policy, and politics. In *Human well-being and the natural world*. Routledge.
Odin, S. (1997). The Japanese concept of nature in relation to the environmental ethics and conservation aesthetics of Aldo Leopold. In M. E. Tucker & D. R. Williams (Eds.), *Buddhism and ecology: The interconnection of dharma and deeds*. Harvard University Press.
Odum, E. P. (1968). Energy flow in ecosystems: A historical review. *American Zoologist, 8*(1), 11–18.
Odum, E. P. (1969). The strategy of ecosystem development. *Science, 164*, 262–270.
Odum, E. P. (1977). The emergence of ecology as a new integrative science. *Science, 195*, 1289–1293.
Odum, W. E. (1982). Environmental degradation and the tyranny of small decisions. *BioScience, 32*(9), 728–729.
Odum. (1983). *Basic ecology*. Saunders Publ.
Odum. (1985). Trends expected in stressed ecosystems. *Bioscience, 35*, 419–422, 499.
Odum, E. P. (1987). Reduced-input agriculture reduces nonpoint pollution. *Journal of Soil and Water Conservation, 42*(6), 412–414.
Odum, H. T. (1988). Self-organization, transformity and information. *Science, 242*, 1132–1139.
Odum, E. P. (1989a). Input management of production systems. *Science, 243*(4888), 177–182.
Odum, H. T. (1989b). *Ecology and our endangered life-support systems*. Sinauer Assoc..
Odum, H. T. (1990). Field experimental tests of ecosystem level hypotheses. *Trends in Ecology & Evolution, 5*, 204–205.
Odum, H. T. (1991). Network ecology: Indirect determination of the life-environment relationship in ecosystems. In M. Higashi & T. P. Burn (Eds.), *Theoretical studies of ecosystems: The network perspectives* (pp. 288–351). Cambridge University Press.
Odum, E. P., & Barrett, G. W. (1971). *Fundamentals of ecology* (Vol. 3, p. 5). Saunders.
Odum, E. P., & Biever, L. J. (1984). Resource quality, mutualism, and energy partitioning in food chains. *The American Naturalist, 124*(3), 360–376.
Odum, H. T., & Odum, E. C. (1981). *Energy basis for man and nature*. McGraw-Hill.
Odum, E. P., Finn, J. T., & Franz, E. H. (1979). Perturbation theory and the subsidy-stress gradient. *Bioscience, 29*, 341–352.
Oelberg, S. (2002). Ecohumanism: So, what's new? God, humanity, and nature caring for creation. In R. B. Tapp (Ed.), *Ecohumanism*. Prometheus Books.
O'Neill, R. V. (1986). *A hierarchical concept of ecosystems*. (No. 23). Princeton University Press.
Orians, G. H. (1990). Ecological concepts of sustainability. *Environment, 32*(9), 10–39.
Orians, G. H., & Kunin, W. E. (1990). Ecological uniqueness and loss of species. In *The preservation and valuation of biological resources*.
Orr, D. W. (1989). Ecological literacy: Education for the twenty-first century. *Holistic Education Review, fall, 1989*, 48–53.
Ortolano, L. (1984). *Environmental planning and decision making*. John Wiley & Sons.
Ottawa Conference on Conservation and Development. (1986). *Conservation with equity: Strategies for sustainable development – implementing the world conservation strategy*.
Our Common Future. (1987). *The World Commission on Environment and Development*. Oxford University Press.
Owen, D. F., & Weigert, R. G. (1976). Do consumers maximize plant fitness? *Oikos, 27*, 489–492.

Panayotou, T. (1990). *The economics of environmental degradation: Problems, causes and responses. Development discussion paper no. 335, A CAER project report, Harvard Institute for International Development*. Harvard University.

Pandey, B., Buytaert, W., Zulkafli, Z., Karpouzoglou, T., Mao, F., & Hannah, D. M. (2016). A comparative analysis of ecosystem services valuation approaches for application at the local scale in data scarce regions. *Ecosystem Services, 22B*, 250–259.

Pappas, S. (2012). *Tipping point? earth headed for catastrophic collapse, researchers warn, live science contributor, June 6, 2012*, Program from http://www.unep.org/greeneconomy/AboutGEI/WhatisGEI/tabid/29784/Default.aspx. Published: October 14, 2016, https://doi.org/10.1371/journal.pone.0164733

Parrot, L., & Kok, R. (2002). *Incorporating complexity in ecosystem modelling*. https://www.researchgate.net/publication/249908876_Incorporating_Complexity_in_Ecosystem_Modelling

Parrott, L. (2002). *Complexity and the limits of ecological engineering*. https://complexity.ok.ubc.ca/publications-by-date/

Parrott, L. (2010). Measuring ecological complexity. *Ecological Indicators, 10*, 1069–1076. https://doi.org/10.1016/j.ecolind.2010.03.014

Parrott, L., & Kok, R. (2000). Incorporating complexity in ecosystem modelling. *Complexity International, 7*, 1–19.

Parrott, L., & Meyer, W. (2012). Future landscapes: Managing within complexity. *Frontiers in Ecology and the Environment, 10*(7), 382–389. https://doi.org/10.1890/110082

Pascual, U., Balvanera, P., Díaz, S., Pataki, G., Roth, E., Stenseke, M., Watson, R. T., Dessane, E. B., Islar, M., Kelemen, E., Maris, V., Quaas, M., Subramanian, S. M., Wittmer, H., Adlan, A., Ahn, S. E., Al-Hafedh, Y. S., Amankwah, E., Asah, S. T., Berry, P., Bilgin, A., Breslow, S. J., Bullock, C., Cáceres, D., Daly-Hassen, H., Figueroa, E., Golden, C. D., Gómez-Baggethun, E., González-Jiménez, D., Houdet, J., Keune, H., Kumar, R., Ma, K., May, P. H., Mead, A., O'Farrell, P., Pandit, R., Pengue, W., Pichis-Madruga, R., Popa, F., Preston, S., Pacheco-Balanza, D., Saarikoski, H., Strassburg, B. B., van den Belt, M., Verma, M., Wickson, F., & Yagi, N. (2017). Valuing nature's contributions to people: The IPBES approach. *Current Opinion in Environmental Sustainability, 26–27*, 7–16. https://doi.org/10.1016/j.cosust.2016.12.006

Pattanayak, S. K. (2004). *Valuing watershed services: Concepts and empirics from Southeast Asia*. https://doi.org/10.1016/j.agee.2004.01.016

Patten, B. C. (1978). System approach to the concept of environment. *The Ohio Journal of Science, 78*, 206–222.

Patten, B. C., & Odum, E. P. (1981). The cybernetic nature of ecosystems. *The American Naturalist, 118*, 886–895.

Patton, W. K. (1972). Studies on the animal symbionts of the gorgonian coral, Leptogorgia virgulata (Lamarck). *Bulletin of Marine Science, 22*(2), 419–431.

Patton, R. J., & Chen, J. (1991). A review of parity space approaches to fault diagnosis. *IFAC Proceedings Volumes, 24*(6), 65–81.

Pearce, D. (1988). Economics, equity and sustainable development. *Futures, 20*(6), 598–605.

Pearce, D., Barbier, E., & Markandya, A. (1990). *Sustainable development: Economics and environment in third world*. Billing & Sons.

Peden, D. G. (1998). Agroecosystem management for improved human health: Applying principles of integrated pest management to people. New directions in animal production systems. In *Proceeding of the Annual Meeting of Canadian Society of Animal Science*, July 5–8, Vancouver, British Columbia, Canada.

Pereira, C. P., Reid, W. V., Sarukhan, J., Scholes, R. J., & Whyte, A. (2009). Science for managing ecosystem services: Beyond the millennium ecosystem assessment. *PNAS, 106*(5), 1305–1312. https://doi.org/10.1073/pnas.0808772106

Perring, C., Folke, C., & Maler, K.-G. (1992). The ecology and economics of biodiversity loss: The research agenda. *Ambio, 21*(3), 201–211.

Pettinger, T. (2017). *List of countries energy use per capita*. https://www.economicshelp.org/blog/5988/economics/list-of-countries-energy-use-per-capita/

Piementel, D., Stachow, U., Takacs, D. A., Brubaker, H. W., Dumas, A. R., Mwaney, J. J., O'Neil, J. A. S., Onsi, D. E., & Corzilius, D. B. (1992). Conserving biological diversity in agriculture/forestry systems. *Bioscience, 42*(5), 354–362.

Piketty, T. (2014). *Capital in the twenty-first century*. Harvard University Press.

Piketty, T., & Saez, E. (2003). Income inequality in the United States, 1913-1998. *The Quarterly Journal of Economics, 118*(1), 1–39.

Pimentel, D. (1968). Population regulation and genetic feedback. *Science, 159*, 1432–1437.

Pinchot, G. (1914). *The training of a forester*. J. B. Lippincott Company.

Pinker, S., & Jackendoff, R. (2005). The faculty of language: what's special about it? *Cognition, 95*(2), 201–236.

Pipenger, N. (1978). Complexity theory. *Scientific American, 238*(6), 114–124.

Pirages, D. C. (1977). Introduction: A social design for sustainable growth. In D. C. Pirages (Ed.), *The Sustainable Society* (pp. 49–72). Praeger, USA: xiv+ 342.

Podgorski, J. S. (2010). Humberto Maturana's view on the theory of evolution from autopoiesis to natural drift metaphor. *Ecological Questions, 13*(2010), 81–88.

Polunin, N. (1972). The biosphere today. In N. Polunin (Ed.), *The environmental future: Proceeding of the First International Conference on Environmental Future* (pp. 49–50).

Pomeroy, L. R. (1974). The Ocean's food web: A changing paradigm. *Bioscience, 24*, 499–504.

Pomeroy, L. R., & Diebel, D. (1986). Temperature regulation of bacterial activity during spring bloom in Newfoundland coastal waters. *Science, 233*, 359–361.

Pomeroy, L. R., & Wiebe, W. J. (1988). Energetics of microbial food webs. *Hydrobiologia, 159*, 7–18.

Ponting, C. (1990). Historical perspectives on sustainable development. *Environment, 32*(9), 4–9.

Popova, M. (2016). Einstein on widening our circles of compassion, *The Marginalian.* https://www.themarginalian.org/2016/11/28/einstein-circles-of-compassion/

Prigogine, I. (1980a). Loi, histoire... et désertion. *Le débat*, (6), 122–130.

Prigogine, I. (1980b). *From being to becoming time and complexity in the physical sciences*. W.H. Freeman.

Prigogine, I. (1984). Irreversibility and space-time structure. In *Fluctuations and sensitivity in nonequilibrium systems: Proceedings of an international conference, University of Texas, Austin, Texas, March 12–16, 1984* (pp. 2–10). Springer Berlin Heidelberg.

Prigogine, I., & Stengers, I. (1984). *Order out of chaos: Man's new dialogue with nature*. Bantam.

Prigogine, I., Nicolis, G., & Babloyantz, A. (1972). Thermodynamics of evolution. *Physics Today, 25*(11), 23–28.

Program from. http://www.unep.org/greeneconomy/AboutGEI/WhatisGEI/tabid/29784/Default.aspx

Pulliam, H. R. (1988). Sources, sink and population regulation. *The American Naturalist, 132*, 652–6661.

Rabesandratana, T. (2023). Scientists urge European Parliament to vote for Nature Restoration Law. *Science*, 10 July 2023. https://www.science.org/content/article/scientists-urge-european-parliament-vote-nature-restoration-law?utm_source=sfmc&utm_medium=email&utm_campaign=DailyLatestNews&utm_content=alert&et_rid=888901169&et_cid=4809038

Rabinowitch, E. I. (1945). *Photosynthesis and related processes, New York.*

Rabinowitz, P., & Conti, L. (2013). Links among human health, animal health and ecosystem health. *Annual Review of Public Health, 34*, 189–204.

Raju, P. T. (1939). Buddhistic conception of dharma. *Annals of the Bhandarkar Oriental Research Institute, 21*(3/4), 1939–1940. https://www.jstor.org/stable/i40079364

Rakic, P. (2009). Evolution of the neocortex: A perspective from developmental biology. *Nature Reviews Neuroscience, 10*(10), 724–735.

Ramade, F. (1984). *Ecology of natural resources*. Wiley.

Rapport, D. J., Regier, H. A., & Hutchinson, T. C. (1985). Ecosystem behavior under stress. *Nature, 125*, 617–640, 1984.

Rapport, D. J., Costanza, R., & McMichael, A. J. (1998). Assessing ecosystem health. *Trends in Ecology & Evolution, 13*(10), 397–402.

Raskin, P. (2002). *Great transition* (pp. 42–43). Stockholm Environment Initiative.

Raskin, P., Banuri, T., Gallopín, G., Gutman, P., Hammond, A., Kates, R., & Swart, R. (2002). *Great transition: The promise and lure of the times ahead.* Stockholm Environment Institute - Boston Tellus Institute.

Raworth, K. (2017). *Doughnut economics: Seven ways to think like a 21st century economist.* Chelsea Green Publishing.

Rees, W. E. (1990). The ecology of sustainable development. *The Ecologist, 20*(1), 18–23.

Reich, C. A. (1970). The Greening of America. Published by Random House, *The New Yorker.* p. 42. Retrieved 2008-07-11.

Reid, W. V., & Miller, K. R. (1989). *Keeping options alive: The scientific basis for conserving biodiversity.* World Resources Institute.

Reis, S., Morris, G., Fleming, L. E., Beck, S., Taylor, T., White, M., Depledge, M. H., Steinle, S., Sabel, C. E., Cowie, H., Hurley, F., McP Dick, J., Smith, R. I., & Austen, M. (2013). Integrating health and environmental impact analysis. *Public Health.*

Renner, M. (1991). *Jobs in sustainable economy* (World-watch paper 104). World-watch Institute.

Repetto, R. (1992). Accounting for environmental assets. *Scientific American, 266*(6), 94–101.

Repetto, R. C., Magrath, W., Wells, M., Beer, C., & Rossini, F. (1989). *Wasting assets: Natural resources in the national income accounts.* (No. INVES-ET P01 R425w). World Resources Institute.

Rifkin, J. (1989). *Entropy into the greenhouse world.* Bantam Books.

Roberts, J. O. M., & Johnson, B. D. G. (1985). "Adventure" tourism and sustainable development: Experience of the Tiger Mountain group's operations in Nepal. In J. A. McNeely & J. W. Thorsel (Eds.), *People and protected areas in the Hindukush-Himalaya* (pp. 81–84). ICIMOD. 250 pp.

Rockefeller, S. C. (1997). Buddhism, global ethics and earth chapter. In M. E. Tucker & D. R. Williams (Eds.), *Buddhism and ecology, the interconnection of dharma and deeds.* Harvard University Press. Buddhism and Ecology by Donald Swearer.

Rockström, J., Steffen, W., Noone, K., Persson, Å., Chapin, F. S., Lambin, E. F., et al. (2009). A safe operating space for humanity. *Nature, 461*(7263), 472–475.

Rode, A., Carleton, T., Delgado, M., et al. (2021). Estimating a social cost of carbon for global energy consumption. *Nature, 598*, 308–314. https://doi.org/10.1038/s41586-021-03883-8

Rolston, H. I. I. I. (1986). *Philosophy gone wild: Essays in environmental ethics.* Prometheus.

Rolston, H., III. (1988). Human values and natural systems. *Society & Natural Resources, 1*(1), 269–283.

Rolston, H., III. (1990). Property rights and endangered species. *The University of Colorado Law Review, 61*, 283.

Rolston, H., III. (1992). Human values and natural system. *Societies and Natural Resources, 1*, 271–283.

Royal Society. (2009). *Geoengineering the climate: Science, governance, and uncertainty.*

Ryan, A. (2015). *On Marx: Revolutionary and Utopian.* National Geographic Books.

Saarikoski, H., Mustajoki, J., Barton, D. N., Geneletti, D., Langemeyer, J., GomezBaggethun, E., Marttunen, M., Antunes, P., & Keune, H. (2016). Multi-criteria decision analysis and cost-benefit analysis: Comparing alternative frameworks for integrated valuation of ecosystem services. *Ecosystem Services, 22B*, 238–249.

Saez, E., & Zucman, G. (2019). *The triumph of injustice: How the rich dodge taxes and how to make them pay.* WW Norton & Company.

Sage, R. F. (2020). Global change biology: A primer. *Global Change Biology, 26*(1), 3–30.

Salisbury, F. B., & Ross, C. W. (1985). *Plant physiology.* Wadsworth.

Sandler, R. (2007). *Character and environment. A virtue-oriented approach to environmental ethics.* Columbia University Press.

Sandler, R. (2012). *The ethics of species: An introduction.* Cambridge University Press.

Sarkar, S. (2005). *Biodiversity and environmental philosophy an introduction*. Cambridge University Press.

Sayre, K. M. (1976). *Cybernetics and the philosophy of mind* (p. 71). Humanities Press.

Schafer, D. (2002). Time is not on our side. In R. B. Tapp (Ed.), *Eco-humanism*. Prometheus Books.

Schelling, T. C. (1978). *Micro motives and macro behavior*. Norton.

Schindler, D. W. (1990). Experimental perturbations of whole lakes as tests of hypotheses concerning ecosystem structure and function. *Oikos, 57*, 25–41.

Schlesinger, W. H. (1978). Community structure, dynamics and nutrient cycling in the Okefenokee cypress swamp-forest. *Ecological Monographs, 48*(1), 43–65.

Schlesinger, W. H., Reynolds, J. F., Cunningham, G. L., Huenneke, L. F., Jarrel, W. M., Virginia, R. A., & Whiteford, W. G. (1990). Biological feedbacks in global desertification. *Science, 247*, 1043–1048.

Schroder, S. A., Toth, S. F., Deal, R. L., & Ettl, G. J. (2016). Multi-objective optimization to evaluate tradeoffs among forest ecosystem services following fire hazard reduction in the Deschutes National forest, USA. *Ecosystem Services, 22B*, 328–347.

Schrodinger, E. (1944). *What is life?* Cambridge University Press.

Schrodinger, E. (1956). *What is life? and other scientific essays*. Garden City, Doubleday.

Schrodinger, E. (1994). *What is life?* Cambridge University Press.

Schröter, D., Cramer, W., Leemans, R., Prentice, I. C., Araújo, M. B., Arnell, N. W., et al. (2005). Ecosystem service supply and vulnerability to global change in Europe. *Science, 310*(5752), 1333–1337.

Schultz, P. W., & Zelezny, L. (1999). Values as predictors of environmental attitudes: Evidence for consistency across 14 countries. *Journal of Environmental Psychology, 19*(3), 255–265.

Schweitzer, A. (1933). *Out of my life and thought: An autobiography*. Holt.

Scott, J. P. (1989). *The evolution of social systems*. Gordon and Breach Science Publishers.

Seppelt, R., Dormann, C., Eppink, F., Lautenbach, S., & Schmidt, S. (2011). *A quantitative review of ecosystem service studies: Approaches, shortcomings and the road ahead: Priorities for ecosystem service studies*. https://www.researchgate.net/publication/331917586

Service, W. M. (1998). An overview of vector-borne diseases associated with irrigation. Page ix to xii 8. In F. Konradsen & W. van der Hoek (Eds.), *Health and irrigation, Proc. Copenhagen workshop on health and irrigation*. IIMI.

Shantz, C. J. (1992). The developing brain. *Scientific American, 267*, 61–67.

Shue, H. (1999). Global environment and international inequality. *International Affairs, 75*(3), 531–545.

Shukla, J., Nobre, C., & Sellers, P. (1990). *Science, 247*, 1322–1325.

Simms, A., Chowla, P., & Johnson, V. (2010). *Growth is not possible: Why we need a new economic direction, Schumacher College Transformative learning for sustainable living*. https://neweconomics.org/uploads/files/f19c45312a905d73c3_rbm6iecku.pdf

Simon, H. A. (1962). The architecture of complexity. *Proceedings of the American Philosophical Society, 106*(6), 467–482.

Simonsen, S. H., Biggs, R., Schlüter, M., Schoon, M., Bohensky, E., Cundill, G., et al. (2014). *Applying resilience system: Seven principles for building resilience in social-ecological systems*. Stockholm Resilience Centre. Research for Biosphere Stewardship and Innovation.

Singer, P. (1977). *Animal liberation*. Paladin.

Skolimowski, H. (1986). In defense of eco-philosophy and of intrinsic values. *The Trumpeter, 3*(4), Fall.

Skolimowski, H. (1990). For the record: On the origin of eco-philosophy. *The Trumpeter, 7*(1), 44.

Smith, K. (1991). The natural Debt. *East West Center Views, 1*(3), 3.

Snyder, G. (1990). Turtle talk: Voices for a sustainable future. In C. Plant & J. Plant (Eds.), *The new catalyst bioregional series*. New Society Publishers.

Snyder, G. (1995). *A place in space: Ethics, aesthetics, and watersheds: New and selected prose*. Counterpoint.

Solow, R. M. (1974). The economics of resources or the resources of economics. *American Economic Review, 64*(1974), 1–14.

Sorell, S., & Ockwell, D. (2020). *Can we decouple energy consumption from economic growth?* Sussex Energy Group (SEG), Policy briefing, Number 7 February 2020.

Soulé, M. E. (1985). What is conservation biology? *Bioscience, 35*, 727–734.

Specktor, B. (2019). *Human civilization will crumble by 2050 if we don't stop climate change now*, New Paper Claims, Live Science Contributor, Senior Writer, June 4.

Spencer, D. F., Alpert, S. B., & Gilman, H. H. (1986). Cool water: Demonstration of clean and efficient new coal technology. *Science, 232*, 609–612.

Speth, J. G. (2008). *The bridge at the edge of the world: Capitalism, the environment, and crossing from crisis to sustainability*. Yale University Press.

Sponberg, A. (1997). Green Buddhism and hierarchy of compassion. In M. E. Tucker & D. R. Williams (Eds.), *Buddhism and ecology: The interconnection of Dharma and Deeds*. Harvard University Press.

Sponberg, A. (2017). The Buddhist conception of an ecological self. In L. Vaughan-Lee (Ed.), *Spiritual ecology – The cry of the EARTH, a collection of essays* (2nd ed.).

Spracklen, D. V., Arnold, S. R., & Taylor, C. M. (2012). Observations of increased tropical rainfall preceded by air passage over forests. *Nature, 489*(7415), 282–285.

Stanley, J., & Loy, D. (2016). At the edge of the roof: The evolutionary crisis of the human spirit. In L. V. Lee (Ed.), *Spiritual ecology* (2nd ed.). Golden Sufi Center.

Stanley, J., & Loy, D. (2020). *Buddhism and the end of economic growth*. https://oneearthsangha.org/articles/buddhism-end-of-economic-growth/

Stapp, H. P. (1971). S-matrix interpretation of quantum theory. *Physical Review D, 3*(6), 1303.

Starr, T. B., & Allen, T. F. H. (1988). *Hierarchy: Perspectives for ecological complexity*. University of Chicago Press.

Statista. (2023, Aug 29). *Statista Research Department*. https://www.statista.com/statistics/262742/countries-with-the-highest-military-spending/

Steffen, W., Broadgate, W., Deutsch, L., Gaffney, O., & Ludwig, C. (2015a). The trajectory of the Anthropocene: The great acceleration. *The Anthropocene Review, 2*(1), 81–98.

Steffen, W., Richardson, K., Rockström, J., Cornell, S. E., Fetzer, I., Bennett, E. M., Biggs, R., Carpenter, S. R., de Vries, W., de Wit, C. A., Folke, C., Gerten, D., Heinke, J., Mace, G. M., Persson, L. M., Ramanathan, V., Reyers, B., & Sörlin, S. (2015b). Planetary boundaries: Guiding human development on a changing planet. *Science, 347*(622). https://doi.org/10.1126/science.1259855

Stephen, C. (1998). *Bridging human health and farming through ecology. New directions in animal production systems*. In Proceeding of the Annual Meeting of Canadian Society of Animal Science, July 5–8, 1998, Vancouver, British Columbia, Canada.

Sterba, J. (2001). *Three challenges to ethics: Environmentalism, feminism, and multiculturalism*. Oxford University Press.

Sterman, J. D. (2000). *Business dynamics: Systems thinking and modeling for a complex world*. Irwin/McGraw-Hill.

Stern, D. (2004). Economic growth and energy. In C. J. Cleveland (Ed.), *Encyclopedia of energy* (p. 43). Academic Press.

Stocker, T. F., Qin, D., Plattner, G.-K., Tignor, M., Allen, S. K., Boschung, J., Nauels, A., Xia, Y., Bex, V., & Midgley, P. M. (Eds.). (2013). *Climate change 2013: The physical science basis*. Contribution of Working Group 1 to the Fifth Assessment Report of the Intergovernmental Panel on Climate Change. Cambridge University Press.

Strogatz, S. (2001). Exploring complex networks. *Nature, 410*(6825), 268–276.

Stubblefield, J. J. (2018). Temporal control of metabolic amplitude by nocturnin. *Cell Reports, 22*(5), 1225–1235.

Stubblefield, J. J., Gao, P., Kilaru, G., Mukadam, B., Terrien, J., & Green, C. B. (2018). Temporal control of metabolic amplitude by nocturnin. *Cell Reports, 22*(5), 1225–1235.

Sumpter, D. J. T. (2006). The principles of collective animal behavior. *Biological Sciences, 361*(1465), 5–22. https://doi.org/10.1098/rstb.2005.1733

Sunkel, O. (1988). Perspectivas democráticas y crisis de desarrollo. *Pensamiento iberoamericano*, (14), 313–317.

Sunkel, O. (1990). *The environmental dimension in development planning*. United Nations Economic Commission for Latin America and the Caribbean.

Surpran, G., & Achakulwisut, P. (2019, November 1). *The F-Word finally enters climate politics*. For the first time, presidential candidates are no longer scared to say "fossil fuels" (okay, that's two f-words).

Sutton, P. (2017). *Nature, environment and society*.

Swearer, D. K. (1997a). The hermeneutics of Buddhist ecology in contemporary Thailand: Buddhadasa Dhammapatika. In M. E. Tucker & D. R. Williams (Eds.), *Buddhism and ecology: The interconnection of dharma and deeds*. Harvard University Press.

Swearer, D. K. (1997b). Theravada Buddhism and ecology: The case of Thailand in Buddhism and ecology. In M. E. Tucker & D. R. Williams (Eds.), *The interconnection of dharma and deeds*. Harvard University Press.

Tackey, C., & Gray, P. M. (2017). Evolutionary GEM: Coevolution of yuccas and yucca moths. *WURJ: Health and Natural Sciences, 8*(1), Article 10. https://doi.org/10.5206/wurjhns.2017-18.6

Tansley, A. G. (1935). The use and abuse of vegetational concepts and terms. *Ecology, 16*(3), 284–307.

Tapp, R. B. (2002). Ecohumanism: Some expansions. In R. B. Tapp (Ed.), *Ecohumanism*. Prometheus Books.

Taylor, P. (1986). *Respect for nature: A theory of environmental ethics*. Princeton University Press.

Thich Nhat Hanh. (1988). *The heart of understanding: Commentaries on the PrajÃ±aapaaramitaa Heart Sutra by Thich Nhat Hanh*. Parallax Press.

Thompson, A. P. (1914). On the relation of pyrrhotite to chalcopyrite and other sulphides. *Economic Geology, 9*(2), 153–174.

Thompson, J. (1990). A refutation of environmental ethics. *Environmental Ethics, 12*, 147–160.

Thompson, P. (1996). Pragmatism and policy: The case of water. In A. Light & E. Katz (Eds.), *Environmental pragmatism* (pp. 187–208). Routledge.

Thompson, E. (2001, May 1). Empathy and consciousness. *Journal of Consciousness Studies, 8*(5–7), 1–32(32), Imprint Academic.

Thompson, E., & Varela, F. J. (2001). Radical embodiment: Neural dynamics and conscious experience. *Trends in Cognitive Sciences, 5*, 418–425.

Thomson, H., Petticrew, M., & Morrison, D. (2001). Health effects of housing improvement: Systematic review of intervention studies. *BMJ, 323*(7306), 187–190.

Tilman, D. (1999). The ecological consequences of changes in biodiversity: A search for general principles. *Ecology, 80*(5), 1455–1474.

Tilman, D., Reich, P. B., & Knops, J. M. H. (2006). Biodiversity and ecosystem stability in a decade-long grassland experiment. *Nature, 441*(7093), 629–632.

Tomasello, M., & Vaish, A. (2013). Origins of human cooperation and morality. *Annual Review of Psychology, 64*, 231–255.

TKP (2023). *The Kathmandu Post*. https://kathmandupost.com/editorial/2023/11/01/do-something-anything-1698801995

Tononi, G. (2016). *Integrated information theory might solve neuroscience's biggest puzzle*.

Tononi, G., Boly, M., Massimini, M., & Koch, C. (2016). Integrated Information Theory: From consciousness to its physical substrate. This article was originally published with the title "What Is Consciousness?". *Scientific American, 318*(6), 60–64, June 2018. https://doi.org/10.1038/scientificamerican0618-60

Torras, M., & Boyce, J. K. (1998). Income, inequality, and pollution: A reassessment of the environmental Kuznets cure. *Ecological Economics, 25*, 147–160.

Toulmin, S., & Goodfield, J. (1982). *The architecture of matter*. University of Chicago Press.

Townsend, K. N. (1993). Steady-State economies and the command economy. In H. E. Daly & N. Kenneth (Eds.), *Townsend edited valuing the earth: Economics, ecology, ethics*. The MIT Press.

Trubetskova, I. L. (2010). *From biosphere to noosphere: Vladimir Vernadsky's theoretical system as a conceptual framework for universal sustainability education*. Doctoral Dissertations. 612. https://scholars.unh.edu/dissertation/612

Tucker, M. E., & Grim, J. (2007). Daring to dream: Religion and the future of the earth. *Reflections – The Journal of Yale Divinity School*, 4.

Tucker, M. E., & Swimme, B. T. (2016). The next transition: The evolution of humanity's role in the universe. In L. Vaughan Lee (Ed.), *Spiritual ecology* (2nd ed.). Golden Sufi Center.

Turner, M. G. (1988). Landscape ecology: The effect of pattern on process. *Annual Review of Ecology and Systematics, 20*, 171–197.

UGCRP. (2017). National Academies of Sciences, Engineering, and Medicine. In *Accomplishments of the US Global Change Research Program*. National Academies Press.

Ulanowicz, R. E. (1986). A phenomenological perspective of ecological development. In T. M. Posten & R. Purdy (Eds.), *Aquatic toxicology and environmental fate: Ninth Volume* (pp. 73–81). American Society for testing and materials, ASTM STP 921.

UNEP. (2015). *Green Economy*. Retrieved November 5, 2015, United Nations Environmental Program from http://www.unep.org/greeneconomy/AboutGEI/WhatisGEI/tabid/29784/Default.aspx

UNEP. (2019). *Global environment outlook (GEO-6): Healthy planet, healthy people*. UNEP.

UNEP. (2021). *State of Finance for Nature*. https://www.unep.org/resources/state-finance-nature

UNESCO. (2013). *UNESCO's medium-the contribution of creativity to sustainable development term strategy for 2014–2021*. http://www.unesco.org/new/fileadmin/MULTIMEDIA/HQ/CLT/images/CreativityFinalENG.pdf

UNFCCC. (2015). *Paris agreement*. United Nations Framework Convention on Climate Change.

UNFCCC. (2016). *Report of the Conference of the Parties on its twenty-first session, held in Paris, United Nations Framework Convention on Climate Change, from 30 November to 13 December 2015*. Retrieved February 15, 2016, from http://unfccc.int/resource/docs/2015/cop21/

United Nations. (1992a). *Rio declaration on environment and development*.

United Nations. (1992b). *Convention on biological diversity*.

United Nations. *Framework convention on climate change, from 30 November to 13 December 2015*.

Uniyal, M. (1993, Jan-Feb 25). *Forest Myths Exploded*. Himal.

Upreti, G. (1987). Ecological problems and conservation needs. *The Rising Nepal, 18*, 2–4.

Upreti, G. (1994). Environmental conservation and sustainable development require a new development approach. *Environmental Conservation, 21*(1), 18–29.

Upreti, G. (1996). *Valuation of biodiversity and environmental services: A basis for sustainable development*. Published in Environment and Biodiversity in the Context of South Asia, pp. 232–241. Proceedings of the Regional Conference on Environment and biodiversity, March 7–9, 1994, Kathmandu, Nepal.

Upreti, G. (2001). Conservation and management of biodiversity resources for sustainable livelihoods. *Conservation Newsletter, 1*, 2, Published by Rural Reconstruction Nepal.

Upreti, G. (2020). *Environmental sustainability requires pragmatic environmental ethics, 2nd NRN global knowledge convention diaspora for innovation and prosperity in Nepal: Post COVID-19 scenario, PROGRAMABSTRACTS*. file:///C:/Users/Microcenter/Documents/NAPA%20Folder/Agri_Newsletter/Program-and-Abstract-Booklet-2nd-NRN-GKC.pdf

Van Dyne, G. M. (1969). In G. M. Van Dyne (Ed.), *The ecosystem concept in natural resource management, based on a symposium, Albuquerque, N. M., Feb. 1968*. Academic Press.

Van Vallen, L. M. (1991). Biotal evolution: A manifesto. *Evolutionary Theory, 10*, 1–13.

Van Zanten, B., Koetse, M. J., & Verburg, P. (2016). Economic valuation at all cost? The role of the price attribute in a landscape preference study. *Ecosystem Services, 22B*, 289–296.

Verhoeff, R. P., Knippels, M. C. P. J., Gilissen, M. G. R., & Boersma, K. T. (2018). The theoretical nature of systems thinking. Perspectives on systems thinking in biology education. *Frontiers in Education, 3*, 40. https://doi.org/10.3389/feduc.2018.00040

Vidal, G. (1984). The oldest eukaryotic cells (in Russian). *V Mire Nauki., 4*, 14–24.

Vidal, C. (2010). *What is the Noosphere?* https://humanenergy.io/projects/what-is-the-noosphere/

Visual Capitalist. (2019). *Global CO2 emissions by Top 15 Countries*. https://www.weforum.org/agenda/2019/06/chart-of-the-day-these-countries-create-most-of-the-world-s-co2-emissions/

Vitousek, P. M., Ehrlich, P. R., Ehrlich, A. H., & Matson, P. A. (1986). Human appropriation of the products of photosynthesis. *BioScience, 36*(6), 368–373.

Vose, R. S., Easterling, D. R., Kunkel, K. E., LeGrande, A. N., & Wehne, M. F. (2017). Temperature changes in the United States. In D. J. Wuebbles, D. W. Fahey, K. A. Hibbard, D. J. Dokken, B. C. Stewart, & T. K. Maycock (Eds.), *Climate science special report: Fourth national climate assessment* (Vol. I, pp. 197–199). U.S. Global Change Research Program. https://doi.org/10.7930/J0N29V45

Wahl, D. C. (2017). *Ecological Footprint, Earth Overshoot Day, and Happy Planet Index, 2017.* https://medium.com/age-of-awareness/ecological-footprint-earth-overshoot-day-and-the-happy-planet-index-8ff5fc19af06

Ward, J. D., Sutton, P. C., Werner, A. D., Costanza, R., Mohr, S. H., & Simmons, C. T. (2016). *Is decoupling GDP growth from environmental impact possible?* Published: October 14, 2016. https://doi.org/10.1371/journal.pone.0164733

Washburn, J. O., Mercer, D. R., & Anderson, J. R. (1991). Regulatory roles of parasites: Impacts on host population shifts with resource availability. *Science, 253*, 185–188.

Webster, K. (2015). *The Circular economy: A wealth of flows*. Ellen McArthur Foundation.

Weiss, P. (1970). Das lebende System: Ein Beispiel fur den Schichtendeterminismus. In A. Koettler & J. R. Smythies (Eds.), *Das neue Menschenbild Wien*.

Weisse, M., & Goldman, E. D. (2018). *Was the second-worst year on record for tropical tree cover loss, 2018.*

Wester, P., Mishra, A., Mukherji, A., & Shrestha, A. B. (Eds.). (2019). *The Hindu Kush Himalaya assessment: Climate change, sustainability and people*. Springer. https://doi.org/10.1007/978-3-319-92288

Weston, A. (1985). Beyond intrinsic value: Pragmatism in environmental ethics. *Environmental Ethics, 7*, 321–339.

White, L. (1967). The historical roots of our ecological crisis. *Science, 155*, 1203–1207.

Whitehead, A. N. (1929). *Process and reality: An essay in cosmology. Gifford lectures delivered in the University of Edinburgh during the session 1927–28*. Macmillan.

WHO (World Health Organization). (1998). *Health promotion glossary*. World Health Organization.

Wilber, K. (1997). *The eye of spirit: An integral vision for a world gone slightly mad* (pp. 71–79). Shambhala.

Willis, K. J., & Whittaker, R. J. (2002). Species diversity—Scale matters. *Science, 295*(5558), 1245–1248.

Wilson, E. O. (1976). Behavioral discretization and the number of castes in an ant species. *Behavioral Ecology and Sociobiology, 1*, 141–154.

Wilson, E. O. (1986). Adaptive indirect effects. In J. Diamond & T. J. Case (Eds.), *Community Ecology* (p. 437–44). Harper and Row.

Wilson, E. O. (1988). The current state of biological diversity. In E. O. Wilson (Ed.), *Biodiversity*. National Academy Press.

Wilson, E. O. (1990). *Success and dominance in ecosystems: The case of the social insects*. Ecology Institute, Oldendorf/Luhe, Federal Republic of Germany.

Wilson, E. O. (1998). *Consilience: The Unity of knowledge* (p. 1998). Random House of Canada.

Wilson, E. (1999). *The two hypotheses of human meaning: Scientific empiricism and religious transcendentalism, address, humanist of the year award.*

Wilson, E. O. (2017, January/February). *A biologist's manifesto for preserving life on earth*, Sierra Club, Magazine.

Wilson, E. O., & Hölldobler, B. (1980). Sex differences in cooperative silk-spinning by weaver ant larvae. *Proceedings of the National Academy of Sciences, 77*(4), 2343–2347.

Woodward, A., & Mcmillan, A. (2015). The environment and climate change. In R. Detels, M. Gulliford, Q. Abdool Karim, & C. Chuan Tan (Eds.), *Oxford textbook of global public health* (p. 2015). Oxford University Press.

World Bank. (1983). *World development report*. The World Bank.

World Bank. (1988). *World development report*. The World Bank.

World Bank. (1991). *The forest sectors – A World bank policy paper*.

World Resource Institute. (2016). *Global tree cover loss reaches a new height in 2016*. https://www.wri.org/insights/global-tree-cover-loss-rose-51-percent-2016#:~:text=Global%20tree%20cover%20loss%20reached,the%20size%20of%20New%20Zealand

World Resource Institute. (2017). *2017 was the second-worst year on record for tropical tree cover loss*. https://www.wri.org/insights/2017-was-second-worst-year-record-tropical-tree-cover-loss

World Wide Fund (WWF) for Nature. (1990, October). *Economic analysis of conservation initiative*. WWF-UK.

Wright, D. H. (1990). Human impacts on energy flow through natural ecosystems, and implications for species endangerment. *Ambio, 19*(4), 189–194.

WWF. (1990). *World Wide Fund (WWF) for nature*. Economic analysis of conservation initiative, WWF-UK.

Yachi, S., & Loreau, M. (1999). Biodiversity and ecosystem productivity in a fluctuating environment: The insurance hypothesis. *Proceedings of the National Academy of Sciences, 96*(4), 1463–1468.

Zhabotinsky, A. M. (1964). Periodic course of oxidation of malonic acid solution. *Bulletin of the Academy of Sciences, USSR Division of Chemical Science, 13*, 145–147.

Zhang, Y., Gao, T., Kang, S., & Sillanpää, M. (2019). Importance of atmospheric transport for microplastics deposited in remote areas. *Environmental Pollution, 254*, 112953.

Index

G. Upreti, *Ecosociocentrism*, https://doi.org/10.1007/978-3-031-41754-2